Using
SmartSketch ®

Ron K. C. Cheng

THOMSON

DELMAR LEARNING

Australia   Canada   Mexico   Singapore   Spain   United Kingdom   United States

THOMSON
™
DELMAR LEARNING

Using SmartSketch®
Ron K. C. Cheng

Vice President, Technology and Trades SBU:
Alar Elken

Editorial Director:
Sandy Clark

Senior Acquisitions Editor:
James De Voe

Senior Development Editor:
John Fisher

Marketing Director:
Dave Garza

Channel Manager:
Dennis Williams

Marketing Coordinator:
Casey Bruno

Production Director:
Mary Ellen Black

Production Manager:
Andrew Crouth

Production Editor:
Stacy Masucci

Technology Project Manager:
Kevin Smith

Technology Project Specialist:
Linda Verde

For more information, contact:
Thomson Delmar Learning
Executive Woods,
5 Maxwell Drive, PO Box 8007
Clifton Park, NY 12065-8007.
Or find us on the World Wide Web at
http://www.delmarlearning.com

For permission to use material from this text or product, contact us by
Tel : 1-800-730-2214
Fax: 1-800-730-2215
www.thomsonrights.com

ISBN: 1-4018-7852-0

NOTICE TO THE READER

About the Author

Ron K. C. Cheng leads the Product Design Unit of the Industrial Center of The Hong Kong Polytechnic University, where he is involved in computer-aided industrial and product design and the development of computer-based learning materials. He is also the developer and manager of the instructional web site *http://pdu.ic.polyu.edu.hk*. Ron is author of many trainng guides, including books on AutoCAD, Mechanical Desktop, Autodesk Inventor, Rhinoceros, and Pro/Desktop. You may contact the author via his email address: ircheng@polyu.edu.hk.

Acknowledgments

This book would never have been realized without the contributions of many individuals. At Delmar/Thomson Learning, special thanks goes to John Fisher, the developmental editor who worked closely with me on this book, as well as publisher Alar Elken, acquisitions editor James De Voe, editorial assistant Tom Best, technical editor Stan Thorton, and production editor Stacy Masucci. Thanks also goes to designer Carol Leyba of Leyba Associates and copyeditor Daril Bentley of Bentley Editorial Services.

Contents

Preface

SmartSketch is a 2D computer-aided design tool for constructing 2D schematic diagrams as well as precision engineering drawings. To help you master SmartSketch, this book delineates various aspects of the application in a concise manner. Each chapter begins with a set of objectives to be achieved and an overview of what you will learn in the chapter. The end of each chapter contains a summary of the subject matter of the chapter and a set of review questions to test your understanding. In addition, Appendix A contains a set of exercises intended to provide you with practice using SmartSketch. The sections that follow describe the book's content broken down by chapter.

■ ■ ■ ■ Chapter 1

Chapter 1 outlines the key functions of SmartSketch, together with illustrations on SmartSketch's user interface and its help system. It also explains the organization of drawing data in a SmartSketch document and how a SmartSketch document can cope with upstream and downstream computerized operations.

■ ■ ■ ■ Chapter 2

This chapter starts with explaining the key concepts of 2D drafting. It then delineates methods of setting up a SmartSketch document and explains how to draw and change drawing objects in a SmartSketch document. While learning how to construct drawing elements, you will learn the concepts of establishing relationships among the elements. By the end of this chapter, you should be able to construct simple geometric drawings and sketches.

■ ■ ■ ■ Chapter 3

This chapter is a continuation of Chapter 2. You will learn how to set up layers and format drawing elements. You will also learn how to insert both driving dimensions and driven dimensions into a document and to add annotations. A brief explanation of system options rounds out the chapter.

Chapter 4

To enhance drawing productivity, you use symbols from the library (as well as symbols you might create and save). In this chapter, you will learn how to incorporate symbols from the library into a document from the symbol library through the use of the Symbol Explorer. You will also learn how to use the Attribute Viewer to modify symbol attributes. Most importantly, you will learn how to construct custom symbols of your own.

Chapter 5

This is the third chapter on drafting methods. You will learn how to integrate images in a document and insert Window objects and hyperlinks. You will also learn how to include detail views, construct isometric views, and manipulate variables.

Chapter 6

As a concluding chapter, you will be introduced to various templates that are organized into two major types of SmartSketch documents: schematic diagrams and precision drawings. Schematic diagrams are further divided into general diagrams, electrical diagrams, and process diagrams. As for precision drawings, they are AEC (architecture, engineering, and construction) solutions and mechanical engineering drawings.

CHAPTER 1

Introduction to SmartSketch

■ ■ ■ ■ ## Objectives

This chapter outlines the key functions of SmartSketch and introduces its user interface and help system. After studying this chapter, you should be able to:

❑ Describe the key functions of SmartSketch

❑ Use the SmartSketch user interface and help system

Overview

SmartSketch is a 2D drafting tool for creating engineering drawings, sketches, and diagrams. You can use it as a tool to construct process diagrams, business diagrams, network diagrams, technical drawings, mappings, AEC drawings, and mechanical drawings. It has a comprehensive help system guiding you as you work along.

■ ■ ■ ■ ## Appreciating SmartSketch Functions

SmartSketch is a 2D computer-aided drafting and sketching system for creating 2D engineering drawings and schematic diagrams.

2D Engineering Drawings and Sketches

In essence, using SmartSketch to create 2D engineering drawing and sketches is to use the computer as an electronic drawing board to construct conventional drawings electronically. In such an electronic

1

medium, you have to think about how an object is to be represented by a number of orthographic views or sketches, further decompose each drawing view into a number of geometric entities, and construct these entities using the tools available in the CAD system. Each drawing view is a set of objects: lines, arcs, circles, splines, ellipses, and elliptical arcs. Figure 1–1 shows a 2D engineering drawing constructed using Smart-Sketch.

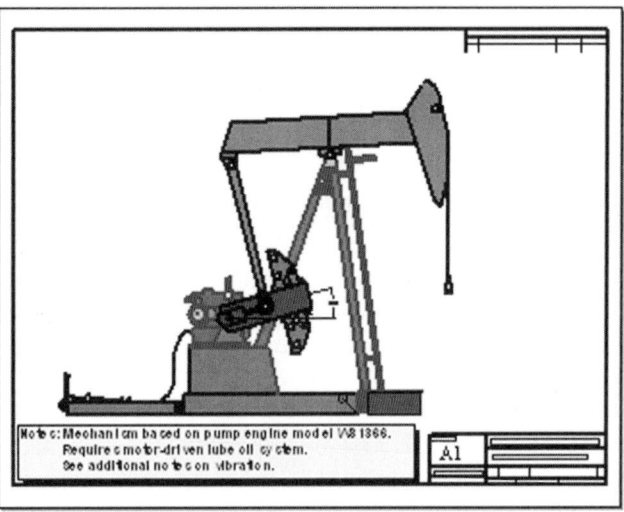

Figure 1–1. 2D engineering drawing (SmartSketch sample document Field_pump.igr).

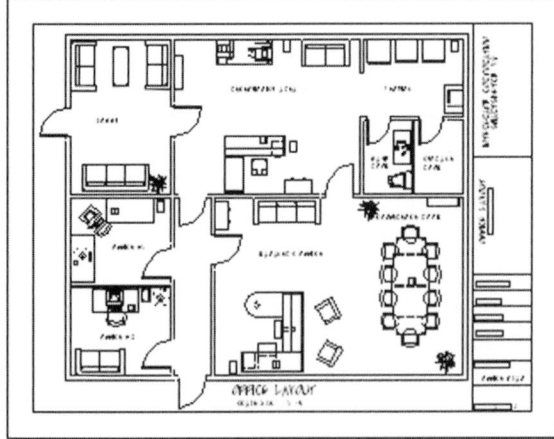

Figure 1–2. Office layout (SmartSketch sample document Office Layout.igr).

Diagrams

Apart from constructing engineering drawings and sketches, you can construct various types of technical sketches, including atlases, flowcharts, network diagrams, office layouts, organizational charts, and workflow diagrams. Figure 1–2 shows an example of an office layout, and Figure 1–3 shows an example of a workflow diagram.

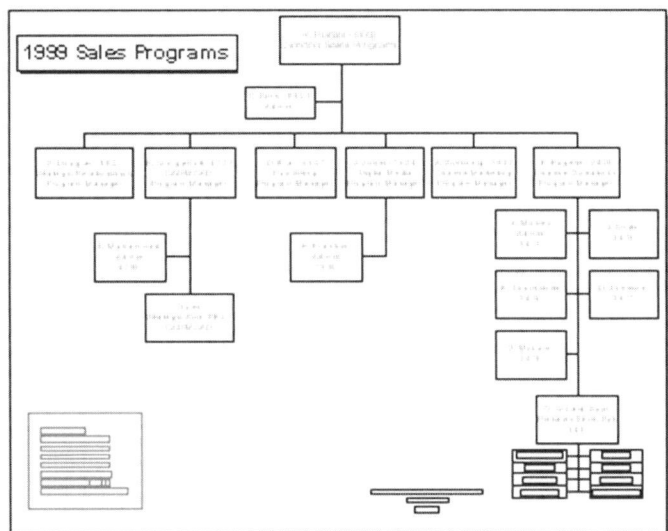

Figure 1–3. Workflow diagram (SmartSketch sample document Org Chart.igr).

■ ■ ■ ■ The SmartSketch User Interface

Figure 1–4. Getting Started dialog box.

Start SmartSketch to familiarize yourself with its user interface. Click on the SmartSketch icon from the desktop or select SmartSketch from the Windows Start menu. In the Getting Started dialog box that automatically appears (shown in Figure 1–4), you will find six options: Product Preview, Exploring the Interface, Office Companion, MicroStation/AutoCAD Users, Learning Center, and Start SmartSketch.

Select Start SmartSketch from the Getting Started dialog to start SmartSketch. Click on the OK button of the Tip of the Day to close the dialog box. The default user interface has a number of areas. At the top of the SmartSketch window you will find the standard Window bar, depicting the application's name and the name of the current SmartSketch document. Figure 1–5 shows the SmartSketch user interface.

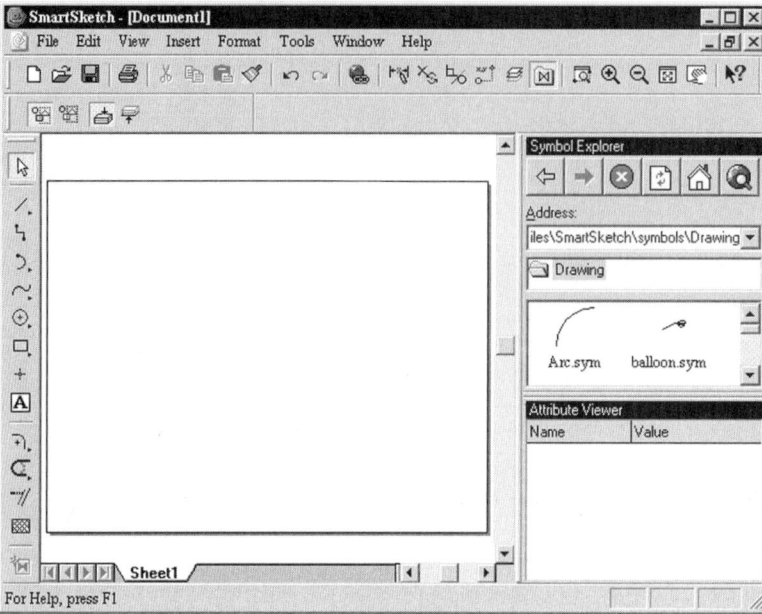

*Figure 1–5.
SmartSketch
user interface.*

Pull-down Menus

Below the standard window bar is the pull-down menu bar containing a number of menu items: File, Edit, View, Insert, Format, Tools, Windows, and Help. Table 1–1 outlines the functions of these menus.

Table 1–1 Main User Interface Pull-down Menus and Their Functions

Pull-down Menu Options	Function
File	For working on files and templates
Edit	For general Windows editing
View	For setting display and toolbars
Insert	For insertion of objects
Format	For setting various styles
Tools	Provides various types of tools
Window	Sets the display of multiple documents
Help	Provides useful help information

Figure 1–6. Toolbars dialog box.

Toolbars

The default configuration of the main interface contains a Main toolbar below the pull-down menus, a Ribbon toolbar below the Main toolbar, and a Draw toolbar at the left side of the SmartSketch window. Other than these, you can access other toolbars by selecting View > Toolbars (selecting Toolbars from the View pull-down menu, which accesses the Toolbars dialog box), checking the checkbox of a toolbar item in the Toolbars selection box, and then clicking on the OK button. (See Figure 1–6.)

Main Toolbar

The default configuration of the main interface contains the Main toolbar, located immediately below the pull-down menus. Figure 1–7 shows the Main toolbar, and Table 1–2 outlines the options of the Main toolbar.

Figure 1–7. Main toolbar.

Table 1–2 Main Toolbar Options and Their Functions

Option	Function
New File	Starts a new document.
Open File	Opens an existing document.
Save	Saves the current document.
Print	Prints the current document.
Cut	Cuts selected objects and places them in the Windows clipboard.

Option	Function
Copy	Copies selected objects and places them in the Windows clipboard.
Paste	Pastes cut or copied objects.
Format Painter	Copies formatting from a selected object and applies it to other objects.
Undo	Undoes the last command.
Redo	Redoes the last undone command.
Hyperlink	Establishes hyperlink to selected objects. (See Figure 1–8.)
Dimension	Displays the Dimension toolbar. (See Figure 1–9.)
Change	Displays the Change toolbar. (See Figure 1–9.)
Relations	Displays the Relation toolbar. (See Figure 1–9.)
PinPoint	Displays the PinPoint ribbon. (See Figure 1–9.)
Layers	Displays the Layer ribbon. (See Figure 1–9.)
Symbol Explorer	Activates the Symbol Explorer window and the Attribute Viewer.
Zoom Area	Zooms the display to fill up a rectangular area defined by two selected points.
Zoom In	Zooms in.
Zoom Out	Zooms out.
Fit	Fits the entire document or selected drawing elements in the display window.
Pan	Pans the display.
Help	Displays help information for a selected command.

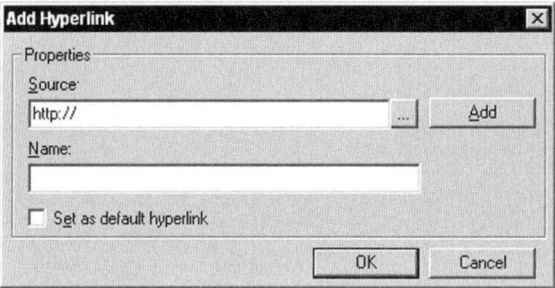

Figure 1–8. Add HyperLink toolbar.

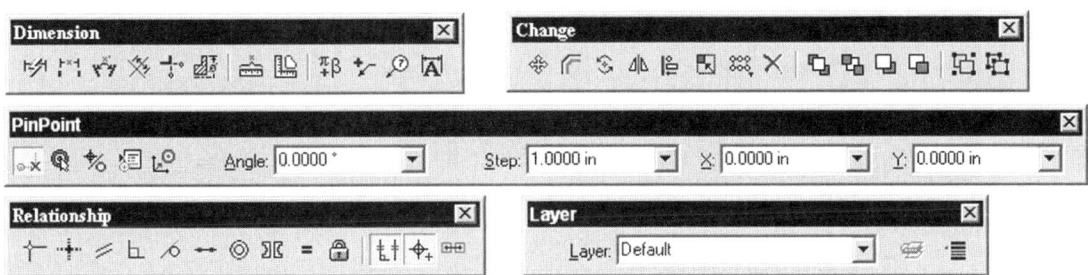

Figure 1–9. From left to right and top to bottom: Dimension toolbar, Change toolbar, PinPoint ribbon, Relation toolbar, and Layer ribbon.

Ribbon Toolbar

Figure 1–10. Ribbon toolbar for selection of objects.

The Ribbon toolbar helps you control various settings when certain commands are activated. Because the Ribbon toolbar is command sensitive, its content varies in accordance with the active command. By default, the Select Tool command of the Draw toolbar is activated and its ribbon is displayed when you first start SmartSketch. (See Figure 1–10.) If the Ribbon toolbar is not displayed, select View > Toolbars, click on Ribbon in the Toolbars selection box of the Toolbars dialog box, and than click on the OK button.

Figure 1–11. Ribbon dialog box for drawing a line.

By default, the Ribbon toolbar is placed under the Main toolbar. However, you may select it and drag it to any position. Figure 1–11 shows the Ribbon dialog box for the Line/Arc Continuous command.

Graphics Area and Display Control

The graphics area is the main working area. You can imagine it as a 2D working area on which you place a drawing sheet and construct 2D objects on the sheet to depict a drawing, sketch, or diagram. You can have more than one working sheet in a document. In addition to *working sheets*, you can include *background sheets*. The concepts of working sheets and background sheets are explained later in this chapter.

Grid and Snap

To help you visualize the actual size of the graphics area, you may display grid points by selecting View > Grid Display. (See Figure 1–12.) To turn off grid and grid snap, you can either select View > Grid Display > View > Grid Snap or uncheck the Grid Display and Grid Snap options on the View tab of the Options dialog box.

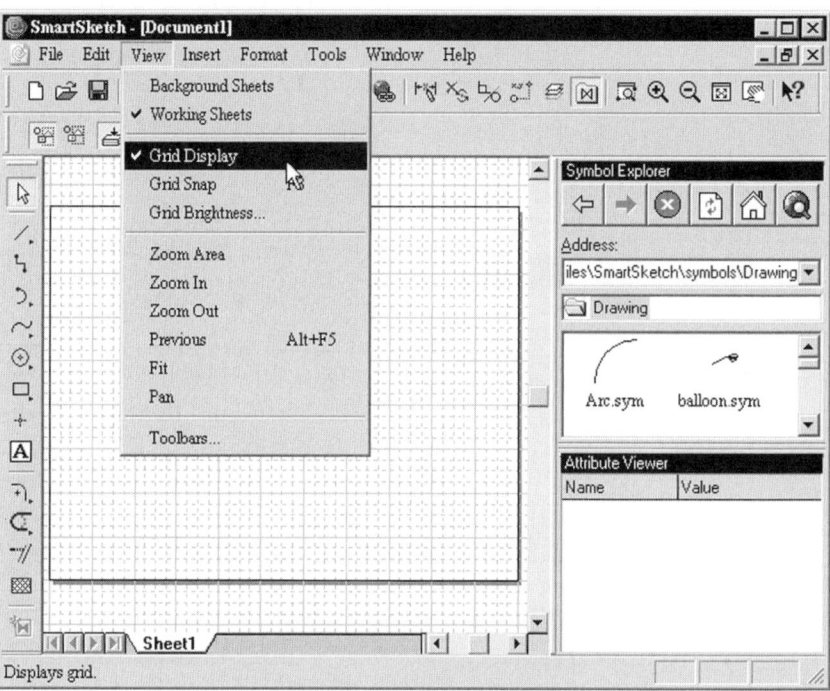

Figure 1–12. Displaying grids in the graphics area.

Grid points are purely visual aids while drafting. Therefore, they will not appear on a printed document and you cannot select the grid point precisely by simply manipulating the cursor. If you wish to move the cursor to specify locations, you need to select View > Grid Snap to activate snapping.

To set the grid display density, select Tools > Options, select the View tab from the Options dialog box, and establish the appropriate settings. (See Figure 1–13.)

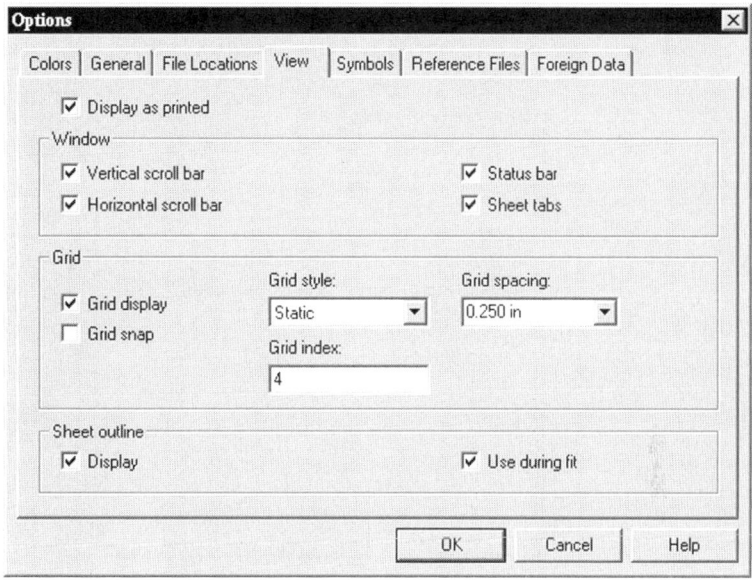

Figure 1–13. View tab of the Options dialog box.

Display Control

To see drawing elements clearly, you use the display control commands: Zoom Area, Zoom In, Zoom Out, Previous, Fit, and Pan from the View pull-down menu or the Main toolbar. Perform the following steps.

1   Select File > Open.

2   In the Open dialog box, select the file Chapter1Display.igr from the companion CD-ROM.

3   Select File > Save As.

4   In the Save As dialog box, click on the Save In pull-down list box and select a folder in your computer.

5   Click on the Save button. The file from the CD-ROM is opened and saved in your computer.

6   Select View > Zoom Area.

7   Click on A and B indicated in Figure 1–14. The rectangular area defined by these two selected points is fitted in the current display window.

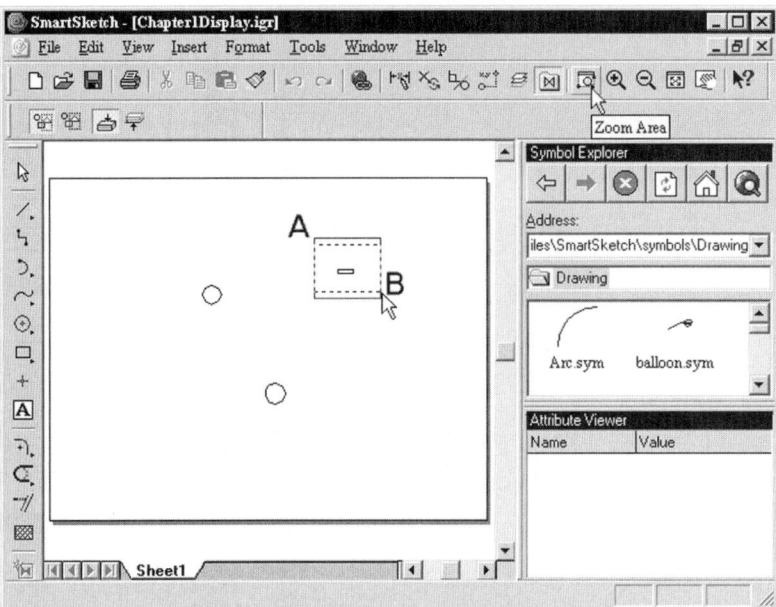

Figure 1–14.
Zooming to
an area.

8   Select View > Zoom In.

9   Click on A and drag the cursor to location B, indicated in Figure 1–15. The display is zoomed in, with the selected point being the center of the display view.

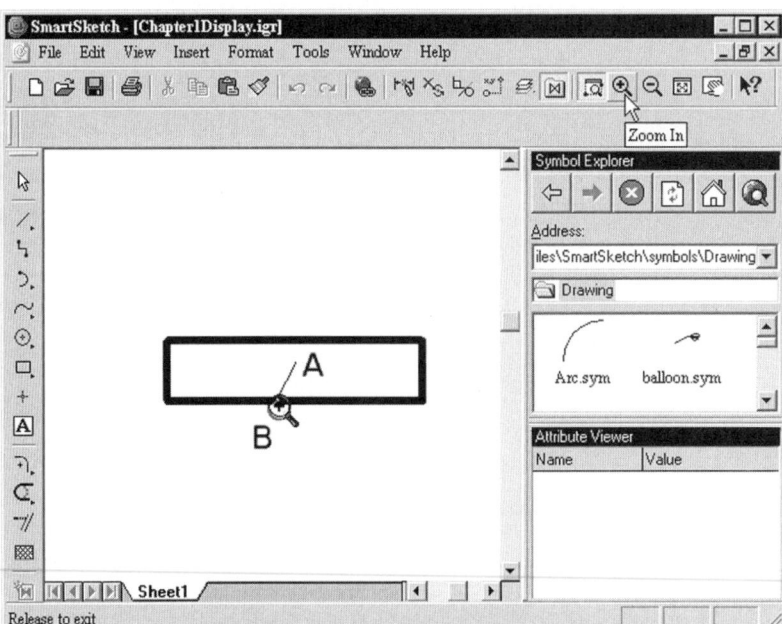

Figure 1–15.
Zooming in.

10 Select View > Zoom Out.

11 Click on A and drag the cursor to location B indicated in Figure 1–16. The display is zoomed out, with the selected point being located at the center of the display window.

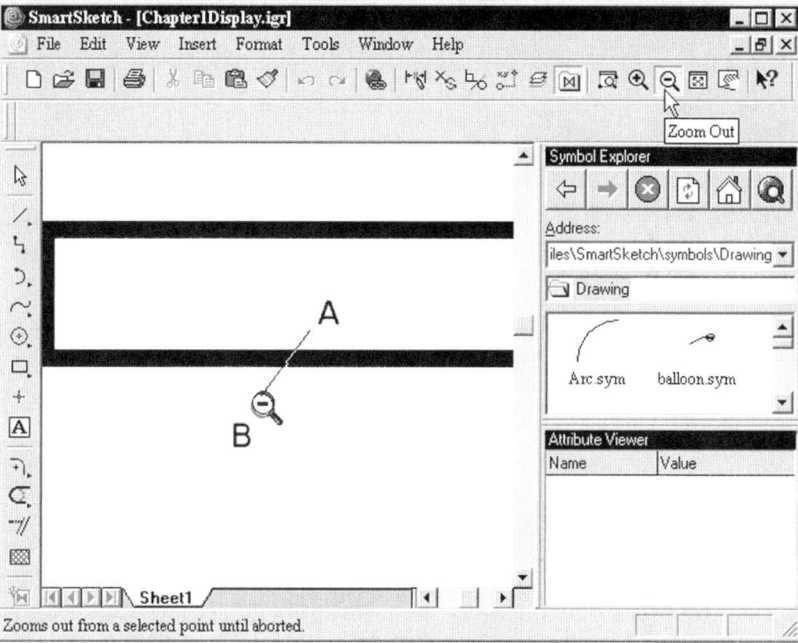

Figure 1–16. Zooming out.

12 Select Zoom > Pan.

13 Click on A (indicated in Figure 1–17), hold down the mouse button, and move the cursor to B. The display is panned.

14 Select View > Previous. The display is reverted to the previous view.

15 Select View > Fit. The document is fitted to the graphics area. (Note: If drawing elements are selected prior to running this command, the selected objects will be fitted to the graphics area rather than the entire document.)

16 Close the file without saving it.

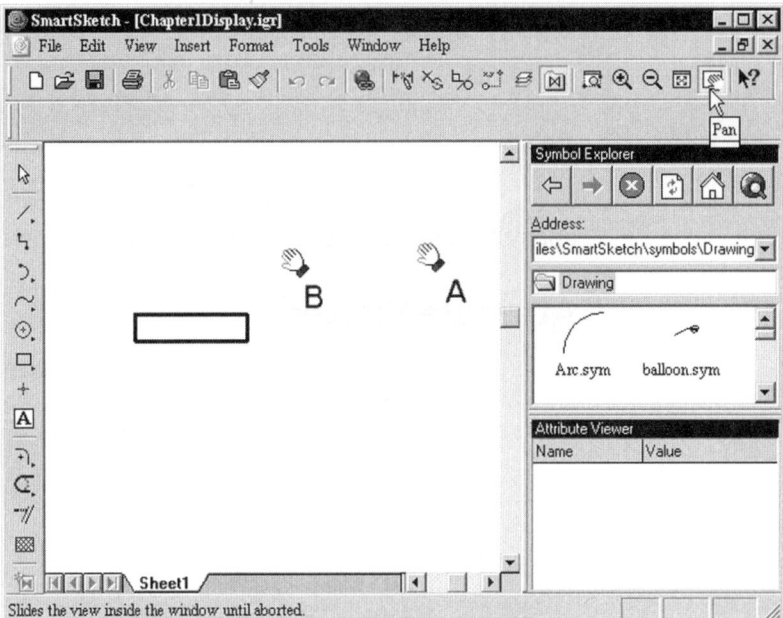

Figure 1–17. Panning.

Symbol Explorer

The Symbol Explorer, as its name implies, is an explorer through which you explore the various types of symbols available. (See Figure 1–18.) By default, the symbol explorer is placed at the right side of the Smart-Sketch window. However, you can select and drag it to any location in the screen.

To close the explorer, click on the Symbol Explorer icon on the Main toolbar. To display the explorer after you close it, click on the Symbol Explorer icon on the Main toolbar again.

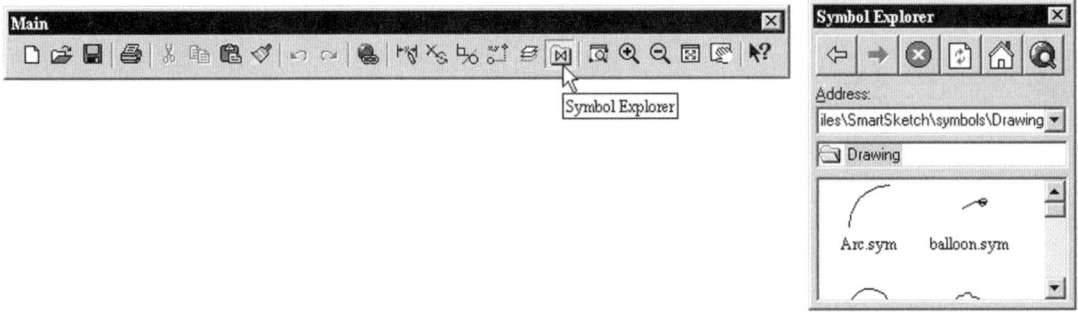

Figure 1–18. Symbol Explorer dialog box.

The Symbol Explorer has four major areas. At the top, there are six buttons. From left to right, they are Back, Forward, Stop, Refresh, Home, and Explore Elsewhere. The second area is the Address area, showing the current location of the explorer. The third area shows the folders available from the address shown, and the fourth area displays the symbols available. Selecting and dragging the icons shown in the fourth area to the Graphics area inserts the selected symbol into the document. To open the symbol document, you double click on it.

Attribute Viewer

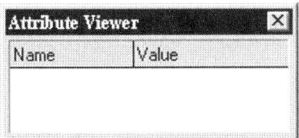

Figure 1–19. Attribute Viewer dialog box.

The Attribute Viewer displays user-defined properties and parameters of an element, symbol, or inserted document. The viewer has two columns: Name and Value. (See Figure 1–19.) The Attribute Viewer is, by default, located below the Symbol Explorer. Again, you can select and drag it anywhere in the screen.

Status Bar

The status bar is located at the bottom of the SmartSketch user interface. The status bar, as in typical interface arrangements, provides prompts and feedback.

Control Keys and Function Keys

In addition to the standard Windows shortcut keys delineated in Table 1–3, SmartSketch provides a set of shortcut keys and key combinations (outlined in Table 1–4) that allow you to perform basic functions quickly.

Table 1–3 Standard Windows Shortcut Keys

Key (s)	Function
Ctrl + A	Select All
Ctrl + C	Copy Text
Ctrl + N	New
Ctrl + O	Open
Ctrl + P	Print

Key (s)	Function
Ctrl + S	Save
Ctrl + V	Paste Text
Ctrl + X	Cut Text
Ctrl + Y	Redo
Ctrl + Z	Undo

Table 1–4 SmartSketch Main Toolbar Shortcut Keys

Key (s)	Function
F1	Help
Shift + F1	What's This? Help
F3	Grid Snap
Alt + F4	Exit
Ctrl + F4	Close File
F5	Update Active View
Alt + F5	Previous Zoom Level
Ctrl + F7	Paste From Clipboard
F9	Toggle PinPoint Display
Ctrl + F9	Cut to Clipboard
F10	PinPoint Lock X axis
Shift + F10	Select First Menu Item
F11	PinPoint Lock Y axis
F12	Reset PinPoint Home

Mouse

A mouse normally has two buttons. You can use the left-hand button to perform a number of tasks: selecting a single object, selecting multiple objects by dragging to fence them, dragging a selected object, clicking or dragging an object, selecting a menu or toolbar item, and activating an embedded or linked object. Using the right-hand button, you can restart a command or display a context-sensitive shortcut menu. If you are using a Microsoft IntelliMouse pointing device, such as the one shown in Figure 1–20, additional tasks that you can perform are outlined in Table 1–5.

Table 1–5 Mouse Button Commands Available via IntelliMouse

Action	Task
Rotate the wheel button forward	Zooms in at the current pointer location
Rotate the wheel button backward	Zooms out at the current pointer location
Drag the wheel button	Pans from one location to another
Press Ctrl and drag the wheel button	Zooms the area of the window you define by dragging the pointer
Press Shift and click the wheel button	Fits the graphics of the document to the window
Press Alt and click the wheel button	Restores the previous view

Figure 1–20. Microsoft IntelliMouse.

■ ■ ■ ■ Understanding the SmartSketch Document

A SmartSketch data file is called a document. In a document, there are two types of drawing sheets: working sheets and background sheets.

Working Sheets

A document's working sheets are the main working areas on which you construct the main constituents of a drawing, sketch, or diagram. Because the number of working sheets is unlimited and because working sheets in a document are independent of one another, in a single SmartSketch document you can construct various types of objects on individual working sheets to depict various drawings, sketches, and diagrams. For example, you might construct the assembly and the individual components of a product or system on separate working sheets of a document.

Background Sheets

Background sheets, as the name implies, serve mainly as a background for working sheets. You can save a lot of time in avoiding duplicative drawing by constructing drawing elements common to a number of working sheets on a background sheet and linking that background sheet to each of the working sheets. The idea is to "draw once and link." You then need only tell the program to show the drawing elements of the background sheet in the working sheets.

Drawing Elements on Working Sheets and Background Sheets

Although you can construct drawing elements on a working sheet or a background sheet, you should construct the main components (those that will not be duplicated) of a document on a working sheet or a number of working sheets and reserve the background sheet for constructing drawing elements that will be common to a number of working sheets.

If you construct drawing elements on a working sheet, the drawing elements will only be displayed on that particular working sheet. If you construct drawing elements on a background sheet and associate it with (link it with) a number of working sheets, drawing elements constructed on the background will be displayed on all associated working sheets.

■ ■ ■ ■ Starting a SmartSketch Document

Starting a SmartSketch document begins with selecting an appropriate template (described further in the following section). You then set up one or more background sheets for constructing drawing elements common to a number of working sheets (if any).

After this preparation work, you set up a number of working sheets as may be required, link the working sheets to selected background sheets if necessary, and start constructing common objects on the background sheets and specific drawing elements on the working sheets. By linking a number of working sheets to a background sheet and showing the drawing elements of the background sheets in the working sheets, you do not have to repeatedly construct the drawing elements common to each individual working sheet in the document.

Selecting a Template

Each time you start a new SmartSketch document, you need to select a template. Depending on the type of document you are going to construct, you select a specific type of template from the SmartSketch *Templates* folder. Perform the following steps.

1 Select File > New.

In the New dialog box, shown in Figure 1–21, you will find a large selection of templates used for various types of work. These are arranged in a number of folders categorically, as shown in Figure 1–21.

Figure 1–21. New dialog box for accessing template folders.

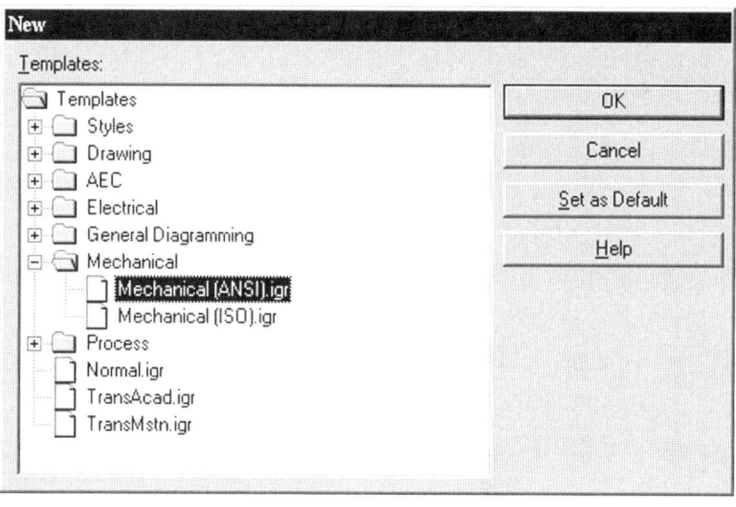

2 Select the file *Mechanical (ANSI).igr* from the *Mechanical* folder and click on the OK button. A new mechanical drawing document is started.

Manipulating Working Sheets and Background Sheets

Several commands from the pull-down menus are used to manipulate working sheets and background sheets in a document. These commands are explained in the sections that follow.

Displaying Working/Background Sheets

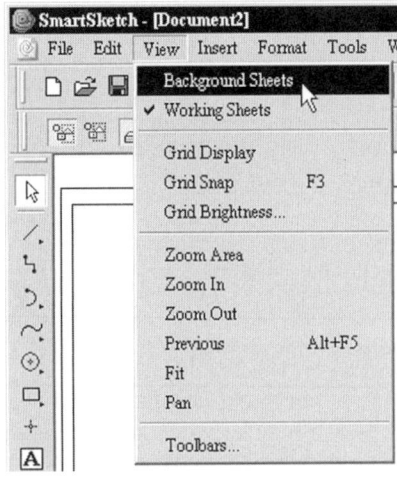

By default, an active working sheet of the document is displayed in the graphics area. To change from displaying working sheets to displaying background sheets, select View > Background Sheets. On the other hand, to change from displaying background sheets to working sheets, select View > Working Sheets.

By default, a working sheet associated with one of the background sheets is displayed in the graphics area. Before we take a closer look at the default working sheet, perform the following steps to view the background sheets in this document.

1   Select View > Background Sheets. (See Figure 1–22.)

Figure 1–22. Viewing background sheets.

Setting Background Sheets

Background sheets in this document are displayed in the graphics area. As can be seen in Figure 1–23, the background sheet (A-Sheet) is displayed and there are four more background sheets available: B-Sheet, C-Sheet, D-Sheet, and E-Sheet (shown as tabs) at the bottom of the graphics area. (Note: Your screen display may not look exactly the same.)

Perform the following steps to explore the options available in the Sheet Setup dialog box, which allows you to establish various sheet properties.

1   Select the B-Sheet tab at the bottom of the graphics area. The background sheet (B-Sheet) will be displayed.

2   Select File > Sheet Setup.

The Sheet Setup dialog box is displayed. (In this case, it is the Sheet Setup dialog box for a background sheet. See Note following.)

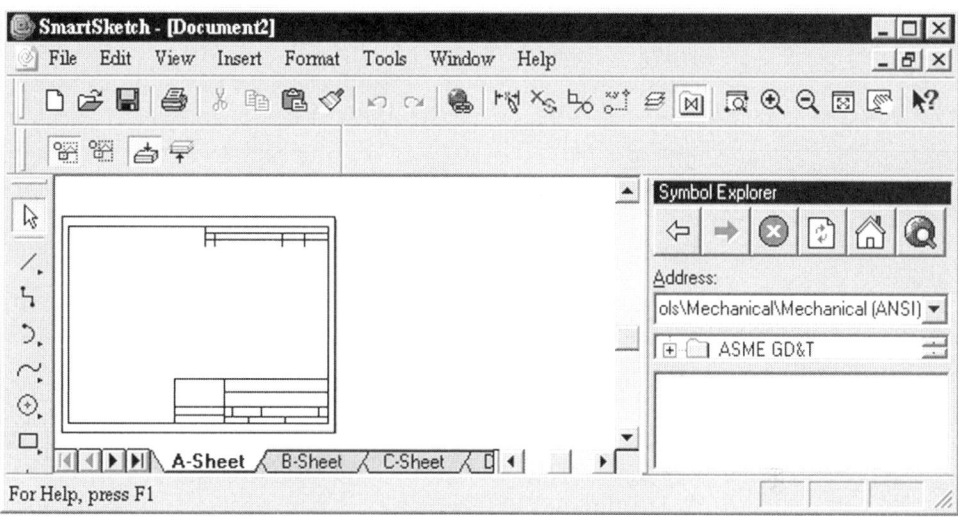

Figure 1–23. Background sheets displayed.

> **NOTE:** *The Select File > Sheet Setup command sets up the displayed drawing sheet. If the displayed sheet is a working sheet, it sets up a working sheet. If the displayed sheet is a background sheet, it sets up a background sheet.*

The Sheet Setup dialog box for setting up a background sheet (shown in Figure 1–24) has two tabs: Name and Size and Scale.

3   Select the Name tab, if it is not already selected. (See Figure 1–24.) The background sheet's name, B-Sheet, is displayed. Change its name to *MY SHEET.*

Figure 1–24. Sheet Setup dialog box for background sheets.

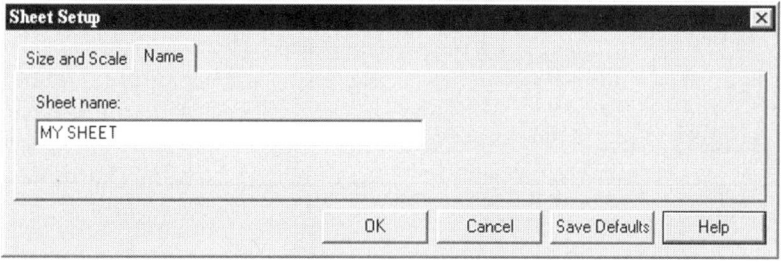

4   Select the Size and Scale tab. (See Figure 1–25.)

The Size and Scale tab of a background sheet has three areas: *Sheet size*, *Drawing scale*, and *Paper units*. There are three ways to define the sheet's size. You can set the sheet's size to be the same as the printer's setup, select a standard sheet size, or specify a custom sheet size. Because the prime purpose of a background sheet is for construction of drawing elements common to a number of working sheets, the *Drawing scale* area is not available. Finally, the *Paper units* area concerns the display of numeric values in dialog boxes and the precision of the display.

5   Leave the settings as they are and click on the OK button to close the dialog box.

Figure 1–25. Size and Scale tab for a background sheet.

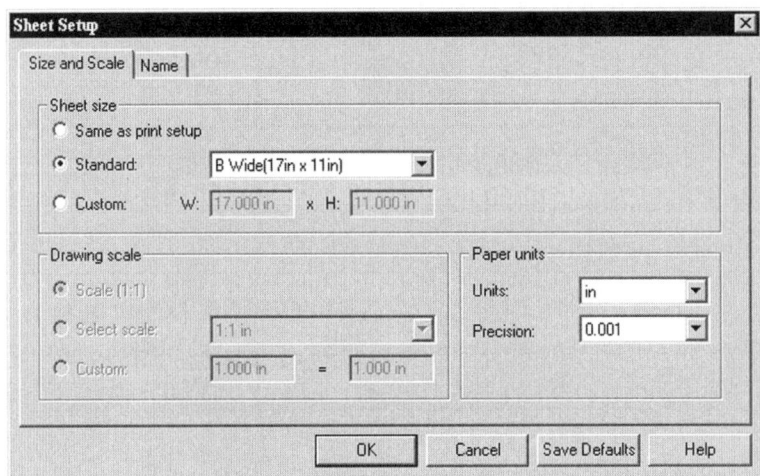

Constructing New Working/Background Sheets

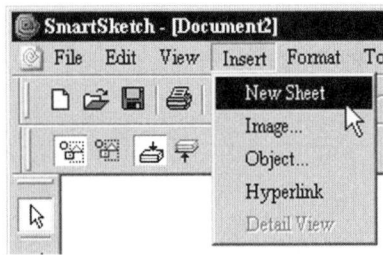

Figure 1–26. Inserting new working/ background sheet.

Depending on which type of sheet you have currently displayed, the Insert > New Sheet command (indicated in Figure 1–26) can be used to add a new working sheet or background sheet to a document. If a working sheet is displayed, selecting Insert > New Sheet adds a new working sheet. If a background sheet is displayed, a new background sheet is added instead.

Perform the following steps to insert and set up a new background sheet.

1   Select Insert > New Sheet.

To reiterate, the type of drawing sheet (background or working) that will be inserted by this command depends on what type of sheet is currently displayed. Because we currently have background sheets displayed, a new background sheet is inserted.

2   Select File > Sheet Setup.

3   In the *Sheet size* area of the Sheet Setup dialog box, select E Tall (34in x 44in). Select the Name tab and change the sheet's name to *MY BACKGROUND*.

4   Click on the OK button. The sheet's size and name are set.

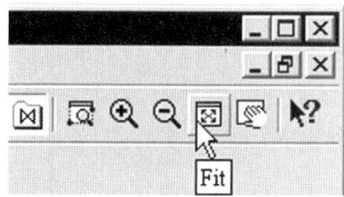

Because the display size of the background sheet is changed, perform the following step to display the entire background sheet.

5   Select Fit on the Main toolbar to fit the entire sheet in the graphics area. (See Figure 1–27.)

Figure 1–27. Entire sheet fitted in the graphics area.

Manipulating Working Sheets

In the previous sections you explored background sheets and set up a new background sheet in a document. Perform the following steps to learn more about the working sheets in a document.

1   Select View > Working Sheets.

As can be seen from the bottom of the graphics area, there is only one working sheet in this document. The default name of this sheet is Sheet 1.

2   Select File > Sheet Setup. (See Figure 1–28.)

The Sheet Setup dialog box for a working sheet has three tabs: Size and Scale, Background, and Name. In the Size and Scale tab, you select a sheet size and decide the drawing scale of the objects in the sheet. Naturally, if you set the sheet to half scale, drawing elements constructed will be shown in half size. Despite setting a scale, any dimension you put in the sheet will not be affected (because dimensions are always shown at full size). Using the Background tab, you can link the active working sheet to a background sheet. You can also display the objects of the background sheet in the working sheet by selecting a background sheet from the pull-down list box and checking the *Show background* box. In the Name tab, you specify the name of the working/background sheet for easy reference.

3 Select the Size and Scale tab, if it is not already selected.

As shown in Figure 1–28, the Size and Scale tab of the Sheet Setup dialog box for a working sheet also has three areas: *Sheet size*, *Drawing scale*, and *Paper units*.

Figure 1–28. Size and Scale tab for a working sheet.

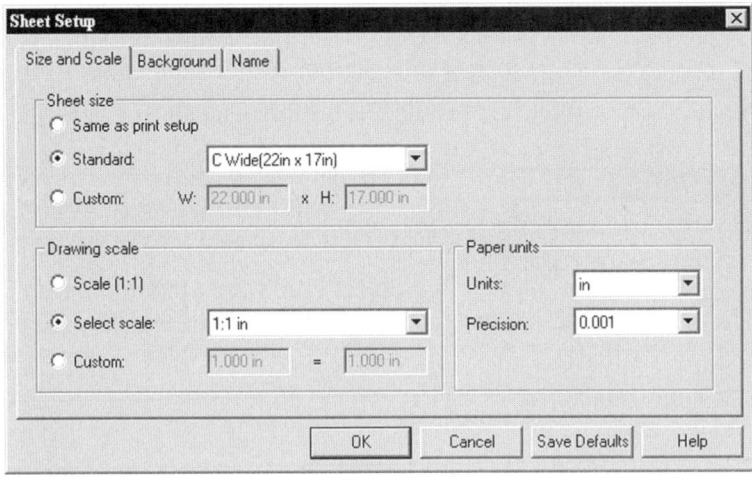

Setting Working Sheet Size

As with setting the sheet size of a background sheet, you can set the working sheet's size to be the same as the printer's setup, select a standard paper size, or specify a custom paper size. However, if you link a background sheet to the working sheet, any sheet size you might specify for the working sheet will be overridden by the sheet size of the associated background sheet. Therefore, if you are going to associate a background sheet to the working sheet, you do not have to specify a sheet size in the Size and Scale tab. Sheet size will change automatically in accordance with the associated background sheet's size.

Setting Working Drawing Scale

As regards drawing scale, it is somewhat important to specify a drawing scale in the Sheet Setup dialog box because this setting affects the displayed size of the drawing elements in relation to the actual size of the working sheet or associated background sheet. To be precise, drawing scale here refers to the scale of display of the drawing elements in relation to the sheet's size, not the scale of the drawing elements themselves.

As we have mentioned, a drawing sheet's working area is unlimited in size. Therefore, you can construct drawing elements of any size and fit them within the graphics area by zooming in, zooming out, or panning around. For the sake of accuracy and consistency in downstream operations (and to avoid confusion), you should specify the drawing elements' true size as you construct them. If they are too large to be placed within the selected working sheet or background sheet, change the drawing scale in the Size and Scale tab of the Sheet Setup dialog box.

Associating Background Sheets and Naming Sheets

To associate a background sheet and change the working sheet's name, perform the following steps.

1 Select the Background tab of the Sheet Setup dialog box.

2 In the *Background sheet* pull-down list box, select the sheet *MY BACKGROUND*. The working sheet's size is changed in accordance with the selected background sheet's size.

3 Check the *Show background* box. Drawing elements constructed on the selected background sheet will be displayed in the working sheet. (See Figure 1–29.)

Figure 1–29. Linking a working sheet to, and showing the content of, a background sheet.

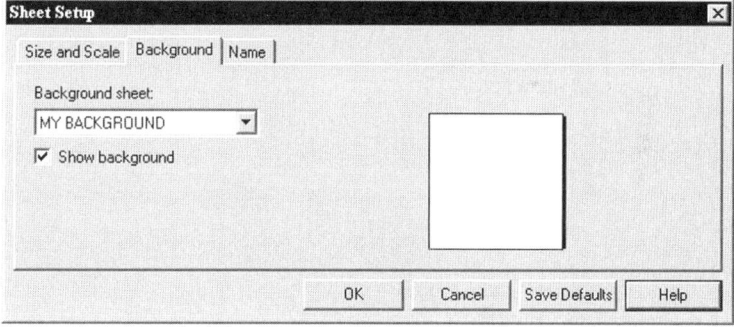

4 Select the Name tab of the Sheet Setup dialog box.

5 Change the name to *MY WORKING SHEET.*

6 Click on the OK button.

Now the working sheet is associated to a selected background sheet and the working sheet's name is changed. If you want to have more than one working sheet, you may continue with selecting Insert > New Sheet to insert as many sheets as you may need. To set up the new working sheet, select File > Sheet Setup.

7 Save the file as *MyDocument.igr.*

Deleting Unwanted Sheets

In addition to inserting new working or background sheets in a document, you can delete unwanted working or background sheets by selecting Edit > Delete Sheet. Because this command deletes the current sheet (working or background), you have to activate the sheet to be deleted before running this command. To activate a sheet, click on it.

▪ ▪ ▪ ▪ Coping with Upstream and Downstream Operations

You can use drawing data constructed in other computer-aided applications in the SmartSketch program. Files saved in a number of different formats can be opened in SmartSketch and saved as SmartSketch documents. On the other hand, you can save SmartSketch documents as other file formats. Both of these capabilities facilitate downstream computerized operations.

SmartSketch-compatible File Formats

File formats you can open in SmartSketch and then save as a SmartSketch document are shown in Figure 1–30.

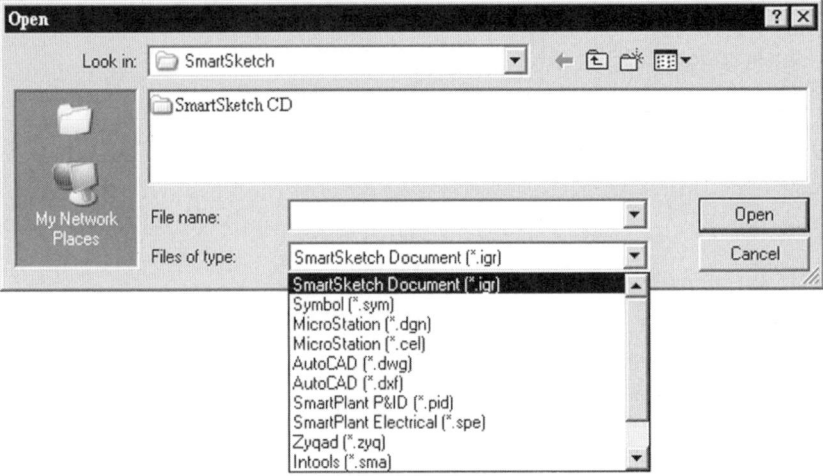

Figure 1–30. SmartSketch-compatible file formats.

"Save As" File Formats

Figure 1–31 shows the file formats a SmartSketch document can be saved as.

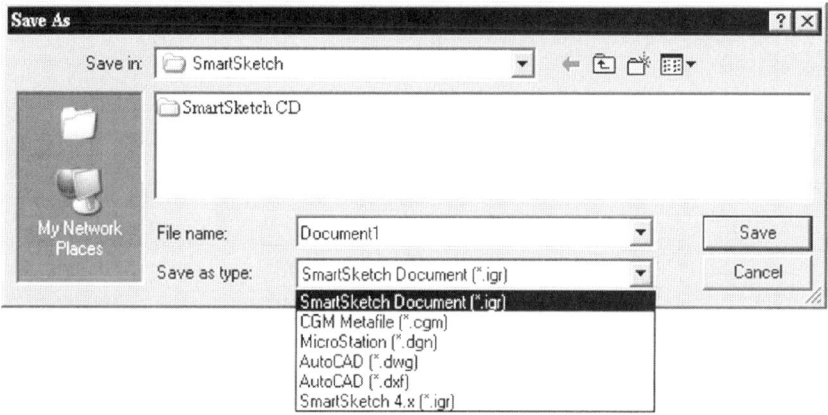

Figure 1–31. "Save As" file formats.

In addition to being able to save a SmartSketch as one of the various file formats indicated in Figure 1–31, you can specify that the SmartSketch document file be saved as an image, a template, or a web page.

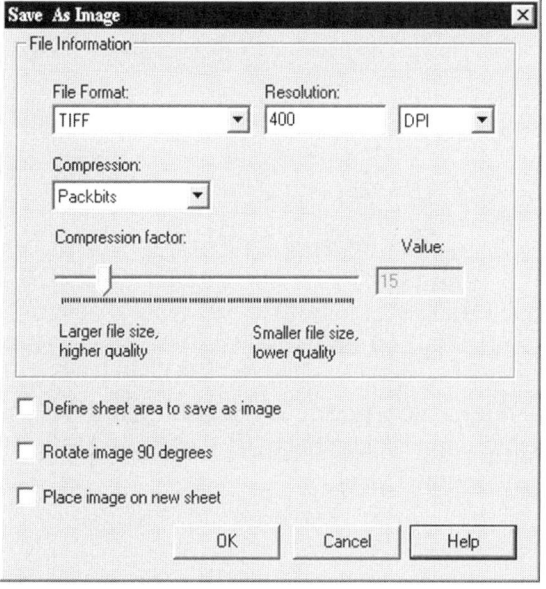

Save As Image

To facilitate insertion to a text document, you can save a SmartSketch document as an image. To save a document as an image, select File > Save As Image. (See Figure 1–32.) In the Save As Image dialog box, select a file format, specify parameters, and then click on the OK button to close the dialog box. After that, specify a file name in the Save As dialog box.

Figure 1–32. Save As Image dialog box.

Saving as Custom Template

You can make use of the settings and parameters in an existing Smart-Sketch document by saving it as a template in SmartSketch's *Template* folder. Perform the following steps.

1 Open the file *MyDocument.igr*, if you already closed it.

2 Select File > Save As Template.

3 In the Save as Template dialog box, select a folder, specify a template file name, and click on the OK button. (See Figure 1–33.)

4 Select File > Close to close the file.

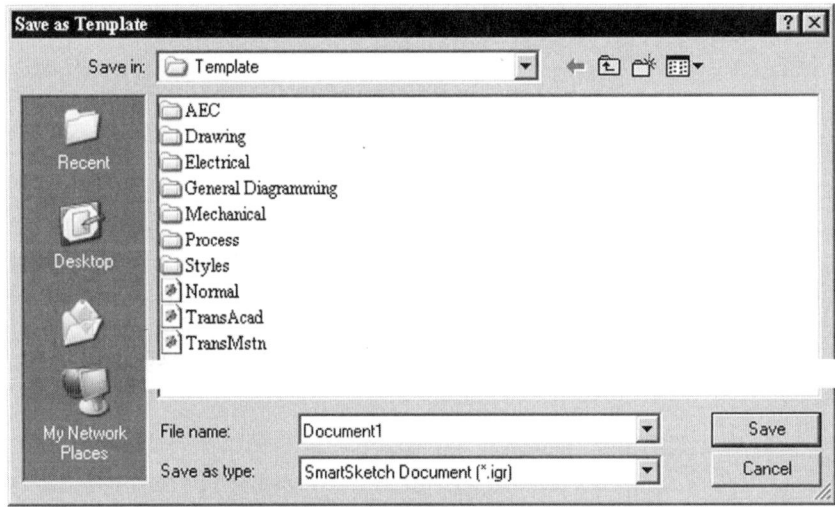

Figure 1–33. Save As Template dialog box.

Save As Web Page

If you want to save a SmartSketch document as a web page, select File > Save As Web Page. (See Figure 1–34.) In the Save As Web Page(s) dialog box, select either All Sheets to save all sheets as web pages or select Active Sheet to save only the current page. To continue, click on the Next button and specify a file name in the Save As Web Page(s) dialog box. Finally, click on the Finish button.

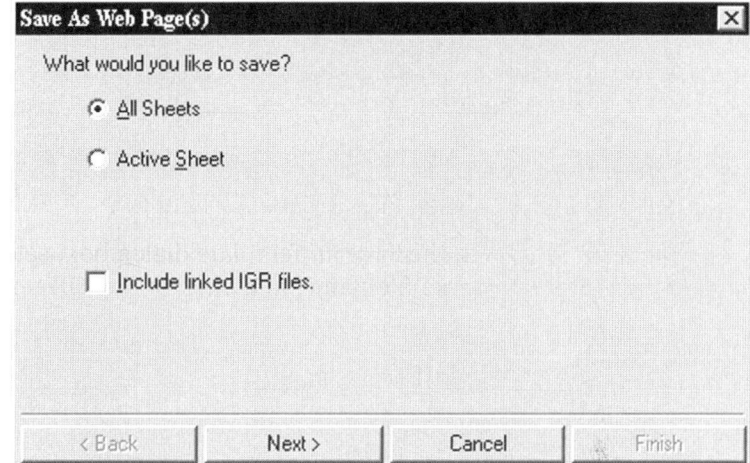

Figure 1–34. Save As Web Page(s) dialog box.

■ ■ ■ ■ Discovering SmartSketch's Help System

The SmartSketch help system consists of a number of elements, including SmartSketch Help, SmartSketch Learning Center, Tip of the Day, SmartSketch on the Web, and SmartSketch Web Forum. These components are explored in the sections that follow.

SmartSketch Help

Help on various aspects of the application can be accessed by selecting Help > SmartSketch Help to display the SmartSketch Help dialog box. (See Figure 1–35.) Selecting Help from the Main toolbar and then selecting an icon from a toolbar or a command from a pull-down menu displays help for the selected command. (See Figure 1–36.)

Figure 1–35. SmartSketch Help dialog box.

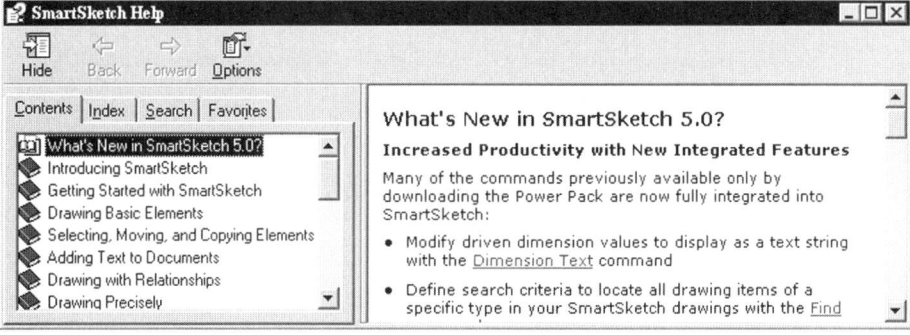

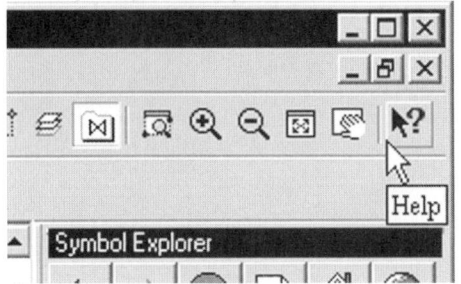

Figure 1–36. Help from the Main toolbar.

Learning Center

To gain access to the tutorials that introduce the key features of Smart-Sketch, you select Help > Learning SmartSketch to display details of the Learning Center. (See Figure 1–37.)

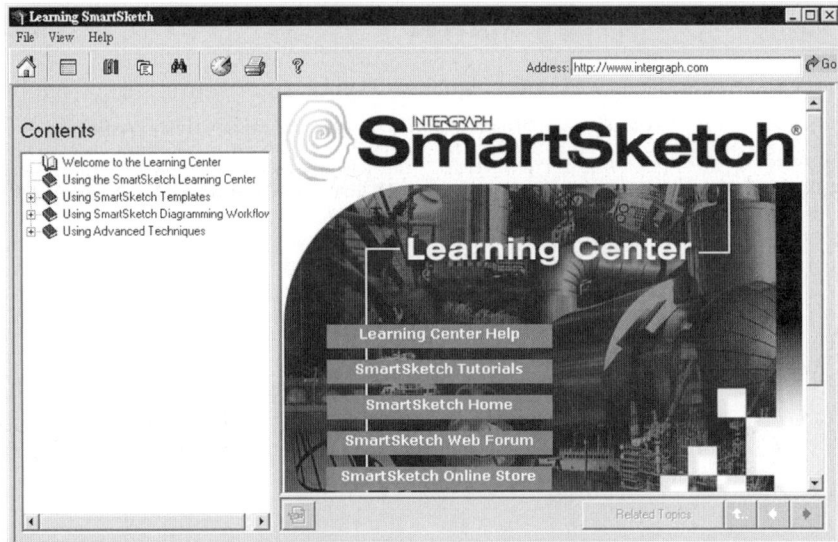

Figure 1–37. Learning Center.

Tip of the Day

The Tip of the Day dialog box shown in Figure 1–38 is accessed by selecting Help > Tip of the Day. If the *Show tips at startup* option of the dialog box is checked, the Tip of the Day dialog box will display each time you start SmartSketch.

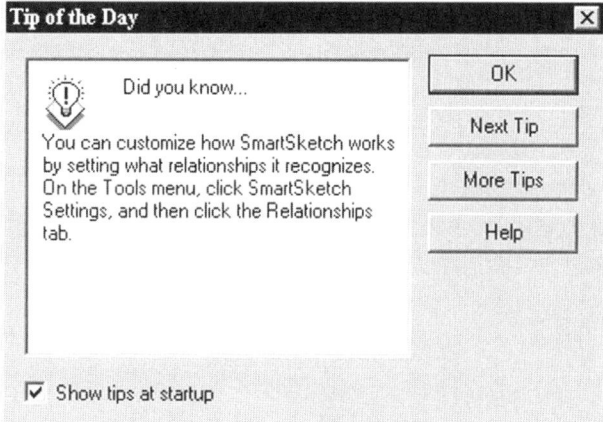

*Figure 1–38.
Tip of the Day
dialog box.*

SmartSketch on the Web

To access the most up-to-date SmartSketch information, select Help > SmartSketch on the Web to browse the SmartSketch web site. (See Figure 1–39.)

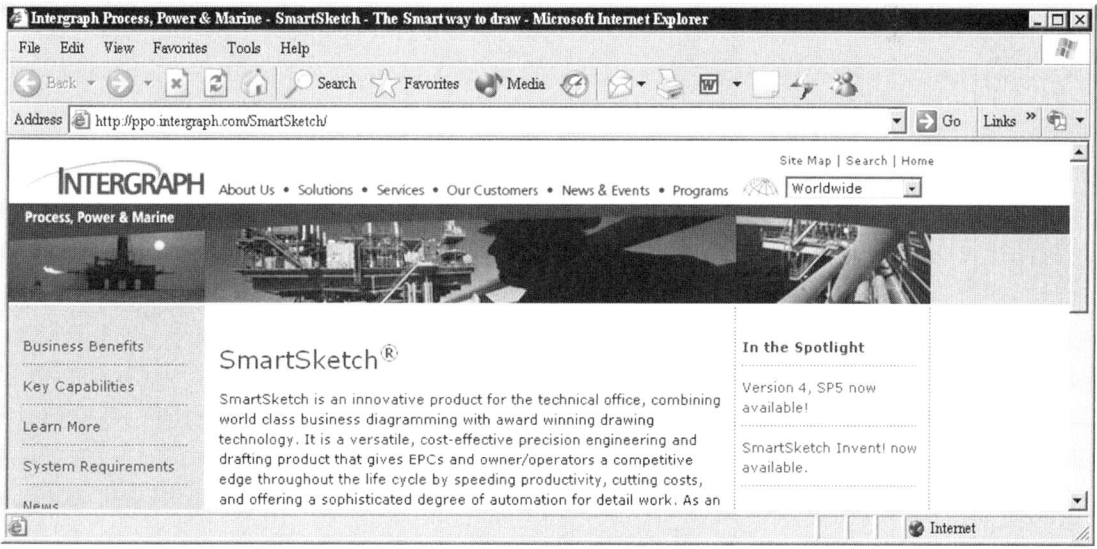

Figure 1–39. SmartSketch web site.

SmartSketch Web Forum

To enter a forum through the Web, select Help > SmartSketch Web Forum. (See Figure 1–40.)

Figure 1–40. Web forum.

■ ■ ■ ■   Summary

SmartSketch is a 2D computer-aided drafting tool for constructing 2D engineering drawings, sketches, and diagrams. The default user interface has a number of areas. From the top, there are the pull-down menus, the Main toolbar, and a context-sensitive Ribbon toolbar (associated with the active command). Below the Ribbon toolbar, from left to right, are the Draw toolbar, a graphics area, the Symbol Explorer, and the Attribute Viewer. A status bar is located at the bottom of the user interface. Like other Windows applications, you can select and drag these objects (except for the pull-down menu, the graphics area, and the status bar) anywhere in the screen.

A SmartSketch data file is called a document. It has an extension of *igr*. In a SmartSketch document, there are two types of drawing sheets on which you construct 2D drawing elements: working sheets and background sheets. Preferably, you should use the working sheets to construct the main constituents of the drawing to depict drawings, sketches, and diagrams. To save time in constructing drawing elements that are common to a number of working sheets, you construct them on a background sheet, link the sheet to respective working sheets, and show the background sheet's drawing elements in the working sheets.

To reuse data constructed by using other computer-aided applications, you open them and save them as SmartSketch documents. On the other hand, you can save SmartSketch documents to other file formats to facilitate downstream computerized operations.

To help use SmartSketch as a tool in drafting, a set of help elements are provided. These include SmartSketch Help, the Learning Center, Tip of the Day, SmartSketch on the Web, and the Web Forum.

■ ■ ■ ■ Review Questions

1 What are the main functions of SmartSketch?

2 Differentiate between a working sheet and a background sheet.

3 State the difference between a toolbar and a ribbon.

4 What help elements are available?

CHAPTER 2

Drafting I

■ ■ ■ ■ **Objectives**

This chapter outlines the key concepts of computer-aided 2D drafting, describes in detail methods of setting up a SmartSketch document, and teaches you how to use SmartSketch as a drafting tool to construct and modify drawing elements. After studying this chapter, you should be able to:

❐ Describe the key concepts of 2D drafting and sketching

❐ Set up a SmartSketch document

❐ Use SmartSketch as a drafting tool to construct and modify drawing elements

Overview

SmartSketch is a 2D computer-aided drafting application, providing a comprehensive set of drafting tools for the construction of 2D engineering drawings, sketches, and diagrams. In this chapter you will learn the basic steps involved in setting up a SmartSketch document and you will practice using SmartSketch as a drafting tool to construct engineering drawings and sketches. As we have explained in Chapter 1, a SmartSketch document can incorporate two types of drawing sheets: working sheets and background sheets. You should construct drawing elements common to a number of working sheets on background sheets and construct the main constituents of drawings, sketches, and diagrams on individual working sheets.

■ ■ ■ ■   Understanding 2D Drafting Concepts

Constructing 2D drawings and sketches in the computer is different from manual drafting using pen and paper. Drawing elements are constructed by selecting appropriate commands from menu and by inputting parameters.

Concepts of Drafting Scale

Before starting to construct a manual drawing, you have to think about the size of the drawing, sketch, or diagram to be constructed and the size of the drawing paper to be used in order to determine the scale of the drawing elements to be constructed on the drawing paper. This thinking process in manual drafting is crucial because if you make a mistake in deciding the scale of the drawing elements you may face difficulties in completing the drawing elements within the confined working area of the selected drawing paper.

Inputting Full Size

In a computer-aided drafting system such as SmartSketch, the drawing elements you construct are defined by mathematical expressions (represented visually in the graphics area) that are stored in the computer's memory in the form of electronic data. As such, drawing elements can be assigned with any numeric value. Different from manual drafting, you do not have to be concerned with the scale of your drawing except when you want to output a hardcopy of it. To preserve original design intent and to avoid confusion, you should always construct the drawing elements in the computer in full size. If a circle is 10,000 units in diameter, you need to specify 10,000 units in the computer. Because the drawing elements are constructed in full size and the graphics area in the SmartSketch window has a fixed viewing size, you may need to zoom in, zoom out, or pan the display to see what you constructed in the computer. Figure 2–1 shows a circle of 10,000 units in diameter fitted to the graphics area.

Drafting Method

Constructing drawing elements in the computer consists of two basic tasks. The first task is to select a command from the menu. The second task is to select a location or a number of locations and specify the parameters. You have to decide what drawing element you want to construct (selecting the command), where to place the drawing element (selecting location), and what size and types you want the drawing element to be (specifying the parameters). The computer-aided application then constructs the drawing elements for you.

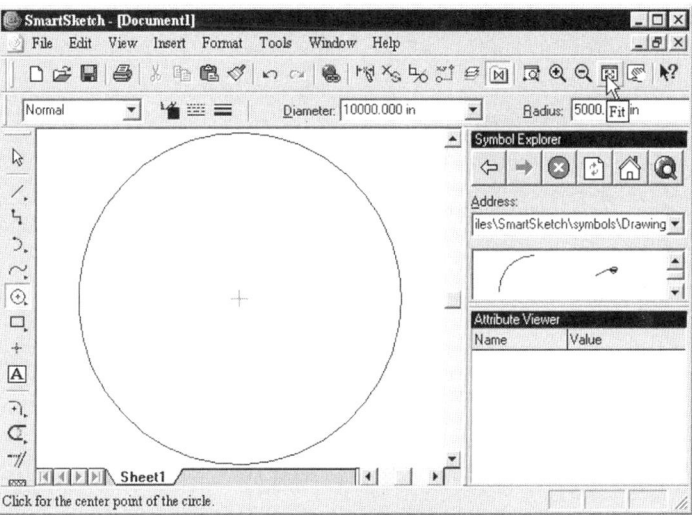

Figure 2–1. Circle of 10,000 units in diameter fitted in the graphics area.

Constructing Drawing Elements

To construct a line, you select a corresponding command and select two points. The system then generates a line joining the two indicated points. To construct a circle, you select a circle command and input the coordinates of the center and a point on the circumference to indicate the radius. In SmartSketch, parameters can be input through the use of command-sensitive ribbons (toolbars or dialogs). Figure 2–2 shows the Circle by Center Point command selected on the Draw toolbar and the Ribbon dialog associated with this command (for inputting the diameter and/or radius of the circle to be created).

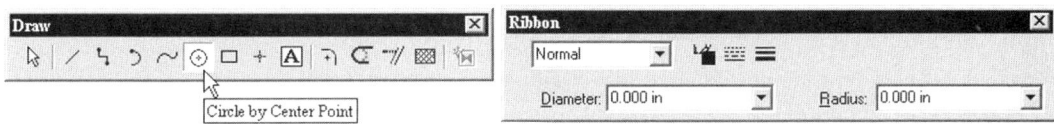

Figure 2–2. Circle by Center Point command and the ribbon associated with this

Modifying Drawing Elements

In manual drafting, modifying an existing drawing can be a tedious job because you have to erase drawing elements drawn on the drawing paper and construct new drawing elements. In addition, the scope and feasibility of modification are in many situations restricted by the drawing scale and paper size already used in making the drawing.

In computer-aided drafting, because both drawing elements and drawing sheets are digital data stored in the database, modifications to drawing elements (as well as to plotting scale and paper size) can be made with little or no problem.

■ ■ ■ ■ Drafting in SmartSketch Documents

In Chapter 1 you learned how to set up a working sheet and a background sheet in a document and how to associate a background sheet to a working sheet. In this chapter you will learn how to construct and modify drawing elements in a SmartSketch document.

To efficiently construct drawing elements in a SmartSketch document you need to first familiarize yourself with the tools available. When you are ready to create a document, you should begin by thinking about how to decompose the engineering drawing or sketch you want to construct into individual drawing elements. You also need to think about what types of tools will be necessary in constructing such drawing elements. You should also consider what tools might be required to modify the drawing elements later in the construction process.

Naturally, the major tools you will use to construct drawing elements are available on the Draw toolbar. To help maintain a proper geometric relationship among the drawing elements, you can use the tools available from the Relationship toolbar and from the Tools pull-down menu. To change the drawing elements you have already constructed, you can use the tools on the Draw toolbar as well as the Change toolbar. Finally, to add dimensions to the drawing elements you use the tools from the Dimension toolbar. Perform the following steps.

1   Select File > New.

2   In the Template area of the New dialog box that appears, select *Technical Drawing (Imperial).igr* from the *Drawing* folder.

3   Click on the OK button. A new technical drawing is started.

The Draw Toolbar

By default, the Draw toolbar is docked vertically at the left side of the SmartSketch window. Like any type of Windows toolbar, you can select and drag the Draw toolbar to place it anywhere you like on the screen. A vertical and a horizontal configuration of the Draw toolbar are shown in Figure 2–3. The toolbar contains 14 icon buttons. Among them, there are a number of fly-outs, denoted by a small black triangular mark at the lower right-hand corner of some of the icons. Commands from the fly-outs are accessible by clicking one of these triangular marks. The functions of the tools of the Draw toolbar are delineated in Table 2–1.

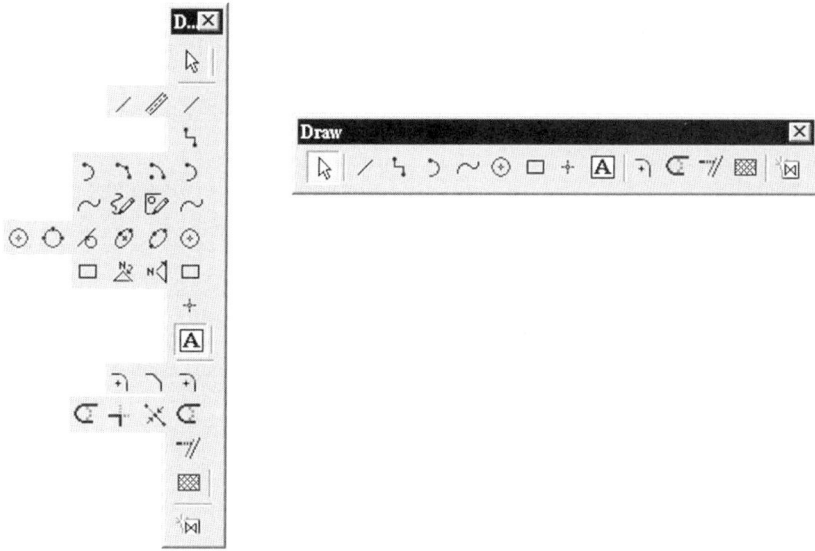

*Figure 2–3.
Draw toolbar.*

Table 2–1 Draw Toolbar Options and Their Functions

Option	Function
Select Tool	Selects objects. The cursor is changed to the arrow-shaped selection cursor with a circle at the end of the cursor depicting the locate zone.
Line/Arc Continuous	Constructs one or a series of connected line or arc segments.
Place Doubleline	Constructs one or a series of connected double lines.
Connector	Draws a series of line segments to connect two elements.
Tangent Arc	Draws an arc tangent or perpendicular to one or two elements.
Arc by 3 Points	Draws an arc by defining three points along the arc.

Option	Function
Arc by Center Point	Draws an arc by defining the center point and the endpoints of the arc.
Curve	Constructs an open or closed curve.
FreeForm	Constructs a free-form curve by sketching.
FreeSketch	Constructs a series of line and arc segments by sketching.
Circle by Center Point	Constructs a circle by specifying the center and the radius.
Circle by 3 Points	Constructs a circle by specifying three points on the circumference of the circle.
Tangent Circle	Constructs a circle tangent to one or two elements.
Ellipse by Center Point	Constructs an ellipse by specifying the center and the axis endpoints.
Ellipse by 3 Points	Constructs an ellipse by specifying two endpoints of an axis and a point on the ellipse.
Rectangle	Constructs a rectangle by specifying two endpoints of an edge and a point on the opposite edge.
Polygon	Constructs a polygon by specifying an edge of the polygon.
Polygon by Center	Constructs a polygon by specifying the center and radius of an inscribed circle.
Point	Constructs a point object.
Text Box	Constructs a text box, which is a rectangular element consisting of text and symbols.
Fillet	Constructs a fillet arc between two elements.
Chamfer	Constructs a chamfer between two elements.
Trim	Trims open and closed elements to the closest intersection in both directions.
Trim Corner	Constructs a corner by extending or trimming two selected elements.
Split	Splits a drawing object into a number of segments of equal length.
Extend to Next	Extends an open element until it meets the nearest element displayed in the graphics window.
Fill	Fills a closed boundary with a selected pattern or solid color.
Create Symbol	Creates a symbol from the selected elements.

To help you understand how to use the Draw toolbar, we will divide the commands here into five groups, as follows.

The first group consists of the Select Tool command only. You click on this command and then select drawing elements from the graphics area.

The second group includes Line/Arc Continuous, Place Doubleline, Connector, Tangent Arc, Arc by 3 Points, Arc by Center Point, Curve, Free-Form, FreeSketch, Circle by Center Point, Circle by 3 Points, Tangent Circle, Ellipse by Center Point, Ellipse by 3 Points, Rectangle, Polygon, Polygon by Center, Point, Text Box, and Fill. Using these commands, you construct various types of drawing elements from scratch.

The third group includes Fillet and Chamfer. Using these commands, you construct new drawing elements from existing elements, as well as modify existing drawing elements.

The fourth group includes Trim, Trim Corner, Split, and Extend to Next. Using these commands, you modify existing drawing elements.

The fifth group consists of the Create Symbol command only. Using this command, you construct a symbol from existing drawing elements.

In conjunction with the Draw toolbar, you need to use the status bar, the cursor, the ribbons associated with some of the commands of the Draw toolbar, the PinPoint ribbon, the Relationships toolbar, the SmartSketch settings, the Relationship handles, and the Alignment Indicator. These elements are explained in the sections that follow.

Using the Status Bar and Cursor

To construct a drawing element using the commands on the Draw toolbar, you select an icon on the Draw toolbar and specify the locations and parameters of the drawing elements to be constructed. After you select a command, you should pay attention to the messages provided in the status bar and respond accordingly. Perform the following steps to construct a rectangle.

1 Select Rectangle on the Draw toolbar. At the status bar, you will see the prompt "Click for the first point."

2 Select location A indicated in Figure 2–4. At the status bar, you will see the prompt "Click for the second point."

3 Select location B indicated in Figure 2–4. At the status bar, you will see the prompt "Click to create the rectangle."

4 Select location C indicated in Figure 2–4. A rectangle is constructed.

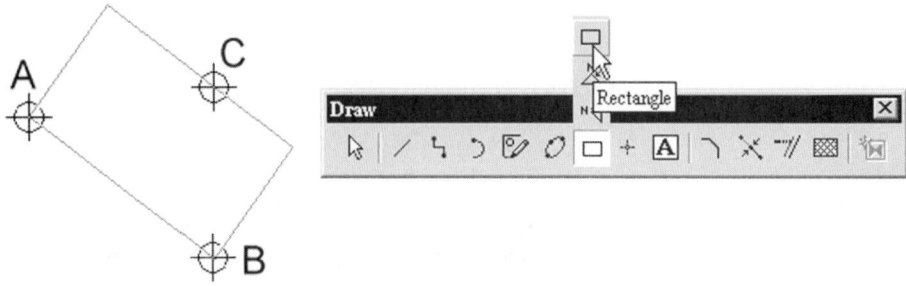

Figure 2–4. Rectangle being constructed.

Using the Associated Ribbons

To construct drawing elements with precise dimensions, you can use the ribbon associated with most of the commands (except the Trim, Trim Corner, Extend to Next, and Create Symbol commands) available from the Draw menu. To display the ribbon, the Ribbon check box of the Toolbars dialog box (accessible from the View > Toolbars menu) has to be selected. Perform the following steps to construct a precise rectangle by using the ribbon.

1   Select View > Toolbars to display the Toolbars dialog box.

2   In the Toolbars dialog box, click the Ribbon option in the Toolbars list (if it is not already selected) and then click on the OK button.

3   Select Rectangle on the Draw toolbar. The Rectangle command's ribbon is displayed. (See Figure 2–5.)

4   In the ribbon, set the rectangle's width to 4 inches, height to 3 inches, and inclination angle to 30 degrees.

Figure 2–5. Using the ribbon associated with the Rectangle command.

5   Select location A, shown in Figure 2–5, to indicate the corner position. A rectangle that measures 4 inches by 3 inches and is inclined at an angle of 30 degrees is constructed.

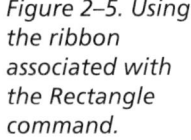

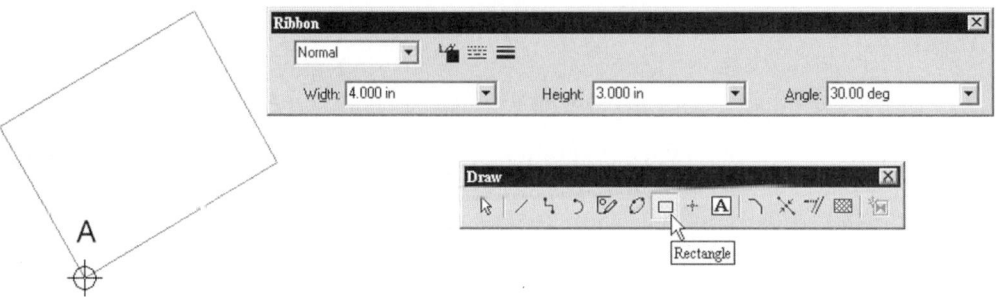

Using the PinPoint Ribbon

PinPoint is a tool that helps you specify a precise location in the graphics area. It shows you the coordinates of the cursor at the cursor and enables you to input a precise location in the dialog box. Let's display the PinPoint dialog box.

1   Click on the Tools pull-down menu and find out whether there is a tick mark prefixing the PinPoint item or not. If there is no tick mark, click PinPoint.

Another way to display the PinPoint ribbon quickly is to click on the PinPoint button on the Main toolbar. The PinPoint ribbon, shown in Figure 2–6, contains a number of buttons and boxes. Table 2–2 outlines the options available.

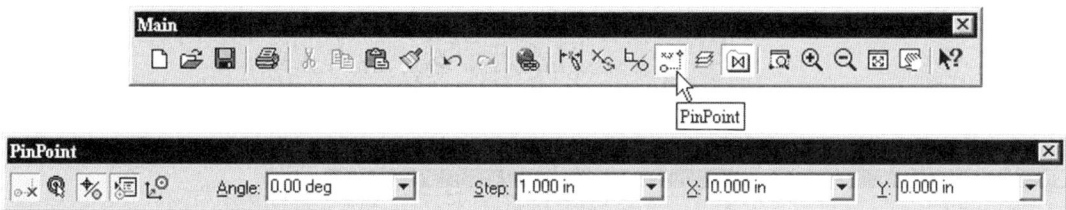

Figure 2–6. Pinpoint ribbon.

Table 2–2 PinPoint Ribbon Options and Their Functions

Option	Function
Display On/Off	Turns on or off coordinate display at the cursor
Reposition Target	Repositions the target point (origin for coordinate measurement) by selecting a new point from the graphics area
Relative Tracking	Sets pinpoint coordinate display relative to the last point clicked during a draw command
Define PinPoint Origin	Displays the Define PinPoint Origin ribbon, enabling you to reposition the target point by inputting coordinate values and saving the target location in the memory
Reposition Target on Origin	Repositions the target point to the saved target point, saved by clicking the Save PinPoint Origin button

Option	Function
Angle	Defines the orientation of the X axis of the pinpoint coordinate system
Step	Defines the pinpoint's step value
X	Defines the X coordinate of point entry for the current draw command
Y	Defines the Y coordinate of point entry for the current draw command

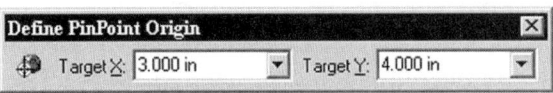

Figure 2–7. Define PinPoint Origin dialog box.

Coordinate measurement at the cursor can be taken relative to either the target point (origin) or the last selected point. By default, the target is located at the lower left-hand corner of the drawing sheet. You can select a target point by clicking on the graphics area. You can define a target point by using the Define PinPoint Origin dialog box, shown in Figure 2–7.

Continue the following steps to construct a circle using the PinPoint ribbon and the ribbon associated with the Circle by Center Point command.

2   In the PinPoint dialog box, click on the Display On/Off button on the PinPoint ribbon, if it is not already selected.

3   Select Insert > New Sheet to insert a new drawing sheet.

4   Click on the Circle by Center Point icon on the Draw toolbar. In the graphics area, you will find the default target point at (0,0) highlighted by a red circle, coordinates displayed at the cursor, and a pair of horizontal and vertical dotted lines connecting the cursor and the target point.

5   In the ribbon associated with the Circle by Center Point command, set the diameter of the circle to 3 inches.

6   In the PinPoint ribbon, set the X value to 4 inches and Y value to 5 inches. This specifies the coordinates of the center point of the circle to be constructed.

7   Left click on the graphics area. A circle of 3 inches in diameter located at 4 inches from the origin in the X direction and 5 inches from the origin in the Y direction is constructed. (See Figure 2–8.)

Continue with the following steps to have a target origin saved in memory and to reposition the target point.

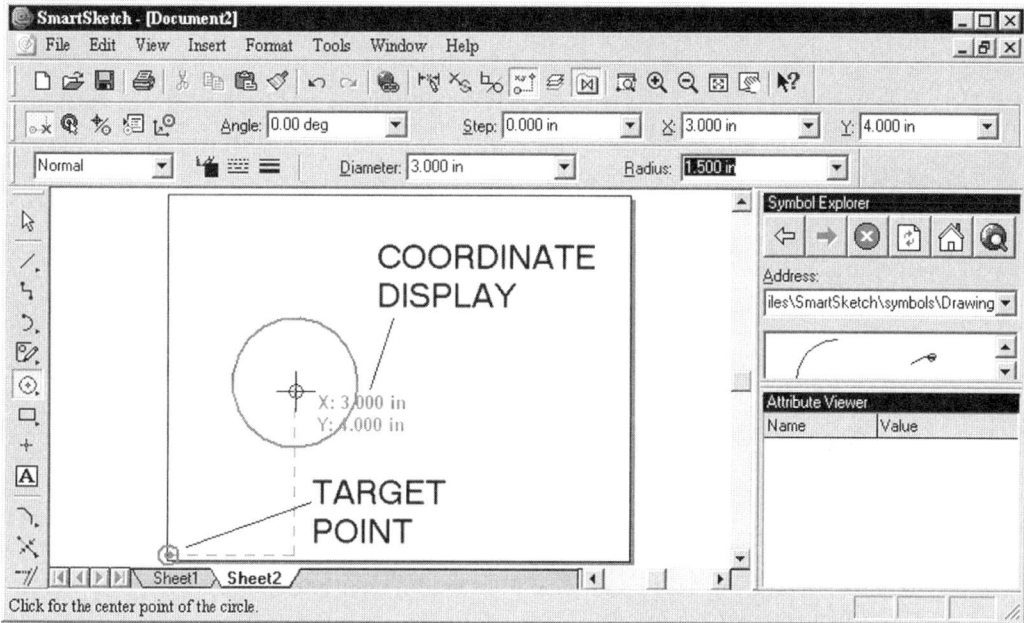

Figure 2–8. Coordinates displayed at the cursor as the Circle command is activated.

8 Select Define PinPoint Origin on the PinPoint ribbon to display the Define PinPoint Origin dialog box.

9 In the Define PinPoint Origin dialog box, set both X and Y values to 0 and click on the Save PinPoint Origin button. The target origin is saved in memory. (See Figure 2–9.)

10 Click on Define PinPoint Origin on the PinPoint ribbon to deselect it, thereby closing the Define PinPoint Origin dialog box.

11 Select Reposition Target on the PinPoint ribbon and move the cursor to center point A indicated in Figure 2–10 until an icon in the shape of a circle is displayed at the cursor, depicting that a center point being selected. (Note: Details regarding relationship will be discussed later in this chapter.)

12 Left click while the circle icon is displayed. The target is repositioned to the center of the circle.

Continue with the following steps to construct a line segment.

13 Click on the Line/Arc Continuous icon on the Draw toolbar.

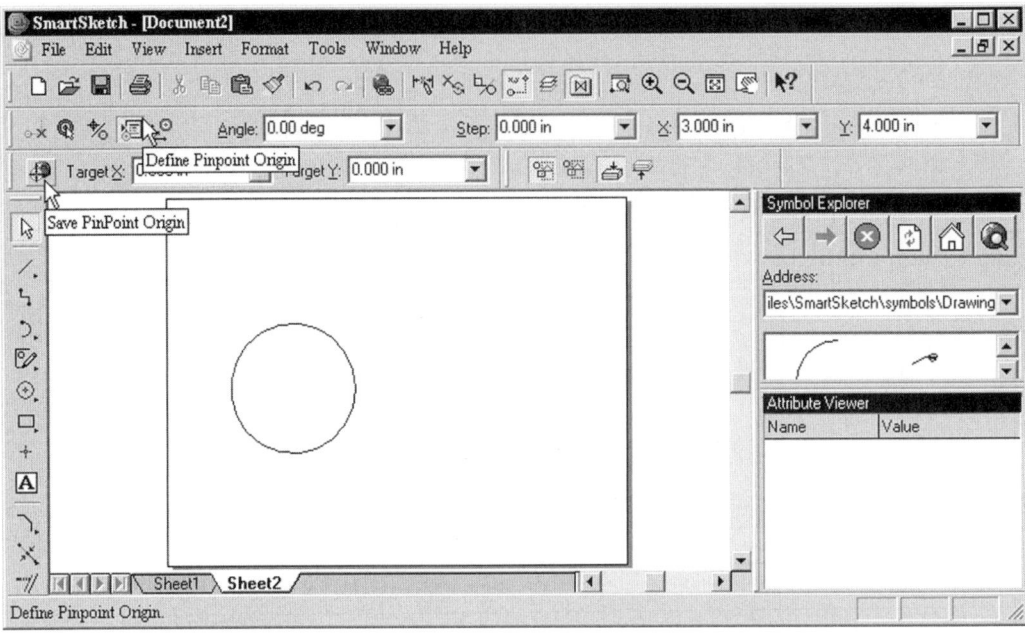

Figure 2–9. Target origin set and saved.

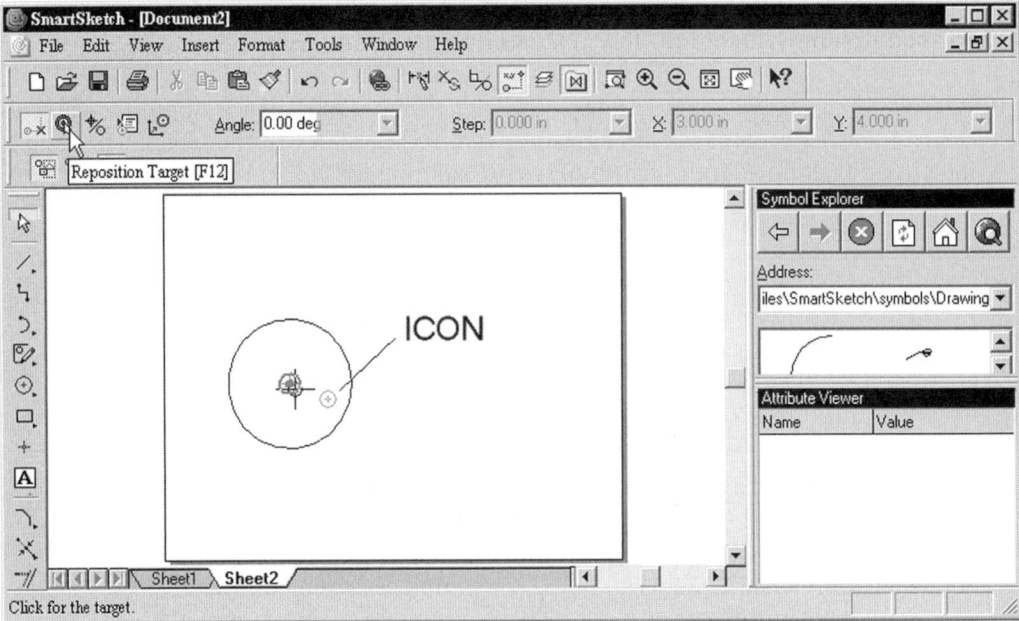

Figure 2–10. Target origin repositioned.

14 Move the cursor to location X:2/Y:2 (inches), as shown in Figure 2–11, and left click.

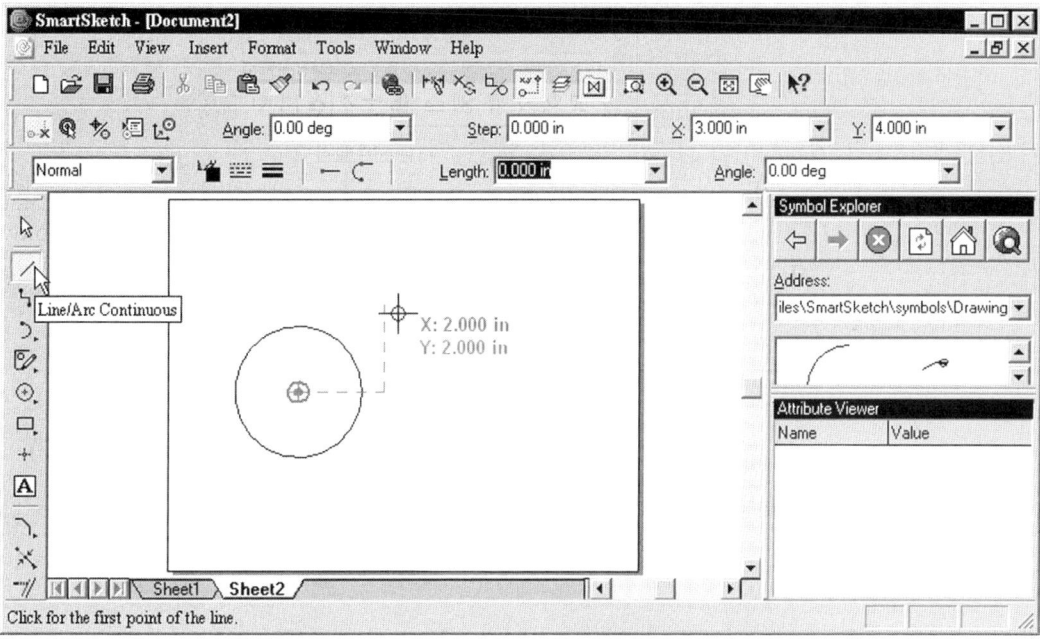

Figure 2–11. First endpoint of a line being selected.

15 With reference to Figure 2–12, move the cursor to location X:4/Y:-1 (inches) and left click. A line segment is constructed. Note that the coordinate display is relative to the target point.

16 Right click to terminate the Draw command.

17 Press the Esc key to exit the Draw command.

18 Click on Relative Tracking on the PinPoint ribbon to activate relative tracking.

19 Click on the Line/Arc Continuous icon on the Draw toolbar.

20 Move the cursor to location A indicated in Figure 2–13 until an icon in the shape of an arrowhead is displayed (depicting an endpoint being selected), and then left click while the icon is displayed. Endpoint A is selected.

21 Move the cursor to location X:-2/Y:-3 (inches) as indicated in Figure 2–14. Note that the coordinate X:-2/Y:-3 is relative to the last selected point, which is the start point of the line segment being constructed.

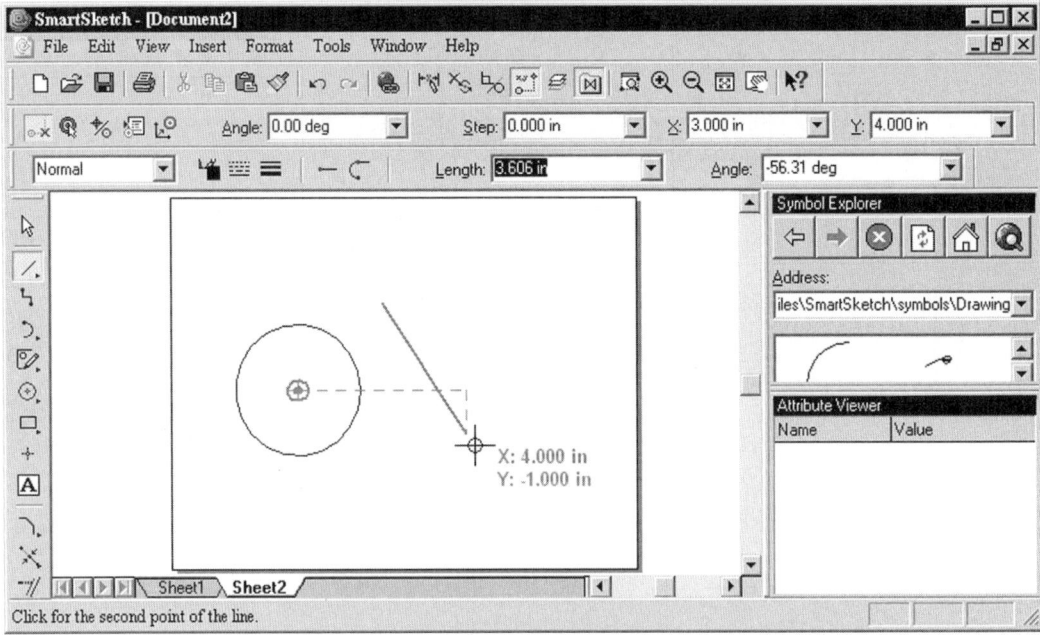

Figure 2–12. Second endpoint of a line being selected.

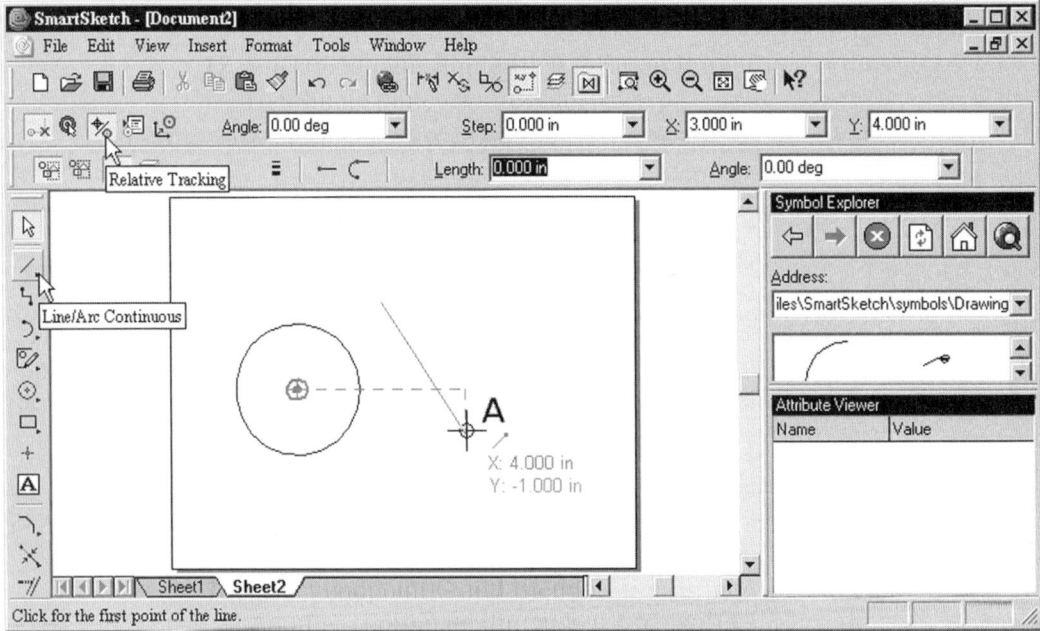

Figure 2–13. Endpoint of a line being selected.

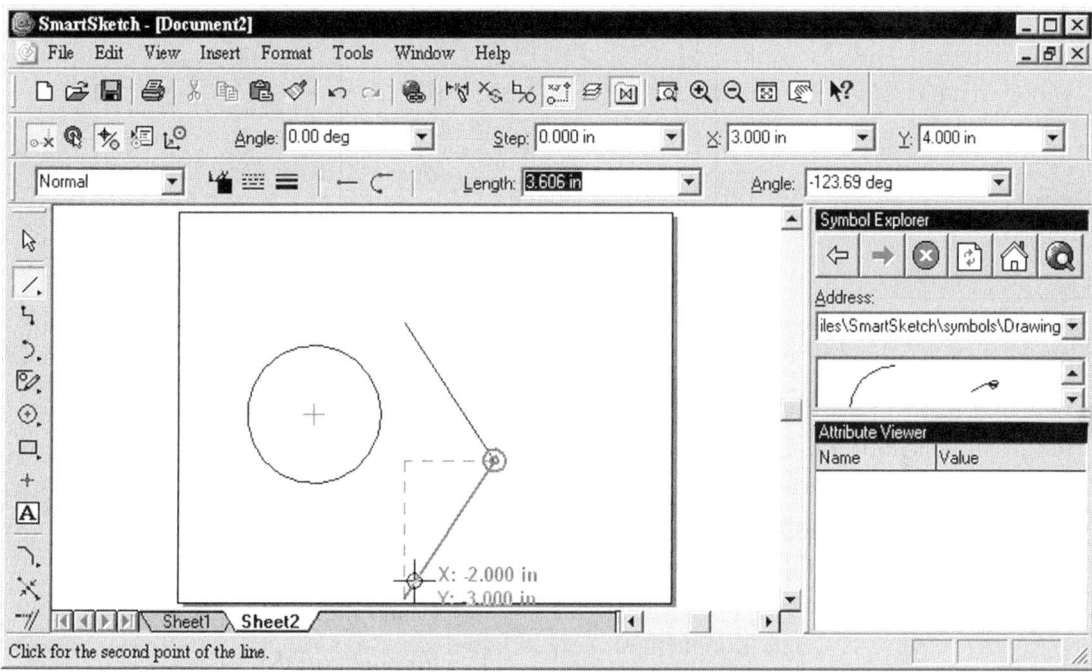

Figure 2–14. Line segment being constructed.

22 Left click to specify the endpoint.

23 Right click to terminate the Draw command.

24 Press the Esc key to exit the Draw command.

25 Save and the file as *TechDraw1.igr* and then close the file.

Moving Sheet Border

As we have mentioned, the default origin is located at the lower left-hand corner of the drawing sheet. To move the location of the sheet's origin, you select File > Move Sheet Border. In the Move Sheet Border dialog box, you can specify a new origin by inputting the coordinates or by clicking on the Interactive Move button of the Move Sheet Border dialog box and selecting a location. Figure 2–15 shows the Move Sheet Border dialog box.

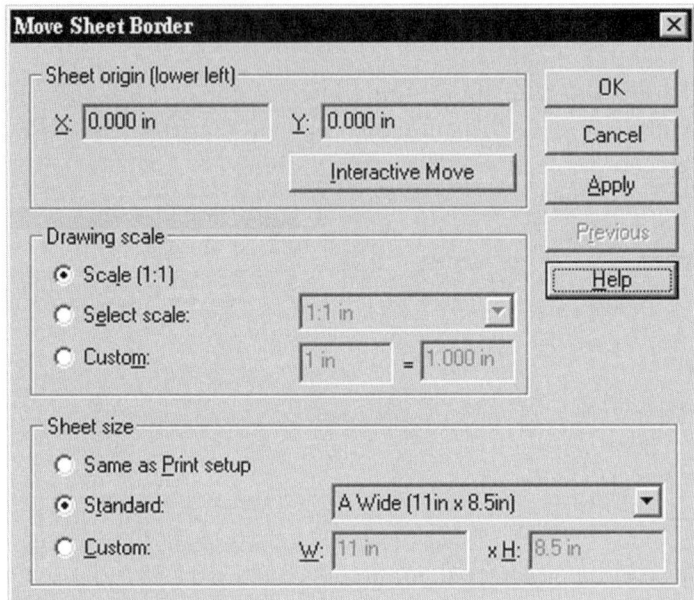

*Figure 2–15. Move
Sheet Border dialog
box.*

Manipulating Relationships and SmartSketch Settings

The term *relationship* refers to how a drawing element's geometric attributes are related to other drawing elements, such as tangency or parallelism with another element. It also refers to the geometric attributes of the drawing element itself, such as whether it is horizontal or vertical.

Relationships can be incorporated in drawing elements in two ways. The first way is to predefine the type of relationship to be incorporated and have the relationship incorporated as you construct the drawing elements. The second way is to add the relationships after the drawing elements are constructed. Perform the following steps to incorporate relationships while constructing drawing elements.

1   Start a new document by selecting File > New.

2   In the New dialog box, select *Technical Drawing (Imperial).igr* from the template list and then click on the OK button.

3   Select Tools > SmartSketch Settings.

4   Select the Relationships tab, if it is not already selected. (See Figure 2–16.)

As shown in the dialog box, there are nine relationship options: Intersection, *End point*, Midpoint, *Center point*, *Point on element*, *Horizontal or vertical*, Parallel, Perpendicular, and Tangent. Each of these relationship options is depicted by a symbol next to the check box. To apply a relationship while you construct a drawing element, you have to click the appropriate check boxes

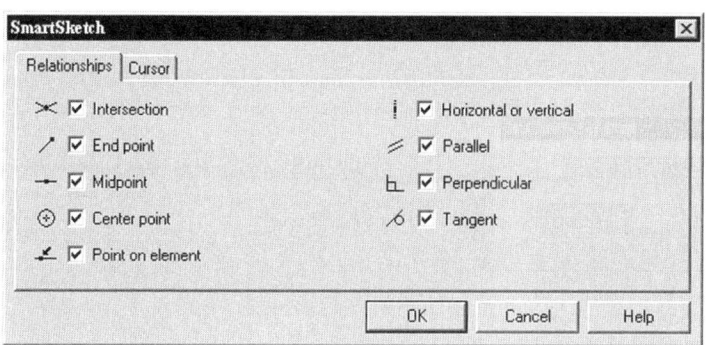

Figure 2–16. Relationships tab of the SmartSketch dialog box.

beforehand. For example, if the *End point* box is checked and you click a point near the endpoint of a drawing element as you construct a line, the endpoint of the constructed line will coincide with the endpoint of the existing drawing element. If the *Horizontal or vertical* box is checked, lines will become horizontal if they are drawn to a nearly horizontal position.

5 Check all boxes on the Relationships tab, if they are not already checked.

You might ask how near the cursor needs to be to an existing drawing element for a relationship to be established between that drawing element and an element to be constructed. The answer is that this is determined by settings you establish in the Cursor tab of the SmartSketch dialog box.

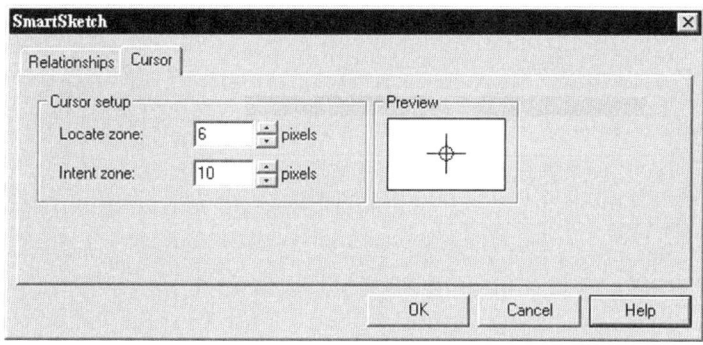

Figure 2–17. Cursor tab of the SmartSketch dialog box.

6 Select the Cursor tab of the SmartSketch dialog box. (See Figure 2–17.)

The Cursor tab contains two settings: *Locate zone* and *Intent zone*. The *Locate zone* setting establishes the radius of a circle around the cursor point in terms of pixel numbers. For example, if the *Locate zone* setting is 6 pixels, the cursor will snap to an endpoint if the cursor is within a distance of 6 pixels from an endpoint. The *Intent zone* setting establishes the size of the zone within which you want the drawing commands to interpret the selected intention as you construct.

7 Set *Locate zone* to 6 pixels and *Intent zone* to 10 pixels, if they are not already set to these values.

8 Click on the OK button.

9 Click on the Line/Arc Continuous icon on the Draw toolbar.

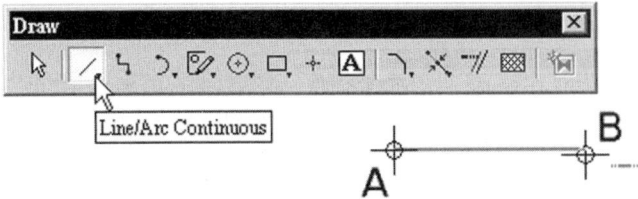

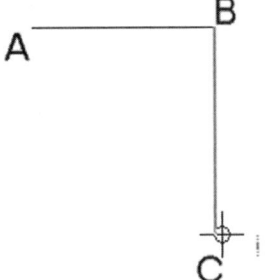

Figure 2–18. Horizontal line being constructed.

10 Click on location A indicated in Figure 2–18. (Note: Exact location is unimportant for this practice session.)

11 Move the cursor slowly to the right in a nearly horizontal position. You will find the horizontal symbol at the cursor. (Refer to the symbol shown in Figure 2–18.)

12 Click a point while the horizontal symbol is displayed. A horizontal line is constructed.

13 Move the cursor slowly to C indicated in Figure 2–19 in a nearly vertical direction until you find a vertical symbol next to the cursor.

Figure 2–19. Vertical line constructed.

14 Click a point while the vertical symbol is displayed. Vertical line BC is constructed. (See Figure 2–19.)

15 With the Line command still activated, move the cursor along line AB. You will find the Point on Element symbol displayed at the cursor. (See Figure 2–20.) Do not click on the line. (Note: If you click while the Point on Element symbol is displayed, the clicked point will lie on the selected element.)

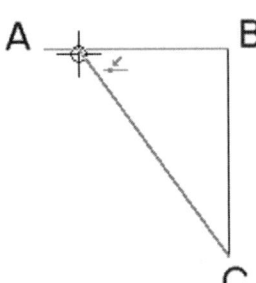

Figure 2–20. Point on Element symbol displayed at the cursor.

16 Move the cursor slowly to endpoint A of line AB until the *End point* symbol is displayed at the cursor. (See Figure 2–21.)

17 While the *End point* symbol is displayed, click on endpoint A. Line segment CA is constructed.

18 With the Line command still activated, move the cursor near the midpoint of line AB until the Midpoint symbol is displayed at the cursor.

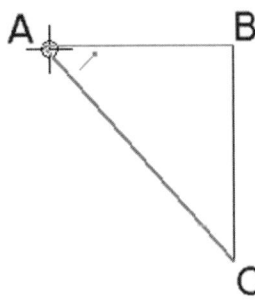

Figure 2–21. End point *symbol displayed at the cursor.*

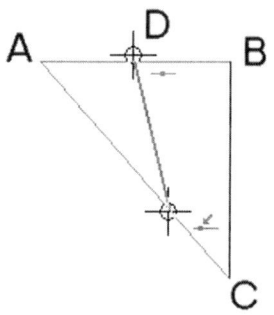

Figure 2–22. Midpoint of line AB selected and line AC being selected.

19 Click on line AB while the Midpoint symbol is displayed.

20 Move the cursor over line AC but do not click on it. (See Figure 2–22.)

21 Move the cursor to location E indicated in Figure 2–23 until the Parallel symbol is displayed at the cursor.

22 Click on the graphics area while the Parallel symbol is displayed. Line DE parallel to line AC is constructed.

23 Move the cursor to location F indicated in Figure 2–24 until the Perpendicular symbol is displayed at the cursor.

24 While the Perpendicular symbol is displayed, click on the graphics area. Line ED perpendicular to line AC is constructed.

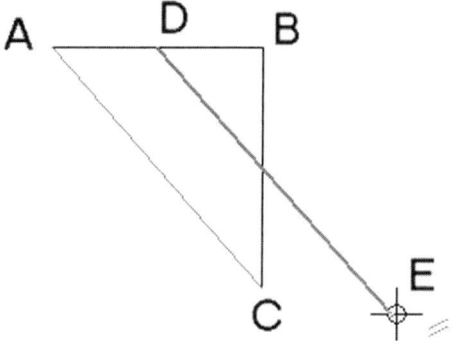

Figure 2–23. Parallel line being constructed.

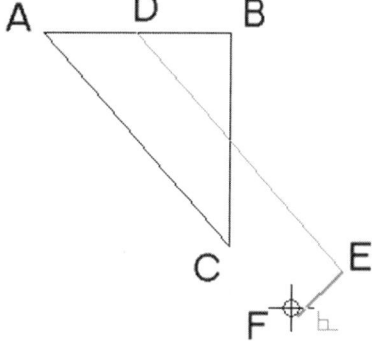

Figure 2–24. Perpendicular line being constructed.

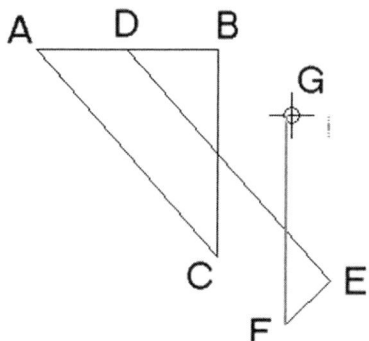

Figure 2–25. Vertical line being constructed.

25 Move the cursor vertically to location G indicated in Figure 2–25 until the Vertical symbol is displayed at the cursor.

26 Click on the graphics area while the Vertical symbol is displayed. A vertical line is constructed. Right click to terminate the command.

27 Click on the Circle by Center Point icon on the Draw toolbar.

28 Move the cursor over line GF near the intersection of the line with line DE, but do not click on the line. (See Figure 2–26.)

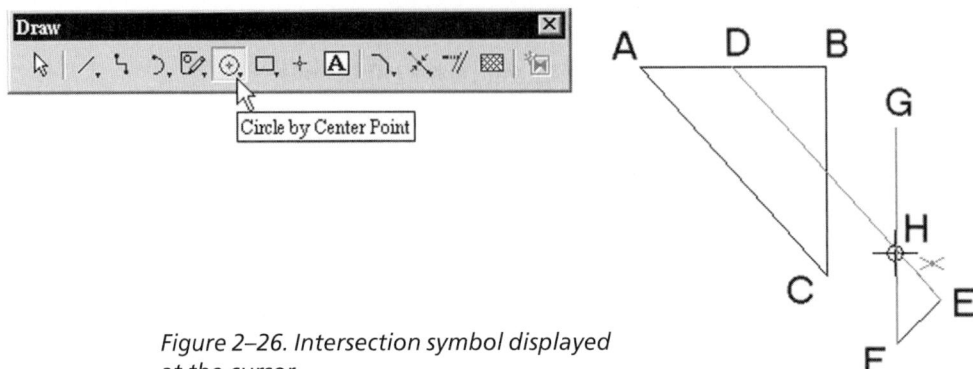

Figure 2–26. Intersection symbol displayed at the cursor.

29 Move the cursor over line DE and then near the intersection until you find the Intersection symbol displayed at the cursor.

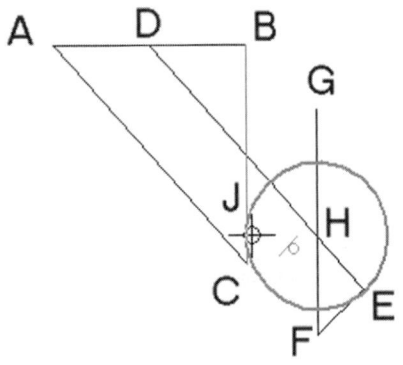

Figure 2–27. Tangent symbol displayed at the cursor.

30 While the Intersection symbol is displayed, left click to select the intersection point. The center of the circle is selected at the intersection of lines DE and GF.

31 Move the cursor near line BC until the Tangent symbol is displayed at the cursor.

32 Click on line BC while the Tangent symbol is displayed. A circle tangent to line BC is constructed. (See Figure 2–27.)

Continue with the following steps to construct two circles concentric to each other.

33 With reference to Figure 2–28, construct circle A. Exact location and size are unimportant.

34 Move the cursor over the center of circle A until the Center Point symbol is displayed at the cursor.

35 While the Center Point symbol is displayed, click on the graphics area. The center point is selected.

36 Select point C indicated in Figure 2–29. A circle is constructed.

37 Save the file as *TechDraw2.igr* and then close the file.

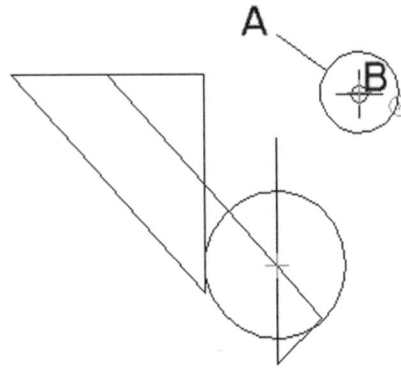

Figure 2–28. Center Point symbol displayed at the cursor.

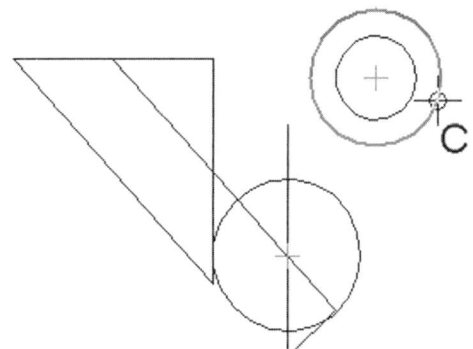

Figure 2–29. Circle being constructed.

Using the Relationship Toolbar

To incorporate relationships after you construct drawing elements, you use the Relationship toolbar. To display the Relationship toolbar, select View > Toolbars to access the Toolbar dialog box, or click on the Relationship icon on the Main toolbar. The Relationship toolbar is shown in Figure 2–30, and the functions associated with its options are outlined in Table 2–3.

Figure 2–30. Relationship toolbar.

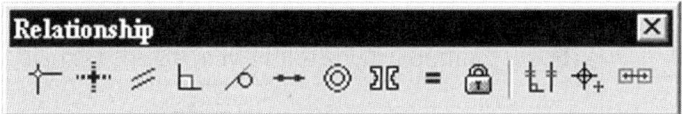

Table 2–3 Relationship Toolbar Options and Their Functions

Option	Function
Connect	Connects two drawing elements.
Horizontal/Vertical	Sets a line horizontal or vertical or aligns two key points horizontally or vertically.
Parallel	Sets two lines parallel to each other.
Perpendicular	Sets a line perpendicular to another line.
Tangent	Sets two drawing elements tangent to each other.
Collinear	Sets two lines collinear to each other.
Concentric	Sets two arcs or circles concentric to each other.

Option	Function
Symmetric	Sets drawing elements symmetric about an axis.
Equal	Sets the length of two lines or the radii of two arcs or circles equal.
Lock	Locks the length, angle, radius, or position of a drawing element.
Relationship Handles	Displays relationship handles on drawing elements. Relationship handles are graphic symbols showing the type of relationship maintained.
Alignment Indicator	Displays dashed horizontal and vertical lines to depict horizontal or vertical alignment when you draw or modify drawing elements.

Perform the following steps to learn how to incorporate relationships after drawing elements are constructed.

1   Start a new document by selecting File > New.

2   In the New dialog box, select *Technical Drawing (Imperial).igr* from the template list and click on the OK button.

3   Click on the Line/Arc Continuous icon on the Draw toolbar.

4   With reference to Figure 2–31, construct three line segments: AB, CD, and EF. The exact length and location of the lines are unimportant.

5   If the Relationship toolbar is not yet displayed, select View > Toolbars, select Relationship from the list of the Toolbar dialog box, and then click on the OK button.

6   Click on the Connect icon on the Relationship toolbar.

7   Move the cursor near endpoint A until you see the *End point* symbol displayed.

8   Click on endpoint A while the *End point* symbol is displayed.

9   Move the cursor near endpoint C until you see the *End point* symbol displayed.

10 Click on endpoint B while the *End point* symbol is displayed.

Line AB is changed, with the endpoint A coincident with endpoint C of line CD. The sequence of selection is important. If you click on endpoint C and then endpoint A, line CD will be changed (with endpoint C moved to endpoint A).

11 Click on the Horizontal/Vertical icon on the Relationship toolbar.

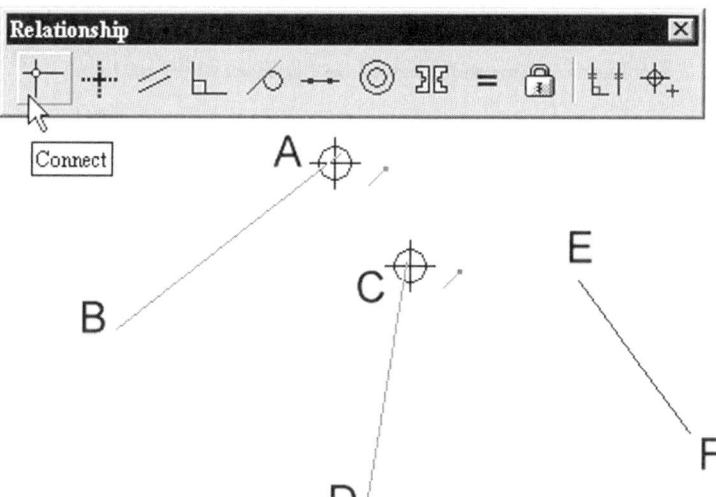

Figure 2–31. Line segments constructed.

12 Select line EF indicated in Figure 2–32. The line becomes vertical. Note that this command will cause a line to be either horizontal or vertical, depending on the inclination angle of the line. If the line is more inclined to vertical (as in this case), the line becomes vertical.

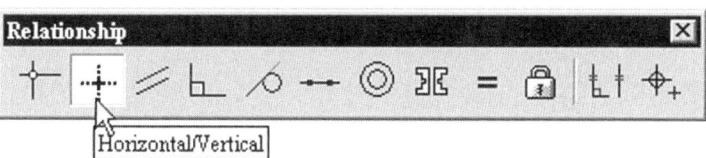

Figure 2–32. Horizontal/vertical relationship being applied to a line.

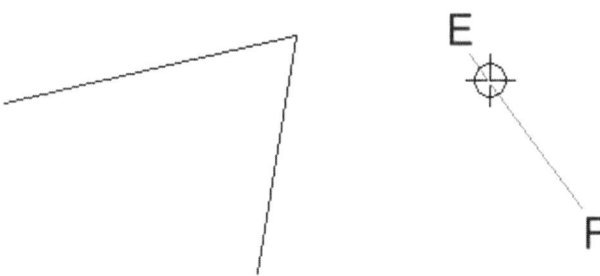

Maintaining Relationships and Displaying Relationship Handles

Relationships incorporated in drawing elements can be either temporary or persistent, depending on the settings established in the Tools pull-down menu. If you select Tools > Maintain Relationships to apply a tick mark prefixing the menu item, relationships incorporated thereafter will be persistently incorporated in the drawing elements. If you select Tools > Relationships again to remove the tick mark, relationships incorporated thereafter are temporary.

To discover whether there are any persistent relationships incorporated in drawing elements, select Tools > Relationship Handles to apply a tick mark prefixing the menu item. Appropriate relationship handles (symbols) depicting the type of relationships persistently incorporated in the drawing elements will be displayed. To remove any of these persistent relationships, select the symbol and press the Delete key. If you wish to hide the relationship symbols, select Tools > Relationship Handles again to remove the tick mark. Perform the following steps.

1   Select Tools > Relationship Handles, if there is no tick mark prefixing the menu item. As there are no persistent relationships incorporated in the drawing elements, there are no relationship handles displayed. Alternatively, you may click on the Relationship Handles icon on the Relationship toolbar.

2   Click on the Parallel icon on the Relationship toolbar.

3   With reference to Figure 2–33, select line AB and then line CD. Line AB is changed. Because the coincident relationship between endpoint A and endpoint C is not persistent, endpoint A of line AB moves away from endpoint C of line CD. (See Figure 2–34.)

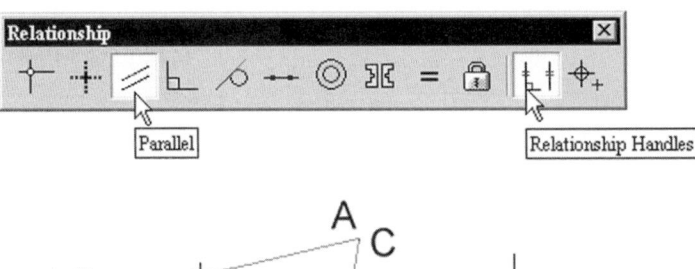

Figure 2–33. Setting line AB parallel to line CD.

Locking causes a drawing element to be fixed in its position. Continue with the following steps to lock a drawing element.

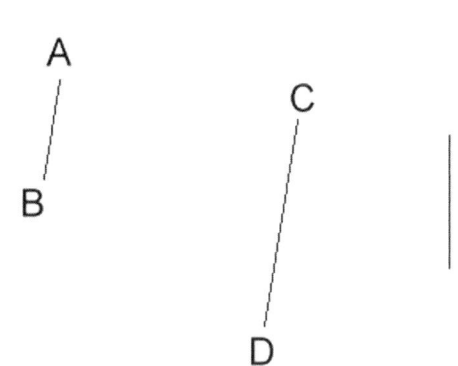

Figure 2–34. Line AB changed.

4 Click on the Lock icon on the Relationship toolbar.

5 Select line AB indicated in Figure 2–35. Line AB is locked. Note the lock symbol attached to the line.

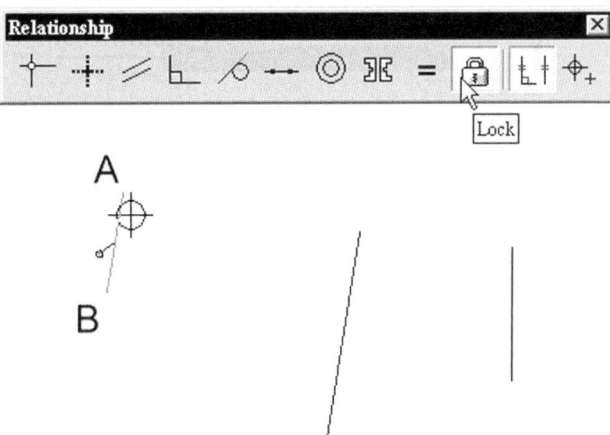

Figure 2–35. Line AB locked.

To apply persistent relationships to drawing elements, continue with the following steps.

6 Select the Tools pull-down menu.

7 Click on Maintain Relationship, if there is no tick mark prefixing it.

8 Click on the Perpendicular icon on the Relationship toolbar.

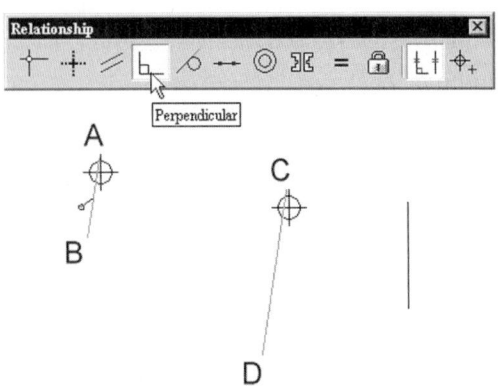

Figure 2–36. Line CD being made perpendicular to line AB.

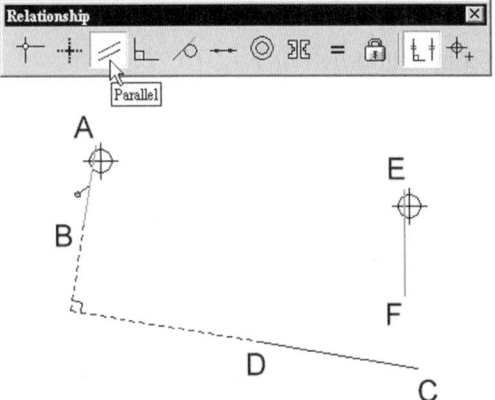

Figure 2–37. Line EF being made parallel to line AB.

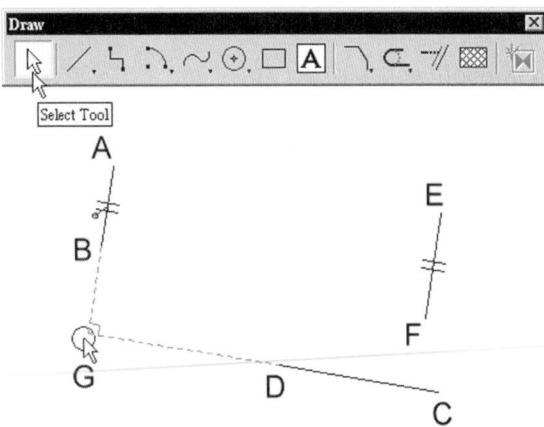

9 Select lines AB and CD indicated in Figure 2–36. Line CD is now perpendicular to line AB. Note the perpendicular symbol and the dashed line connecting lines AB and CD in Figure 2–37.

10 Click on the Parallel icon on the Relationship toolbar.

11 Select lines AB and EF indicated in Figure 2–37. Line EF is now parallel to line AB. Note the parallel symbols on lines AB and EF in Figure 2–38.

Now you will remove a persistent relationship. Continue with the following steps.

12 Click on the Select Tool icon on the Draw toolbar and select the perpendicular symbol G shown in Figure 2–38.

13 Press the Delete key. The perpendicular relationship between lines AB and CD is removed.

Now that the perpendicular relationship is removed, continue with the following steps to add a horizontal/vertical relationship.

14 Click on the Horizontal/Vertical icon on the Relationship toolbar.

15 Select line CD indicated in Figure 2–39. Line CD becomes horizontal. (See Figure 2–40.)

16 Save the file as *TechDraw3.igr* and then close the file.

Figure 2–38. Perpendicular relationship being removed.

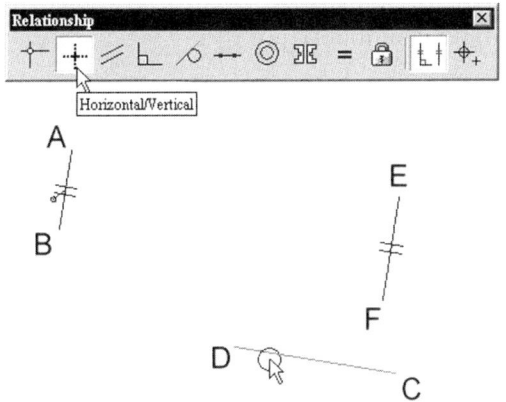

Figure 2–39. Perpendicular relationship
removed and horizontal/vertical
relationship being added.

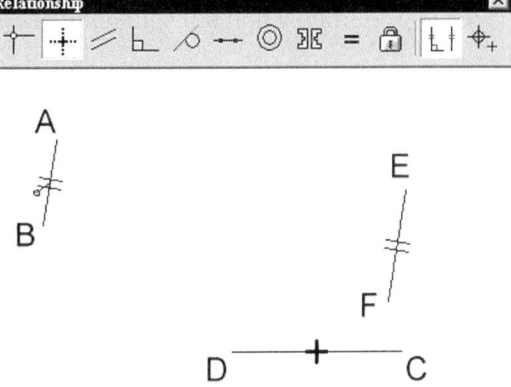

Figure 2–40. Line made horizontal.

Using the Alignment Indicator

To help visualize and indicate horizontal or vertical alignment while you are constructing drawing elements, you can activate the Alignment Indicator, which is a set of horizontal and vertical dashed lines. To activate the Alignment Indicator, select Tools > Alignment Indicator. Because this menu item is a toggle button, selecting Tools > Alignment Indicator again will turn off the Alignment Indicator. Perform the following steps.

1   Start a new document by selecting File > New.

2   In the New dialog box, select *Technical Drawing (Imperial).igr* from the template list and then click on the OK button.

3   Select the Tools pull-down menu. If there is no tick mark prefixing the menu item Alignment Indicator, click on it to activate the Alignment Indicator.

4   Click on the Rectangle icon on the Draw toolbar.

5   With reference to Figure 2–41, click on the graphics area at A, move the cursor slowly to the right until the Horizontal/Vertical symbol is displayed at the cursor, and click on B while the Horizontal/Vertical symbol is displayed.

6   Click on location C indicated in Figure 2–42 to complete the rectangle's definition.

7   Click on the Circle by Center Point icon on the Draw toolbar.

Figure 2–41. Endpoint positions of an edge of a rectangle being selected.

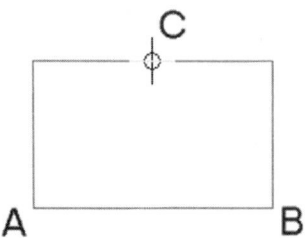

Figure 2–42. Rectangle constructed.

8 Move the cursor near midpoint A of the lower edge of the rectangle until the Mid Point symbol is displayed at the cursor. (See Figure 2–43.) Do not click on the rectangle.

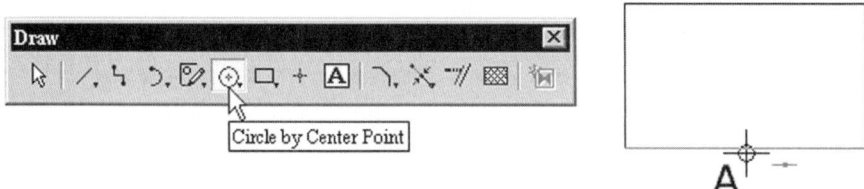

Figure 2–43. Mid Point symbol displayed at the cursor.

9 Move the cursor vertically upward. You will find a vertical dashed line joining the cursor to the midpoint of the lower edge of the rectangle. (See Figure 2–44.) Again, do not click on the graphics area.

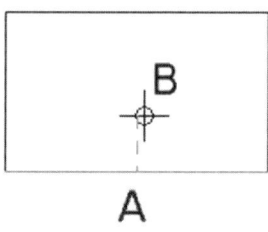

Figure 2–44. Vertical dashed line connecting the cursor and the midpoint of an edge of the rectangle.

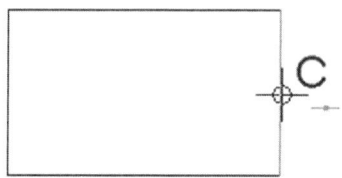

Figure 2–45. Mid Point symbol displayed at the cursor.

10 With reference to Figure 2–45, move the cursor near midpoint C of one of the vertical edges of the rectangle. Do not click on the rectangle.

11 Move the cursor slowly to the left in a horizontal direction. You will see a dashed horizontal line connecting the cursor and the midpoint of a vertical edge of the rectangle. (See location D in Figure 2–46.) Do not click on the graphics area.

12 Move the cursor to location E until you see both the vertical dashed line and the horizontal dashed line. (See Figure 2–47.)

13 Click on the graphics area while the dashed lines are displayed. A point vertical to the midpoint of a horizontal edge of the rectangle and horizontal to the midpoint of a vertical edge of the same rectangle is selected.

14 Click on location F indicated in Figure 2–48 to specify a point on the circumference of the circle. A circle is constructed.

15 Save the file as *TechDraw4.igr* and then close the file.

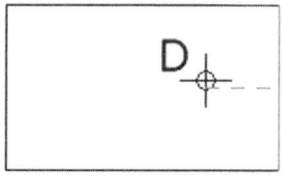

Figure 2–46. Horizontal dashed line connecting the cursor and the midpoint of an edge of the rectangle.

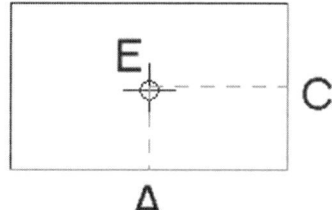

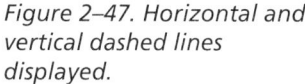

Figure 2–47. Horizontal and vertical dashed lines displayed.

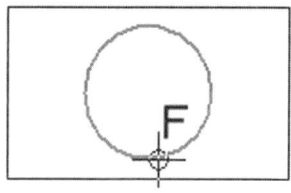

Figure 2–48. Circle completed.

Using the Draw Toolbar

As we have mentioned, the Draw toolbar contains five groups of commands, including those for selection, construction of new drawing elements, construction of new drawing elements from existing drawing elements, modification of existing drawing elements, and construction of a symbol.

Constructing New Drawing Elements

This group of commands includes Line/Arc Continuous, Place Double-line, Connector, Tangent Arc, Arc by 3 Points, Arc by Center Point, Curve, FreeForm, FreeSketch, Circle by Center Point, Circle by 3 Points, Tangent Circle, Ellipse by Center Point, Ellipse by 3 Points, Rectangle, Text Box, and Fill. Perform the following steps to learn these commands.

1 Start a new document by selecting File > New.

2 In the New dialog box, select *Technical Drawing (Imperial).igr* from the template list and then click on the OK button.

3 Select Tools > SmartSketch Settings.

4 In the SmartSketch dialog box, check all boxes on the Relationship tab and then click on the OK button.

5 Select the Tools pull-down menu, and look at the PinPoint, Alignment Indicator, and Maintain Relationships menu items. If there is no tick mark prefixing them, click them one by one. Once these options are all activated, click anywhere on the graphics area to exit.

The Line/Arc Continuous command can be used to construct a line or an arc tangent to an existing drawing element. Continue with the following steps.

6 Click on the Line/Arc Continuous icon on the Draw toolbar.

7 Move the cursor to location A (coordinates X:1/Y:7) and click on the graphics area.

8 Move the cursor to location B (coordinates X:5/Y:7) and click on the graphics area. A line segment is constructed. (See Figure 2–49.)

9 While the Line/Arc Continuous command is active, press the Shift key in conjunction with the A key to change the command to arc construction mode. Move the cursor to location C (coordinates X:7/Y:5) and click on the graphics area. An arc BC tangent to line AB is constructed. (See Figure 2–50.) (Note: If you already exited the Line/Arc Continuous command, activate the command, press Shift + A, move the cursor to endpoint B until you see the *End point* symbol displayed at the cursor, click on the graphics area while the *End point* symbol is displayed, and click on location C.)

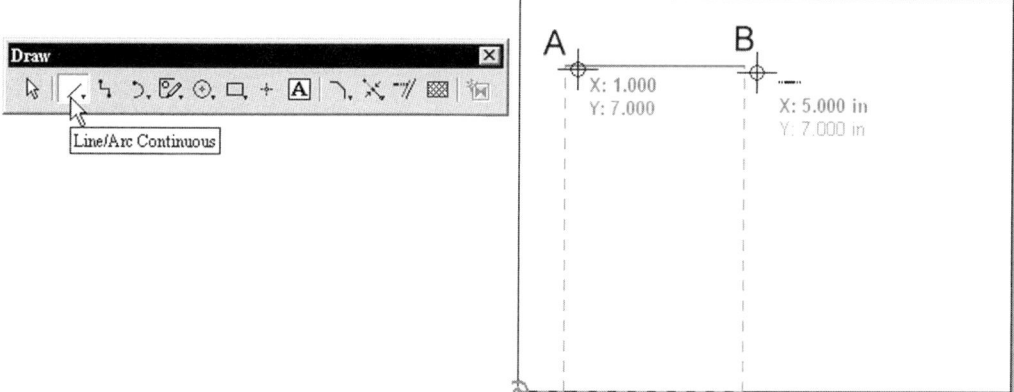

Figure 2–49. Line segment constructed.

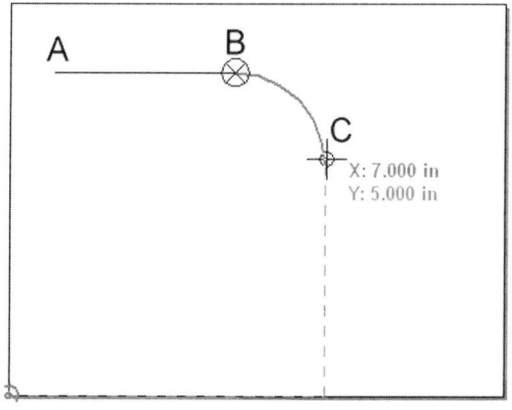

Figure 2–50. Arc segment constructed.

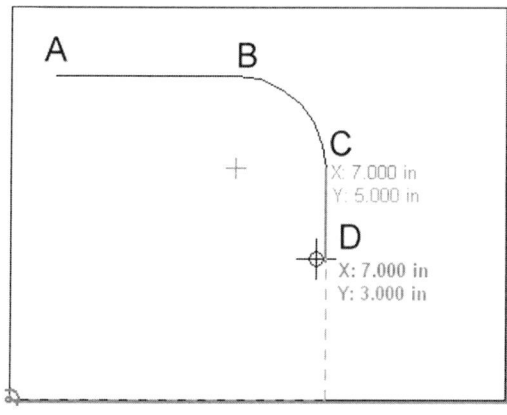

10 While the Line/Arc Continuous command is active, press Shift + L to change back to line construction mode, move the cursor to location D (Figure 2–51, coordinates X:7/Y:3), and click on the graphics area. Line segment CD is constructed.

Apart from using the Line/Arc Continuous command, there are three other ways to construct an arc. Perform the following steps.

1 Click on the Tangent Arc icon on the Draw toolbar.

2 Move the cursor to location D (Figure 2–52) until you see the *End point* symbol displayed at the cursor. Click on the graphics area while the *End point* symbol is displayed.

3 Move the cursor to location E (Figure 2–52, coordinates X:3/Y:3) and click on the graphics area. Arc ED tangent to line CD is constructed.

Figure 2–51. Line segment constructed.

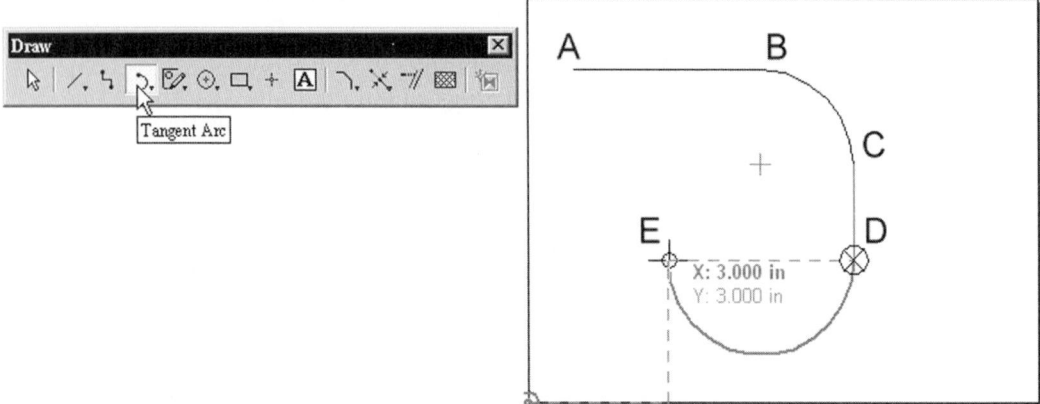

Figure 2–52. Tangent arc constructed.

4 Click on the Arc by 3 Points icon on the Draw toolbar. (If you could not find the Arc by 3 Points command, click on the small triangle at the lower right-hand corner of the Tangent Arc button.)

5 Move the cursor to location E (Figure 2–53) until you see the *End point* symbol displayed at the cursor and click on the graphics area while the *End point* symbol is displayed.

6 Move the cursor to location A (Figure 2–53) until you see the *End point* symbol displayed at the cursor and click on the graphics area while the *End point* symbol is displayed.

7 Move the cursor to location F (Figure 2–53, coordinates X: 3/Y:6) and click on the graphics area. Arc EFA is constructed.

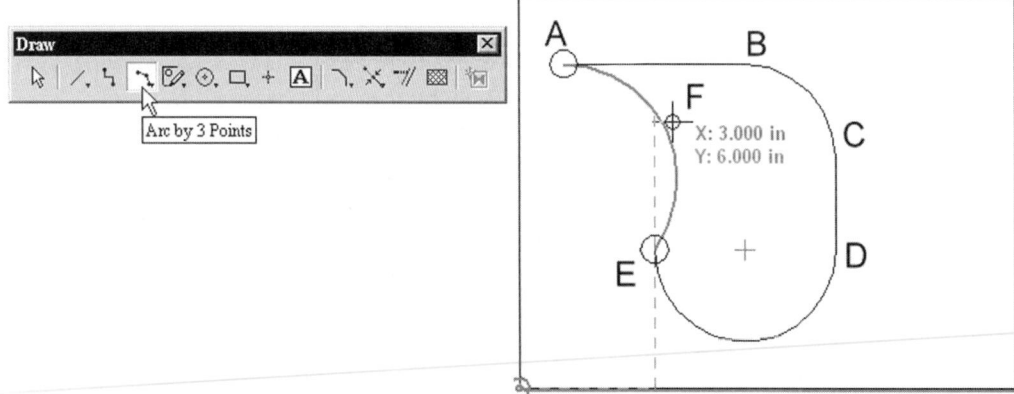

Figure 2–53. Arc by three points constructed.

8 Click on the Arc by Center Point icon on the Draw toolbar.

9 Move the cursor to location C (Figure 2–54) until you see the *End point* symbol displayed at the cursor and click on the graphics area while the *End point* symbol is displayed.

10 Move the cursor to the right and then to location D until you see the *End point* symbol displayed at the cursor, and then click on the graphics area while the *End point* symbol is displayed. (Note: This command can draw a clockwise or a counterclockwise arc. If you move the cursor to the left and then to location D, the result will be different.)

11 Move the cursor to location G (Figure 2–54, coordinates X:5/Y:5) and click on the graphics area. Arc GD is constructed.

12 Save the file as *TechDraw5.igr.*

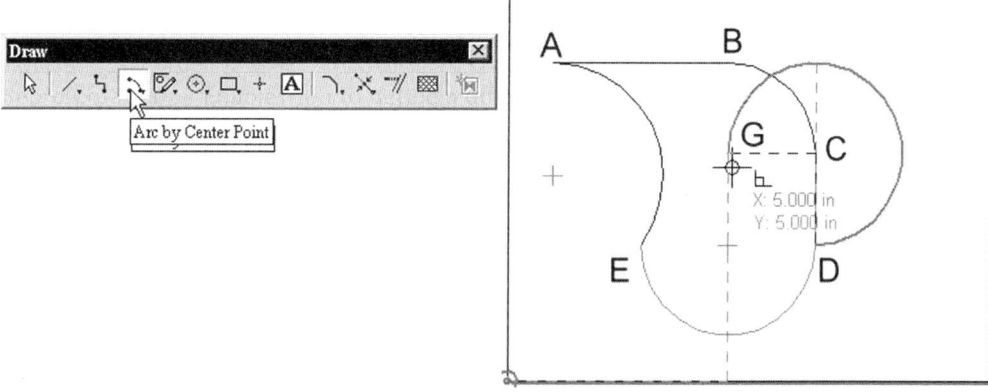

Figure 2–54. Arc by center point constructed.

Perform the following steps to construct a number of circles and ellipses.

1 Select Insert > New Sheet. Because you are working on a working sheet, a new working sheet is constructed.

2 Click on the Circle by Center Point icon on the Draw toolbar.

3 Move the cursor to location A (Figure 2–55, coordinates X:4/Y:5) and click on the graphics area to specify the center point.

4 Move the cursor to location B (Figure 2–55, coordinates X:7/Y:5) and click on the graphics area to specify a point on the circumference of the circle. A circle is constructed.

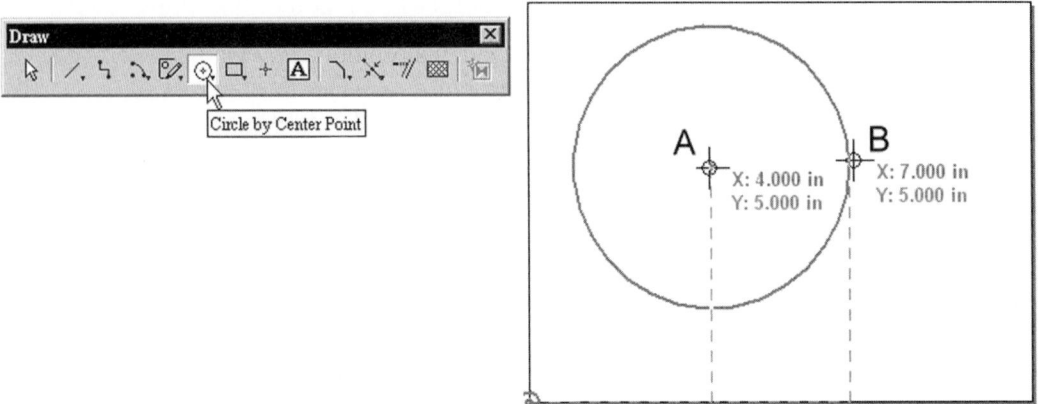

Figure 2–55. A circle constructed on a new working sheet.

5   Click on the Circle by 3 Points icon on the Draw toolbar.

6   Move the cursor near the center of the last circle.

7   Click on the graphics area at location A (Figure 2–56, coordinates X:4/Y:5).

8   Move the cursor to location B (Figure 2–56, coordinates X:7/Y:5) and click on the graphics area.

9   Move the cursor to location C (Figure 2–56, coordinates X:4/Y:2) and click on the graphics area. A circle is constructed.

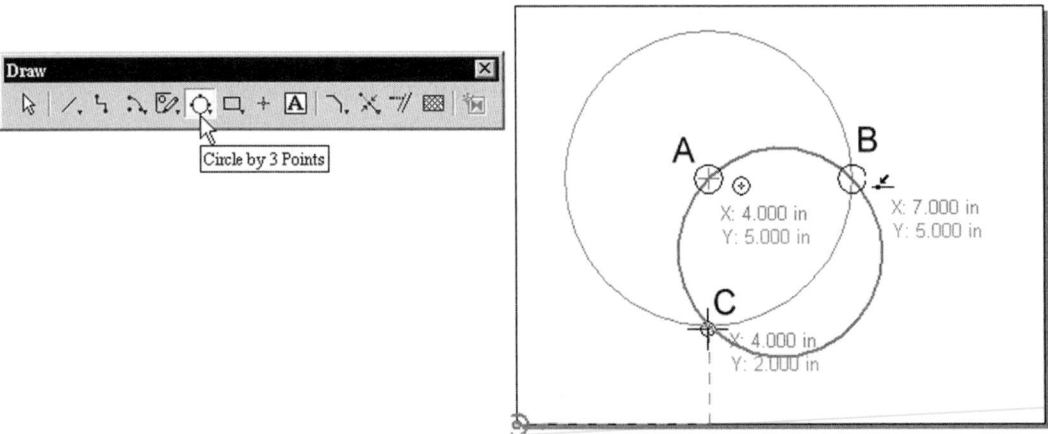

Figure 2–56. Circle by three points constructed.

10 Click on the Tangent Circle icon on the Draw toolbar.

11 Move the cursor to location D (Figure 2–57, coordinates X:4/Y:8) and click on the graphics area.

12 Move the cursor to location E (Figure 2–57) until you see the tangent symbol displayed at the cursor (coordinate display is unimportant here) and click on the graphics area while the tangent symbol is displayed. A tangent circle is constructed.

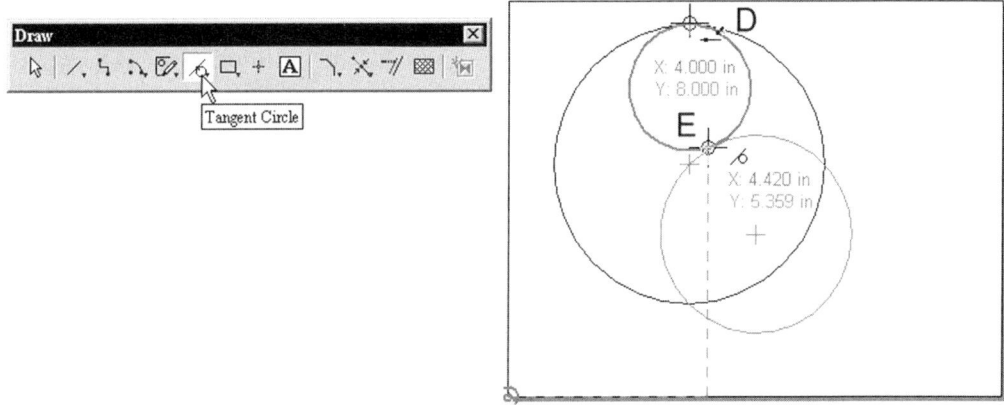

Figure 2–57. Tangent circle constructed.

Perform the following steps to construct two ellipses.

1 Select Insert > New Sheet. A new working sheet is constructed in the document.

2 Click on the Ellipse by Center Point icon on the Draw toolbar.

3 Move the cursor to location A (Figure 2–58, coordinates X:4/Y:2) and click on the graphics area to specify the center point of the ellipse.

4 Move the cursor to location B (Figure 2–58, coordinates X:6/Y:2) and click on the graphics area to specify the endpoint of the primary axis of the ellipse.

5 In the associated ribbon, set the secondary dimension to 1 inch and press the Enter key. An ellipse is constructed.

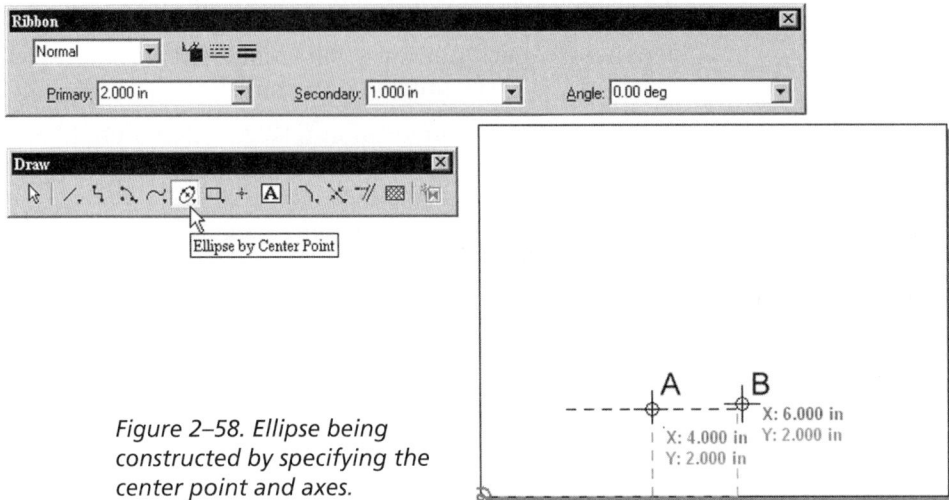

Figure 2–58. Ellipse being constructed by specifying the center point and axes.

6   Click on the Ellipse by 3 Points icon on the Draw toolbar.

7   Move the cursor to location C (Figure 2–59, coordinates X:5/Y:5) and click on the graphics area to specify the first endpoint of the primary axis of the ellipse.

8   Move the cursor to location D (Figure 2–59, coordinates X:8/Y:5) and click on the graphics area to specify the second endpoint of the primary axis of the ellipse.

9   In the associated ribbon, set the secondary dimension to 1 inch and press the Enter key. An ellipse is constructed.

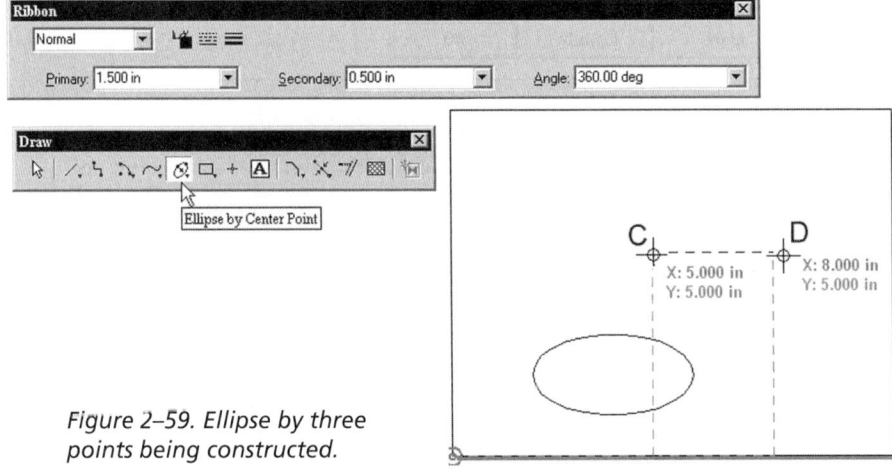

Figure 2–59. Ellipse by three points being constructed.

Perform the following steps to fill the ellipses.

1 Click on the Fill icon on the Draw toolbar.

2 Click on the Solid Color icon on the ribbon. (See Figure 2–60.)

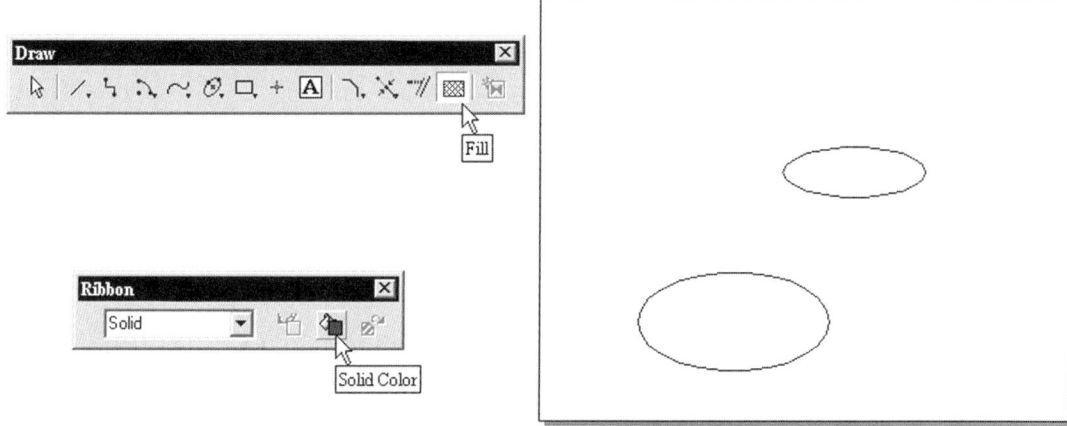

Figure 2–60. Fill command selected.

3 Set the color to yellow and select ellipse A indicated in Figure 2–61.

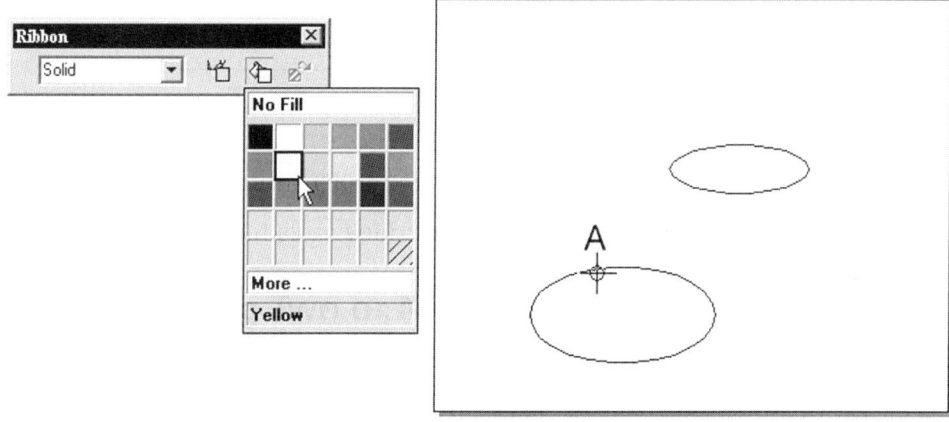

Figure 2–61. An ellipse being filled.

4 Click on the Solid Color icon on the ribbon, set the color to green, and select ellipse A indicated in Figure 2–62.

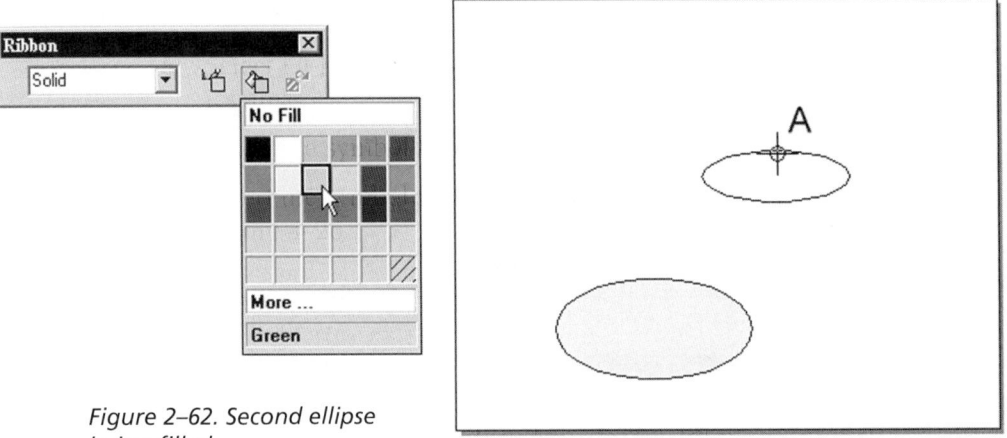

Figure 2–62. Second ellipse being filled.

Perform the follow steps to construct a connector and then manipulate it.

1 Click on the Connector icon on the Draw toolbar.

2 Move the cursor to location A (Figure 2–63) and click on it.

3 Move the cursor slowly upward and then to the left.

4 Move the cursor to location B (Figure 2–63) and click on it. A connector is constructed.

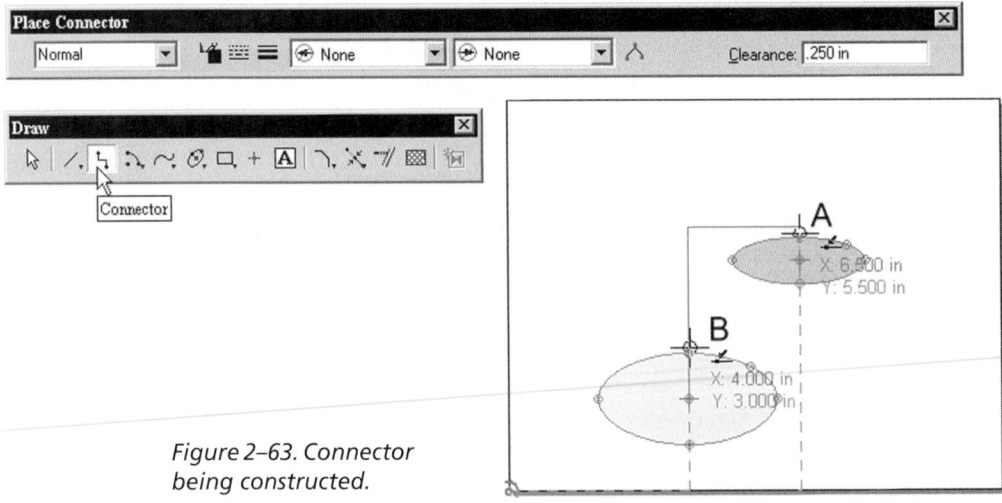

Figure 2–63. Connector being constructed.

5 Click on the Select Tool icon on the Draw toolbar.

6 Click on the connector to highlight it.

7 Select the connector at A (Figure 2–64) and drag it upward. The connector is moved. However, the endpoints of the connector are still attached to the ellipses.

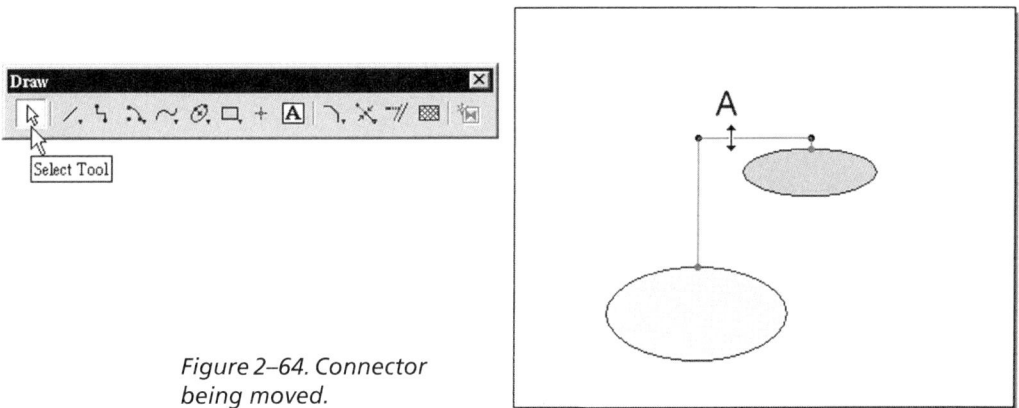

Figure 2–64. Connector being moved.

8 Select ellipse A (Figure 2–65) and drag it to location B (Figure 2–65). Upon releasing the mouse button, the connector's shape is changed but its endpoints are still attached to the ellipses.

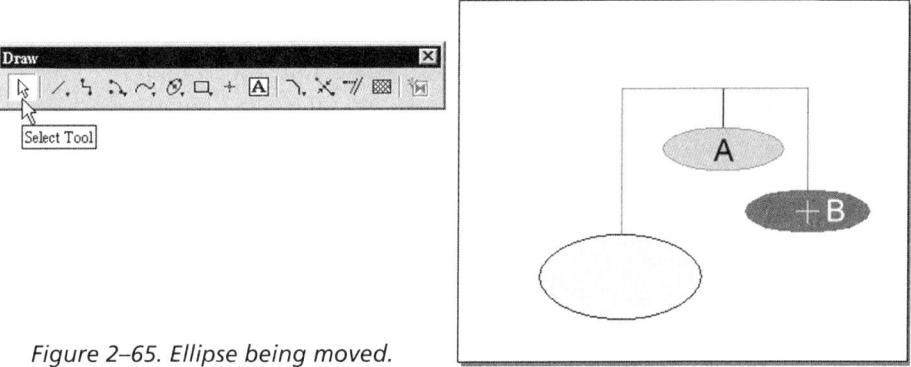

Figure 2–65. Ellipse being moved.

Perform the following steps to construct a series of double lines.

1 Select Insert > New Sheet. A new working sheet is constructed in the document.

2 Click on the Place Doubleline icon on the Draw toolbar.

3 In the Place Doubleline ribbon, click on the Center Primary Line icon and set thickness to 0.5 inch.

4 Move the cursor to location A (Figure 2–66, coordinates X:2/Y:5) and click on the graphics area.

5 Move the cursor to location B (Figure 2–66, coordinates X:6/Y:5) and click on the graphics area. A double line segment is constructed.

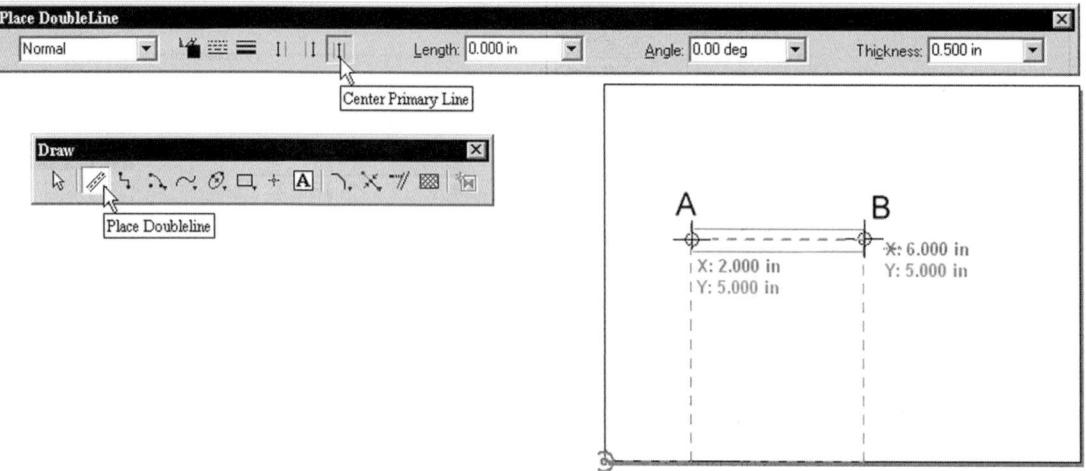

Figure 2–66. A double line segment constructed.

6 Move the cursor to location C (Figure 2–67, coordinates X:6/Y:2) and click on the graphics area. Segment BC is constructed.

7 Move the cursor to location D (Figure 2–68, coordinates X:8/Y:2) and click on the graphics area. Segment CD is constructed.

8 Right click to exit the command.

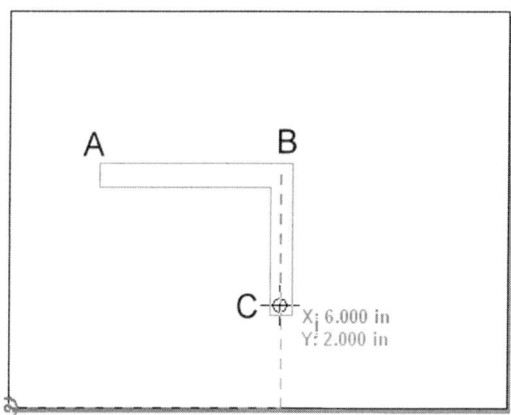

Figure 2–67. Second double line segment constructed.

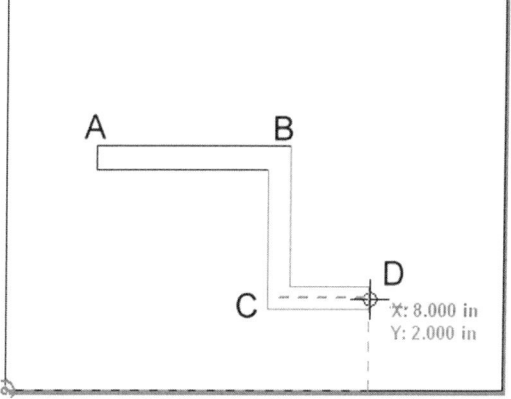

Figure 2–68. Third double line segment constructed.

Perform the following steps to construct free-form curves.

1   Select Insert > New Sheet. A new working sheet is constructed in the document.

2   Click on the Curve icon on the Draw toolbar.

3   Click on the Open icon on the associated ribbon.

4   Move the cursor to location A (Figure 2–69, coordinates X:2/Y:5) and click on the graphics area.

5   Move the cursor to location B (Figure 2–69, coordinates X:5/Y:3) and click on the graphics area.

6   Move the cursor to location C (Figure 2–69, coordinates X:8/Y:6) and click on the graphics area.

7   Right click to exit the command. A curve is constructed.

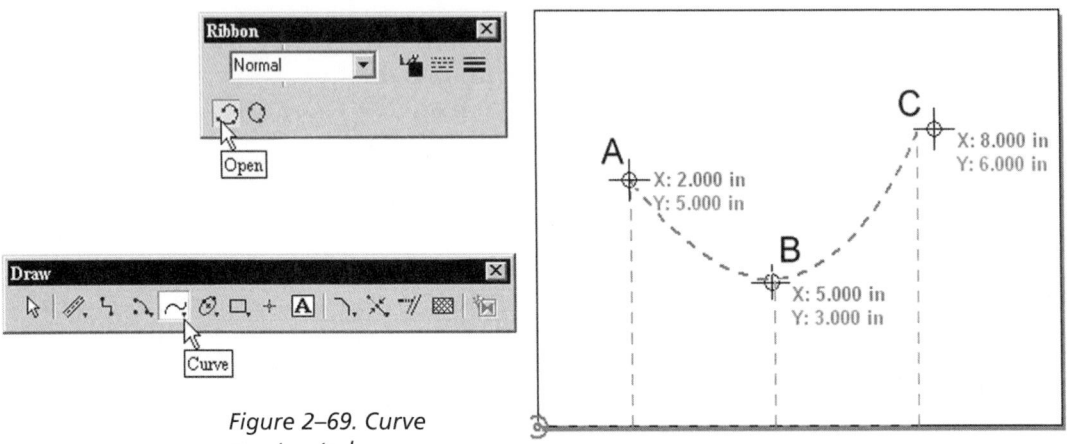

Figure 2–69. Curve constructed.

8   Click on the FreeForm icon on the Draw toolbar.

9   Click on the Smoothing On icon on the associated ribbon.

10 Move the cursor to location D (Figure 2–70). (Exact location is unimportant.)

11 Hold down the mouse button and drag the mouse to location E (Figure 2–70). (Again, exact location is unimportant.)

12 Release the mouse button. A curve is constructed approximating the path moved by the mouse.

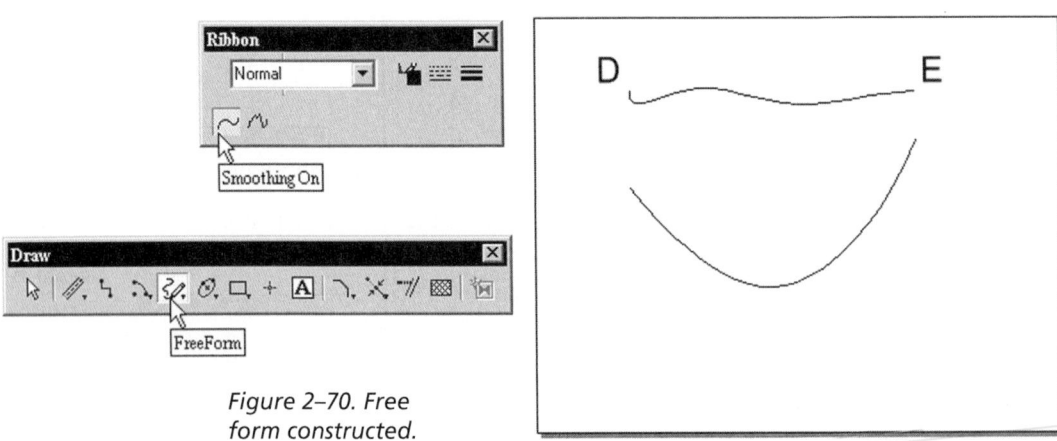

Figure 2–70. Free form constructed.

13 Click on the FreeSketch icon on the Draw toolbar.

14 Click on the Adjust On icon on the associated ribbon.

15 Move the cursor to location F (Figure 2–71). (Exact location is unimportant.)

16 Hold down the mouse button and drag the mouse to location G (Figure 2–71). (Again, exact location is unimportant.)

17 Release the mouse button. A set of line and arc segments is constructed to approximate the path moved by the mouse.

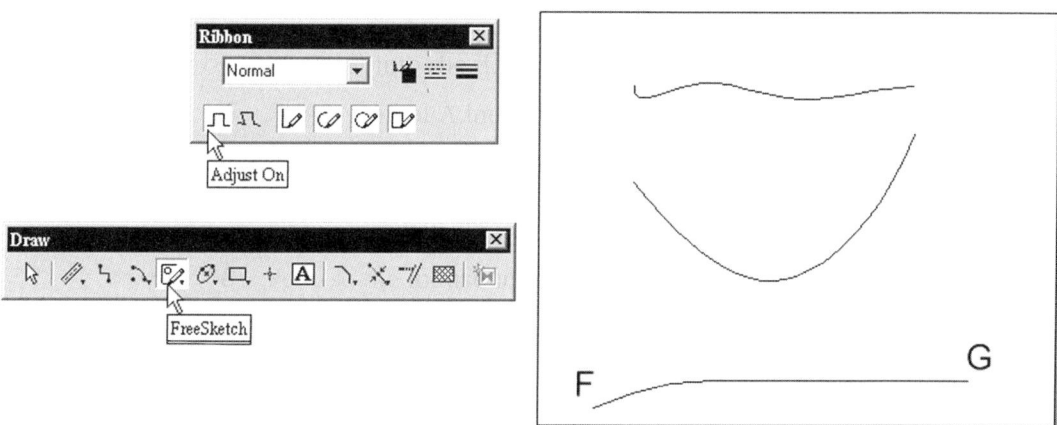

Figure 2–71. Free sketch constructed.

Perform the following steps to construct a text box and a rectangle.

1 Select Insert > New Sheet. A new working sheet is constructed in the document.

2 Click on the Text Box icon on the Draw toolbar.

3 In the associated ribbon, set text height to 0.5 inch.

4 Move the cursor to location A (Figure 2–72, coordinates X:3/Y:4) and click on the graphics area to specify the starting point of the text box.

5 Type *TEXT* and right click. A text box is constructed.

6 Click on the Rectangle icon on the Draw toolbar.

7 Move the cursor to location A (Figure 2–73, coordinates X:2/Y:3) and click on the graphics area to specify the endpoint of an edge of the rectangle.

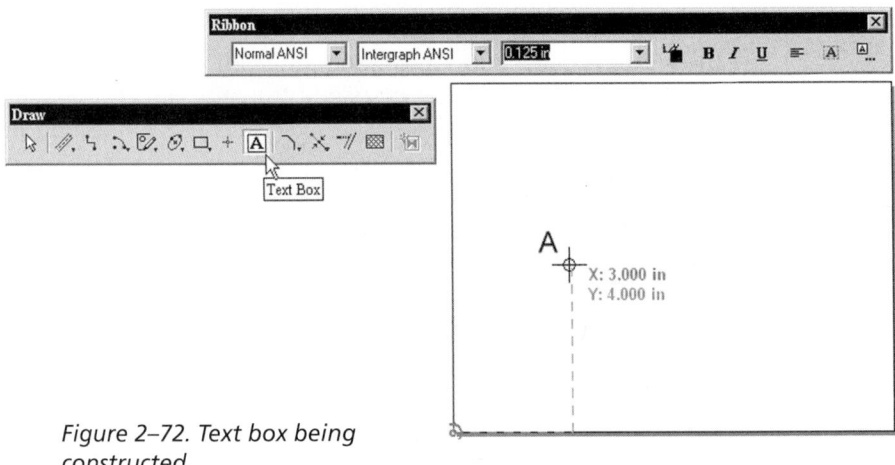

Figure 2–72. Text box being constructed.

8 Move the cursor to location B (Figure 2–73, coordinates X:5/Y:3) and click on the graphics area to specify the other endpoint of the rectangle's first edge.

9 Move the cursor to location C (Figure 2–73, coordinates X:5/Y:5) and click on the graphics area to specify the location of the opposite edge of the rectangle. A rectangle is constructed.

10 Save and close your file.

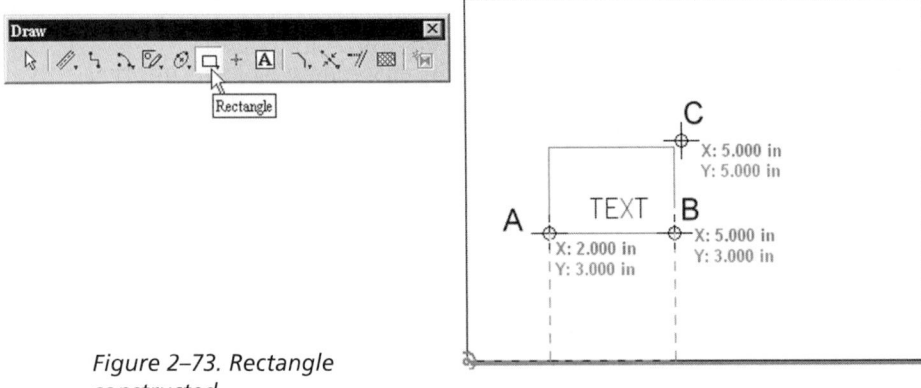

Figure 2–73. Rectangle constructed.

Perform the following steps to construct two polygons.

1 Select Insert > New Sheet. A new working sheet is constructed in the document.

2   Click on the Polygon icon on the Draw toolbar.

3   Set the number of sides to 5 in the associated ribbon.

4   Click on location A (Figure 2–74, 4 inches, 4 inches) and then location B (Figure 2–74, 6 inches, 4 inches). A polygon is constructed.

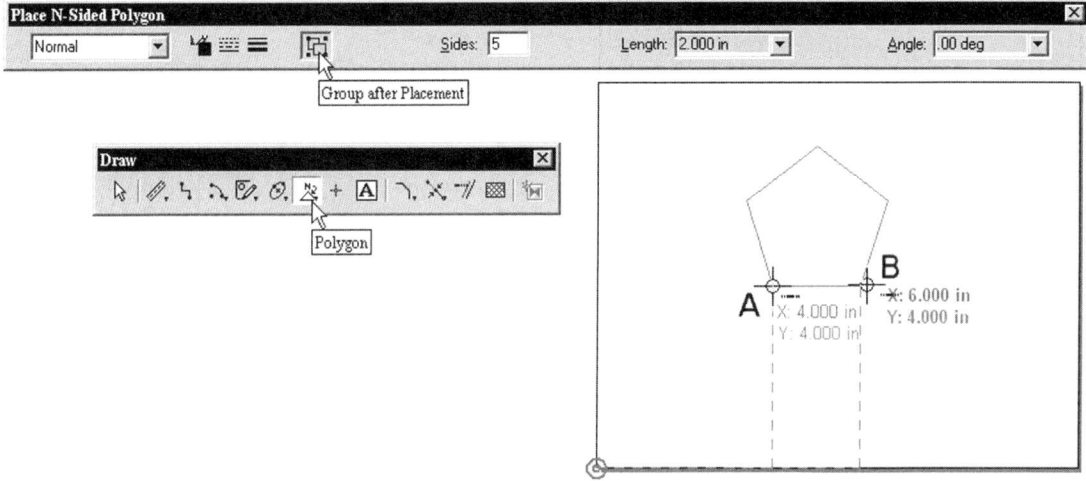

Figure 2–74. Polygon being constructed by specifying an edge of the polygon.

5   Click on the Polygon by Center icon on the Draw toolbar.

6   Click on location A (Figure 2–75, 6 inches, 4 inches) and then location B (Figure 2–75, 3 inches, 5 inches). Another polygon is constructed.

7   Save your file.

Perform the following steps to construct a number of point objects.

1   Select Insert > New Sheet. A new working sheet is constructed in
the document.

2   Click on the Point icon on the Draw toolbar.

3   Click on locations A, B, and C indicated in Figure 2–76. Three
point objects are constructed.

4   Turn off the PinPoint function.

5   Click on the Curve icon on the Draw toolbar.

6   Select points A, B, and C and right click. A curve is constructed.
Save and close your file.

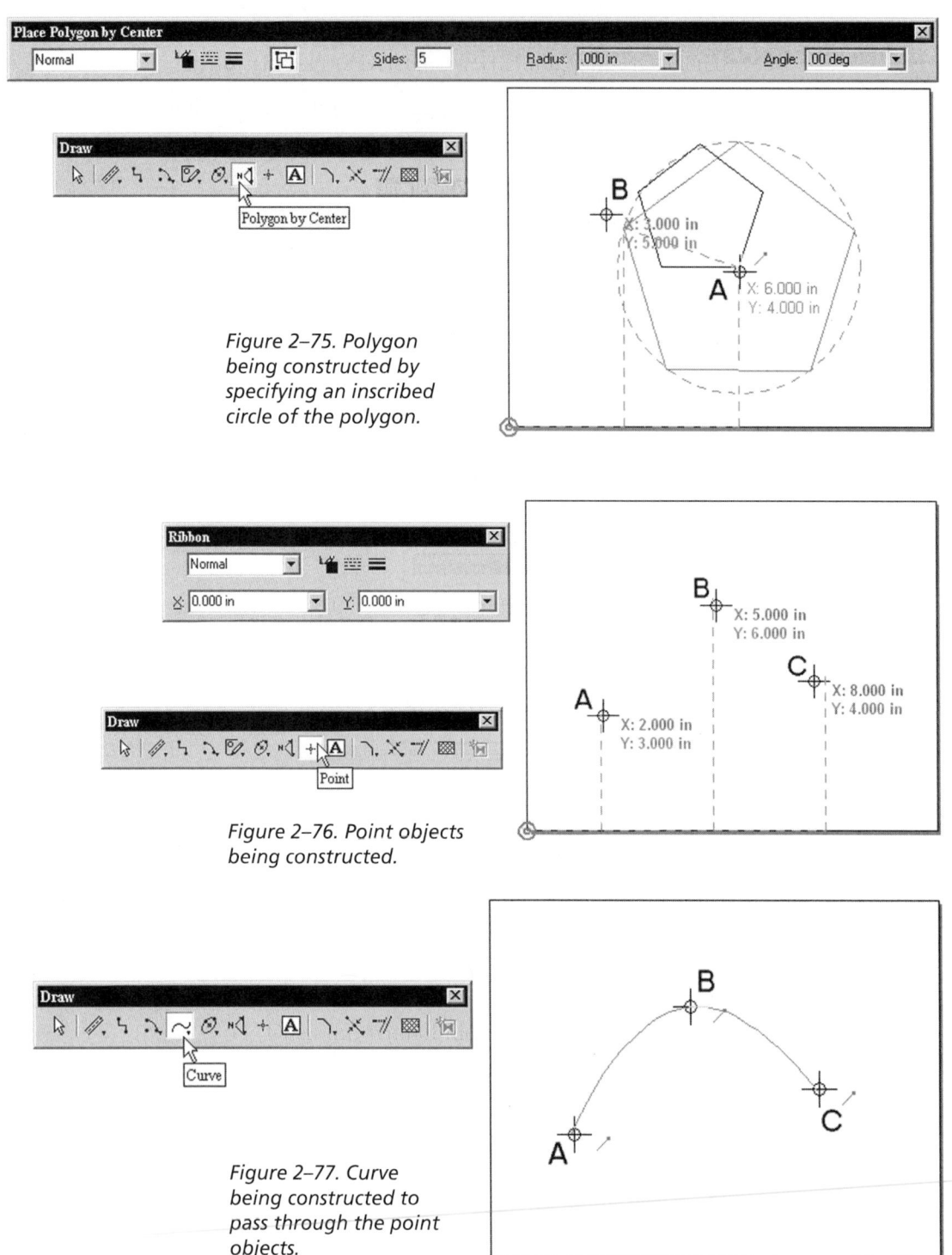

Figure 2–75. Polygon being constructed by specifying an inscribed circle of the polygon.

Figure 2–76. Point objects being constructed.

Figure 2–77. Curve being constructed to pass through the point objects.

Constructing Drawing Elements from Existing Drawing Elements

Two commands (Fillet and Chamfer) on the Draw toolbar can be used to create new elements and to modify existing drawing elements. Perform the following steps.

1 Open the file *Chapter2Draw.igr* from the *Chapter 2* folder of the companion CD-ROM.

2 This document has four working sheets. Sheet 1 should be displayed. If it is not, click on Sheet 1 to activate it.

3 Click on the Fillet icon on the Draw toolbar.

4 In the associated ribbon, click on the Trim icon and set the radius to 1 inch.

5 Select A and B indicated in Figure 2–78. A fillet is constructed and the lines are trimmed.

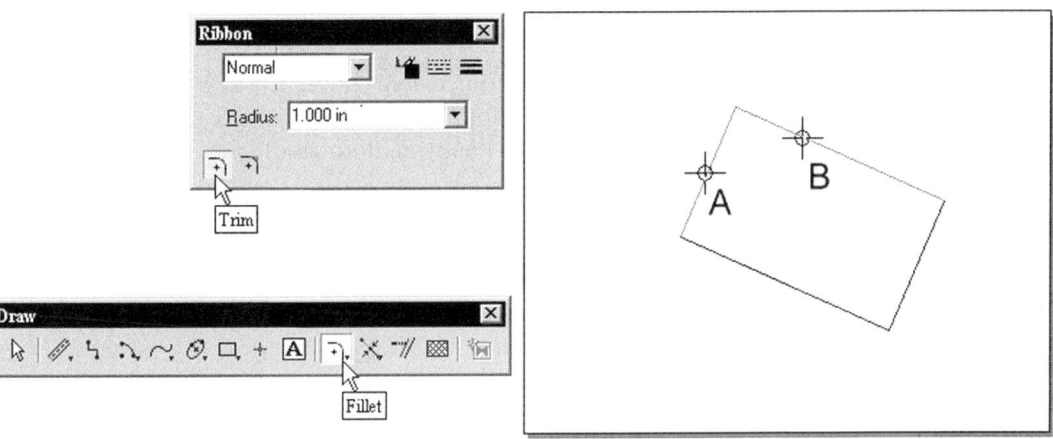

Figure 2–78. Fillet being constructed.

6 Click on the Chamfer icon on the Draw toolbar.

7 In the associated ribbon, specify "set back" distances of 1 inch and 2 inches.

8 Select X and Y indicated in Figure 2–79. A chamfer is constructed. (See Figure 2–80.)

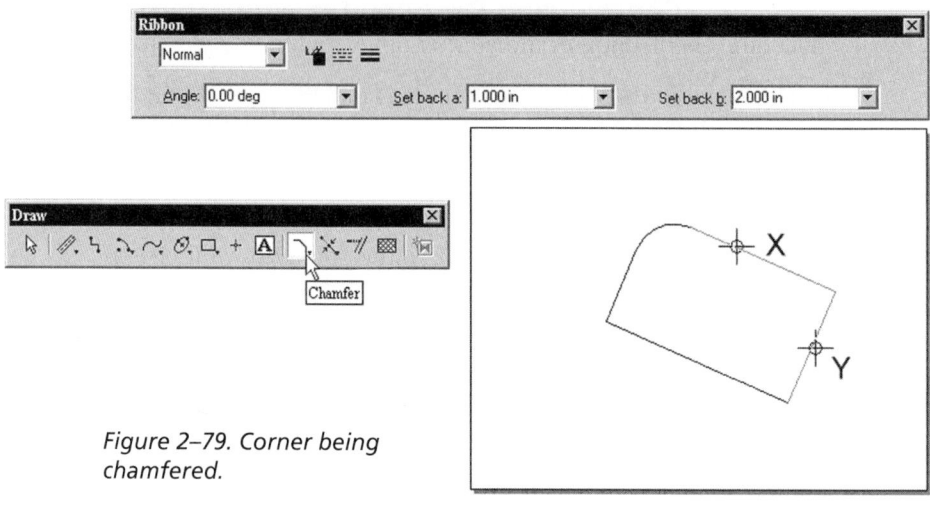

Figure 2–79. Corner being chamfered.

Figure 2–80. Chamfer orientation being selected.

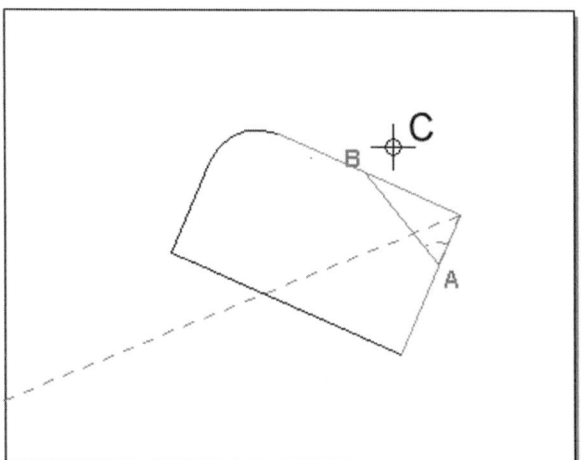

Modifying Existing Drawing Elements

This group has three commands: Trim, Trim Corner, and Extend to Next. These commands modify existing drawing elements. Perform the following steps.

1   Activate Sheet 2.

2   Click on the Trim icon on the Draw toolbar.

3   Select A indicated in Figure 2–81. A portion of the line is trimmed. (Note: the selected portion of the line is trimmed by this command.)

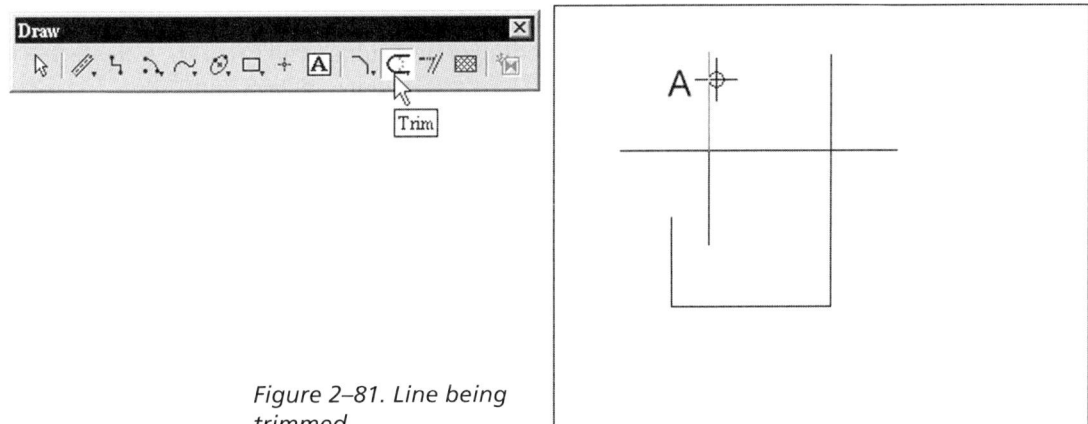

Figure 2–81. Line being trimmed.

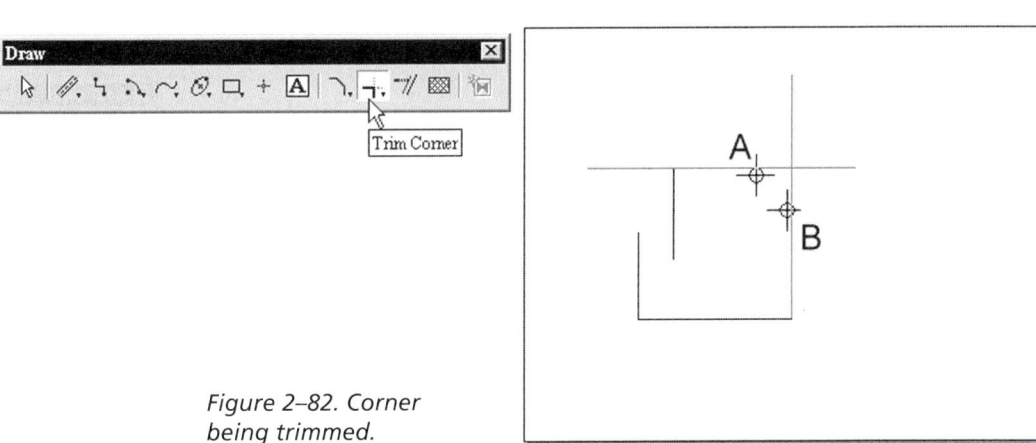

Figure 2–82. Corner being trimmed.

4 Click on the Trim Corner icon on the Draw toolbar.

5 Select A and B indicated in Figure 2–82. The corner is trimmed. (Note: the selected portion of the lines is retained.)

6 Click on the Extend to Next icon on the Draw toolbar.

7 Select A indicated in Figure 2–83. The line is extended to the next drawing element.

Now you will split a curve into segments of equal length. Perform the following steps.

1 Activate Sheet 4.

2 Click on the Split icon on the Draw toolbar.

3 Set the number of segments to 5 in the associated ribbon.

4 Select curve A indicated in Figure 2–84. The curve is split into five equal segments. Save your file.

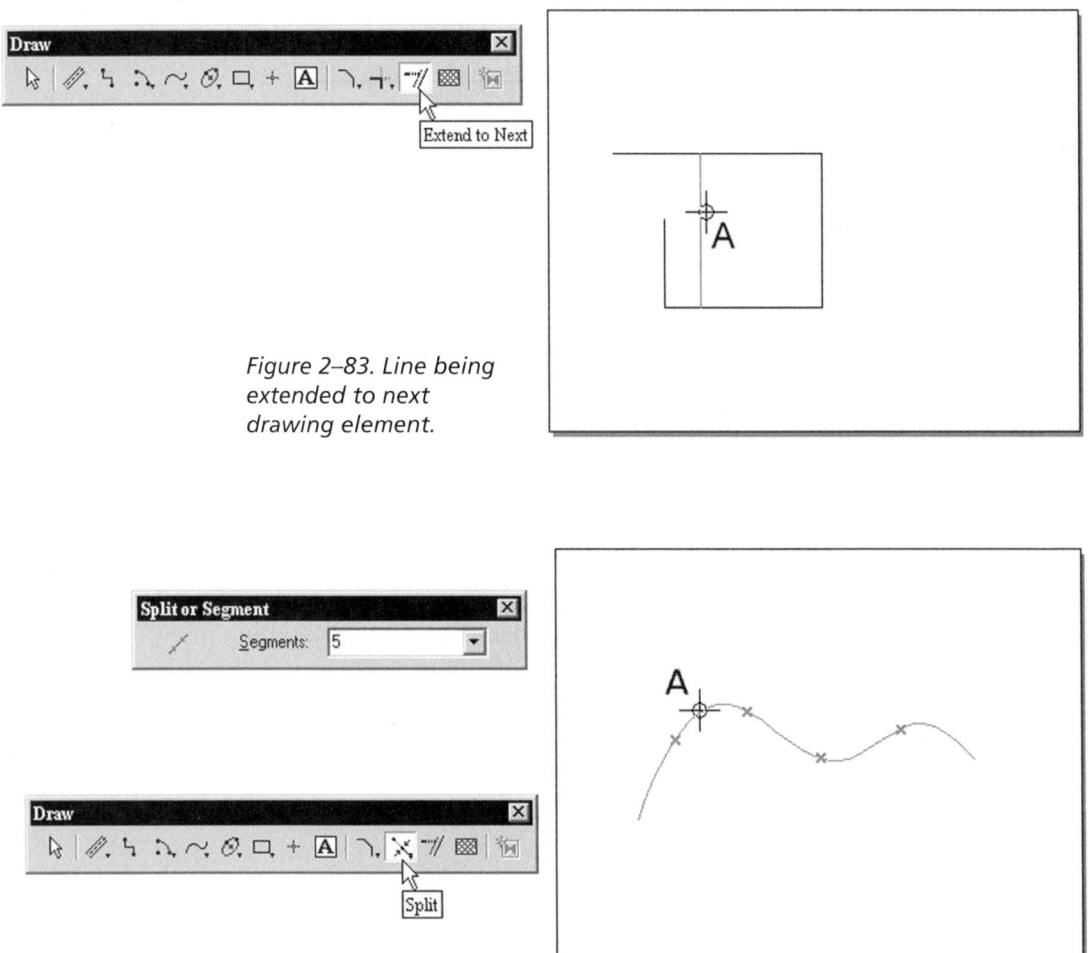

Figure 2–83. Line being extended to next drawing element.

Figure 2–84. A curve being split into five segments of equal length.

Constructing a Symbol from Existing Drawing Elements

If you want to reuse drawing elements already constructed in another document, you can construct a symbol from the elements to form a symbol and then insert the symbol in other documents. Perform the following steps.

1 Activate Sheet 3.

2 With reference to Figure 2–85, click on A, hold down the mouse button, drag the mouse to B, and then release the mouse button.

3 Click on the Create Symbol icon on the Draw toolbar. (Note: This command is not available if no drawing elements are selected. In other words, you have to select drawing elements before clicking on this command.)

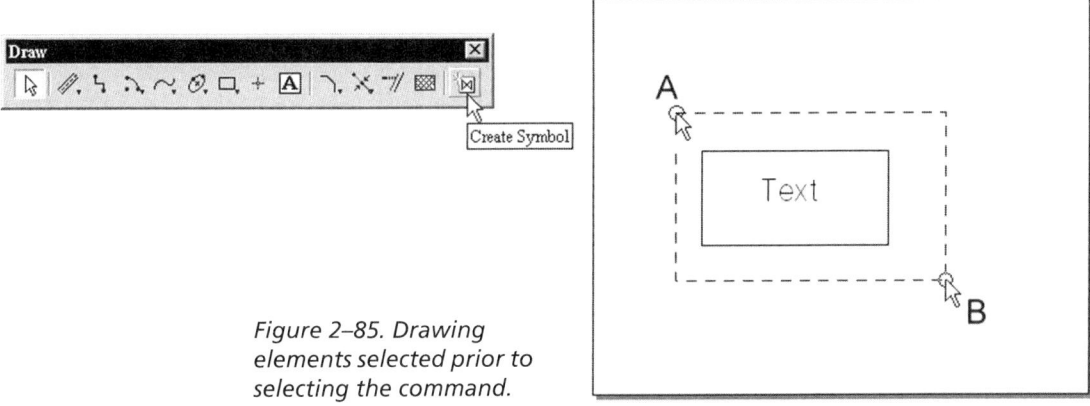

Figure 2–85. Drawing elements selected prior to selecting the command.

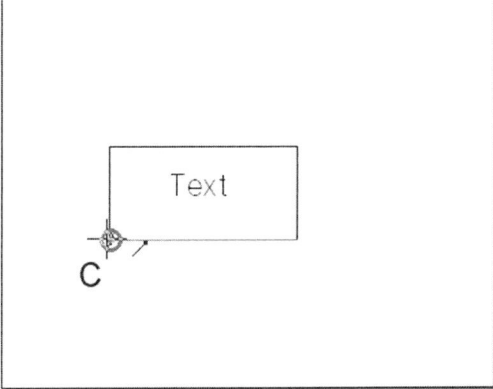

Figure 2–86. Origin being selected.

4 Select endpoint C indicated in Figure 2–86 (to specify the origin of the symbol).

5 In the Save and Symbol dialog box, specify a file name and then click on the OK button. (Note: The file extension is *.sym.*)

6 Save your file in your computer and close the file.

Chapter 4 explores the use of symbols in drafting further.

Using the Change Toolbar

To change drawing elements you have already constructed, you use the tools available from the Change toolbar. (See Figure 2–87.) The Change toolbar can be displayed by using the Toolbar dialog box (accessible from the View > Toolbars menu) or by clicking on the Change button on the Main toolbar.

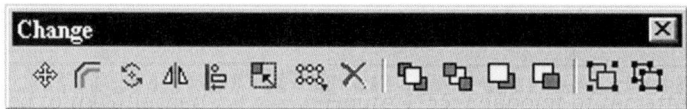

Figure 2–87. Change toolbar.

Details of the options available from the Change toolbar are outlined in Table 2–4.

Table 2–4 Change Toolbar Options and Their Functions

Option	Function
Move/Copy	Moves or copies selected drawing elements. To copy drawing elements, either click on the Copy button on the associated ribbon or hold down the Ctrl key while specifying the distance and direction of move/copy.
Offset	Constructs an offset copy of a drawing element or a set of contiguous drawing elements.
Rotate	Rotates selected drawing elements about a center of rotation. To keep the original drawing elements while rotating, either click on the Copy button on the associated ribbon or hold down the Ctrl key while specifying the start and endpoint of rotation.
Mirror	Mirrors selected drawing elements about a mirror axis. To keep the original drawing elements while mirroring, either click on the Copy button on the associated ribbon or hold down the Ctrl key while specifying the mirror line.
Align	Aligns selected drawing elements in several ways.

Option	Function
Scale	Scales selected drawing elements with reference to a center point. To keep the original drawing elements after scaling, either click on the Copy button on the associated ribbon or hold down the Ctrl key while scaling.
Rectangular Pattern	Repeats selected drawing elements in a rectangular pattern.
Circular Pattern	Repeats selected drawing elements in a circular pattern.
Delete	Deletes selected drawing elements.
Bring to Front	Moves selected drawing elements to the front of the display order.
Send to Back	Moves selected drawing elements to the back of the display order.
Pull Up	Moves selected drawing elements up one position in the display order.
Push Down	Moves selected drawing elements down one position in the display order.
Group	Groups selected drawing elements so that they can be manipulated as a single unit.
Ungroup	Ungroups grouped drawing elements.

Basically, the commands in the Change toolbar can be classified in four major groups. The first group concerns transformation of drawing elements. This group includes the Move/Copy, Rotate, Mirror, Align, and Scale options. You move, rotate, mirror, and scale selected existing drawing elements. While moving, rotating, mirroring, and scaling, you can keep the original drawing elements.

The second group includes the Offset, Rectangular Pattern, and Circular Pattern options. You construct drawing elements from existing drawing elements. You construct offset drawing elements and repeat drawing elements in a rectangular or circular pattern.

The third group includes the Bring to Front, Send to Back, Pull Up, and Pull Down options. You rearrange the order of display of drawing elements that overlap.

The fourth group includes the Group and Ungroup options. You manipulate drawing elements by grouping them so that you can treat them as a single unit. You can also ungroup drawing elements that are already grouped.

Transforming Drawing Elements

Perform the following steps to copy a drawing element.

1 Open the file *Chapter2Change.igr* from the *Chapter 2* folder of the companion CD-ROM and save it to the working folder of your computer.

2 This document has ten working sheets. Sheet 1 should be displayed. If it is not, click on Sheet 1 to activate it.

3 If PinPoint is not activated, select Tools > PinPoint to activate it.

4 If the Change toolbar is not displayed, select View > Toolbars, select Change in the Toolbar list, and then click on the OK button to display it.

5 Click on the Move/Copy icon on the Change toolbar.

6 In the associated ribbon, click on the Copy icon. If you do not click on this button, this command moves the selected drawing elements to a new location.

7 Select the rectangle A (Figure 2–88). Where you select is unimportant.

8 Move the cursor near endpoint B (Figure 2–88) until the *End point* symbol displays at the cursor. Click on the *End point* symbol while it is displayed.

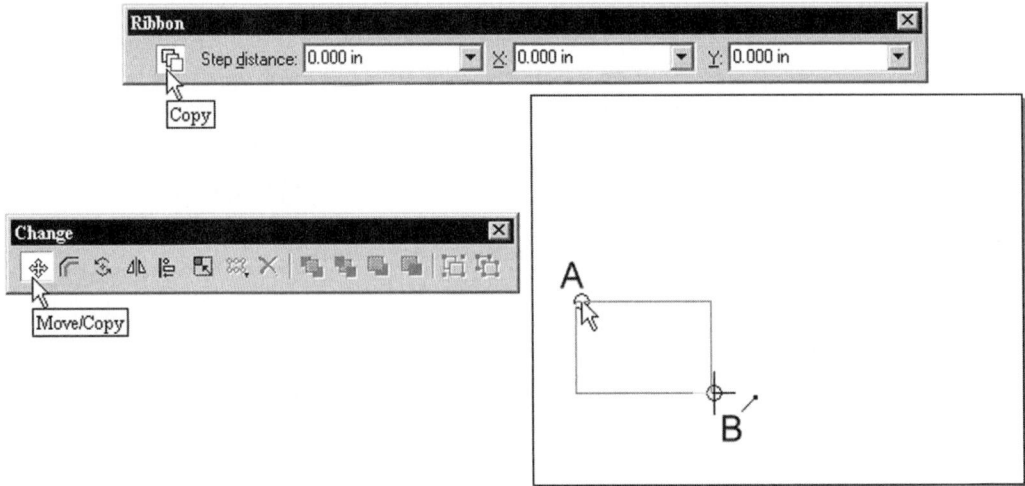

Figure 2–88. Drawing element being copied.

9 Move the cursor to location C (Figure 2–89, coordinates X:7/Y:5).

10 Right click to terminate the command.

Perform the following steps to rotate a drawing element.

1 Select Sheet 2 to activate it.

2 Click on the Rotate icon on the Change toolbar.

3 In the associated ribbon, click on the Copy icon. The original drawing element will be retained after rotating.

4 Select A (Figure 2–90), the drawing element to be rotated.

5 Move the cursor over circle B (Figure 2–90) and then to a location near its center until the center symbol is displayed at the cursor.

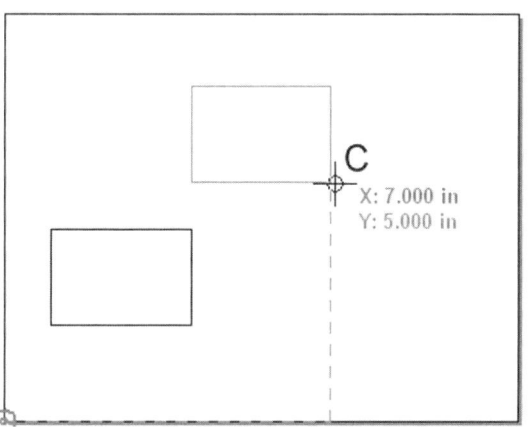

Figure 2–89. Drawing element copied to new location.

6 Click on the graphics area while the center symbol is displayed. The center of circle B (Figure 2–90) is used as the center of rotation.

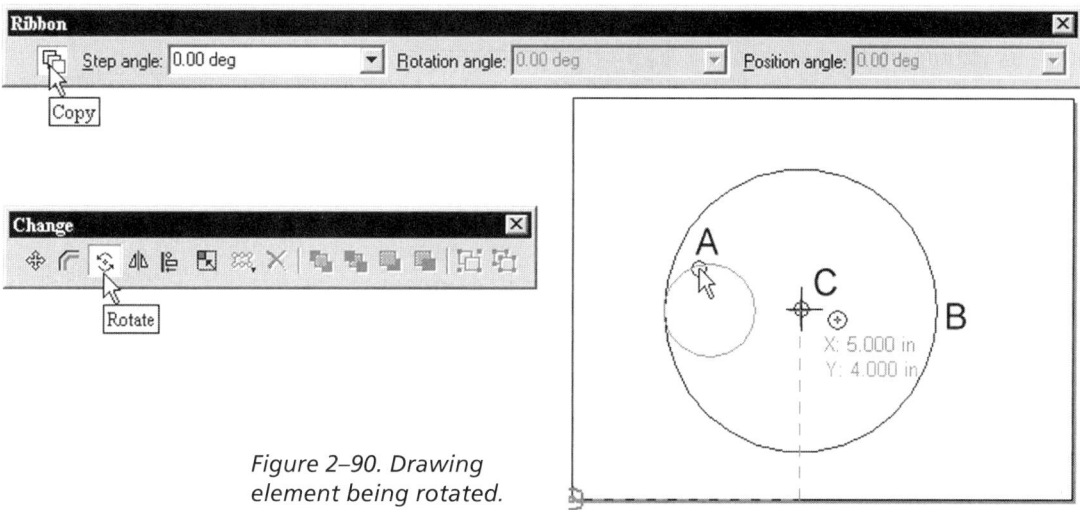

Figure 2–90. Drawing element being rotated.

7 Move the cursor to location D (Figure 2–91) until the horizontal symbol is displayed at the cursor.

8 Click on the graphics area while the symbol is displayed. The position angle is defined.

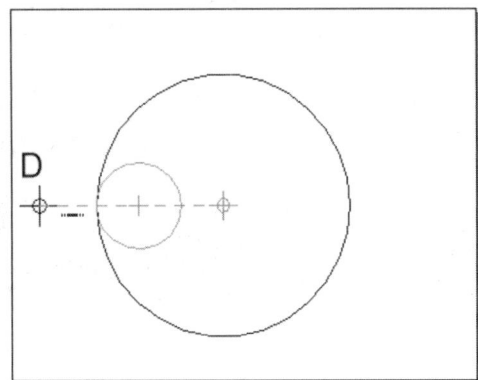

Figure 2–91. Defining the position angle.

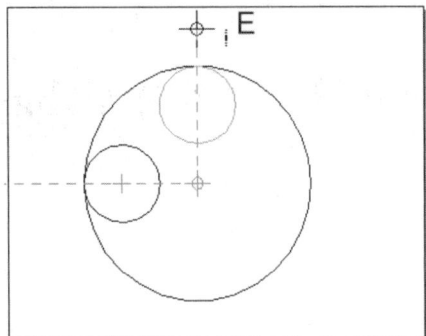

Figure 2–92. Drawing element rotated.

9 Move the cursor to location E (Figure 2–92) until the vertical symbol is displayed at the cursor.

10 Click on the graphics area while the symbol is displayed. The drawing element is copied rotated.

11 Right click to exit the command.

Perform the following steps to mirror a drawing element.

1 Activate Sheet 3.

2 Click on the Mirror icon on the Change toolbar.

3 Click on the Copy icon on the associated ribbon. This way, the original drawing element is retained after mirroring.

4 Select A (Figure 2–93).

5 Move the cursor to location B (Figure 2–94) until the *End point* symbol is displayed at the cursor.

6 Click on the graphics area while the *End point* symbol is displayed. The second point of the mirror line is selected.

7 Move the cursor to location C (Figure 2–95) until the End point symbol is displayed at the cursor.

8 Click on the graphics area while the *End point* symbol is displayed. The first point of the mirror line is selected.

9 Right click to exit the command.

Perform the following steps to scale a drawing element.

1 Activate Sheet 4.

2 Click on the Scale icon on the Change toolbar.

3 Click on the Copy icon on the associated ribbon. The original drawing element will be retained after scaling.

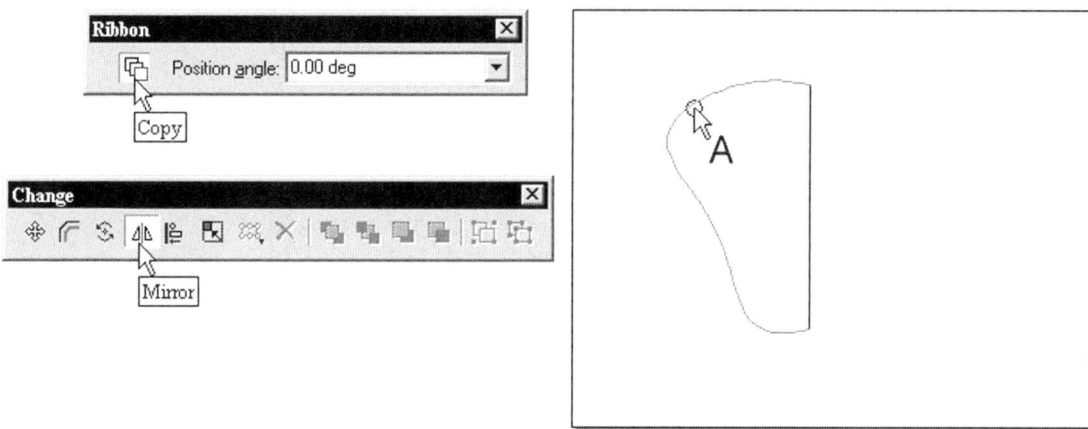

Figure 2–93. Drawing element selected.

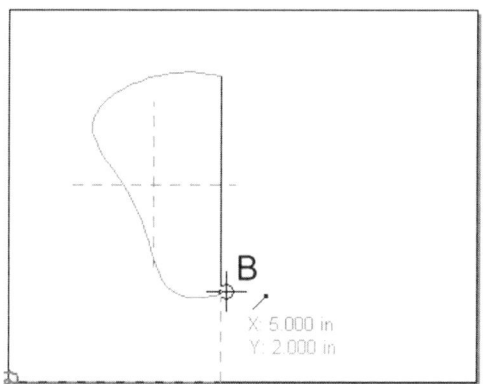

Figure 2–94. First point of the mirror line being selected.

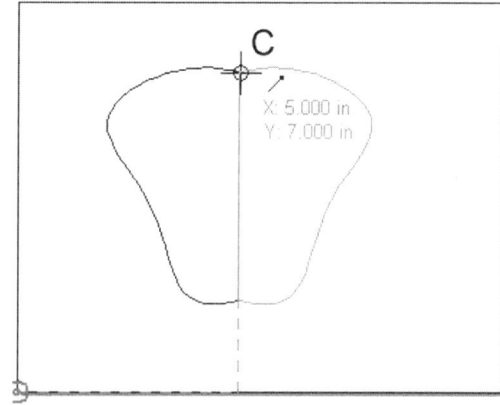

Figure 2–95. Second point of the mirror line selected.

4 Select drawing element A indicated in Figure 2–96.

5 Move the cursor near endpoint B indicated in Figure 2–97 until you see the *End point* symbol displayed at the cursor.

6 Click on the graphics area while the *End point* symbol is displayed. The center point of scaling is selected.

7   Type a value of *2* in the scale factor box of the ribbon and press the Enter key. The drawing element is scaled. (See Figure 2–98.)

8   Right click to exit the command.

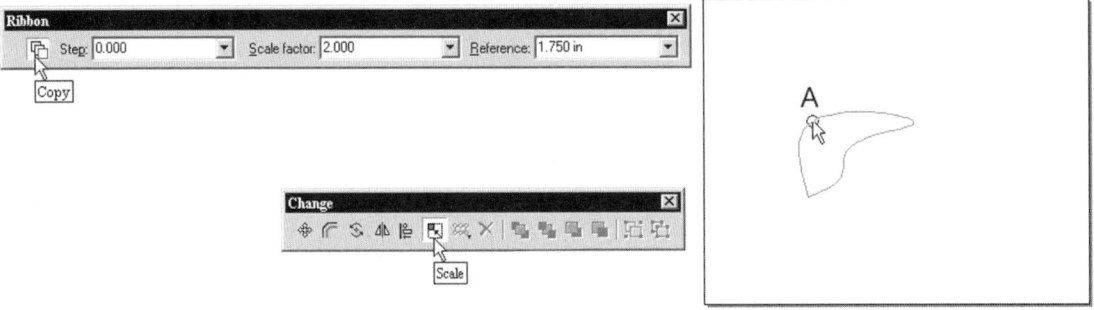

Figure 2–96. Drawing element selected for scaling.

Figure 2–97. Scale center point selected.

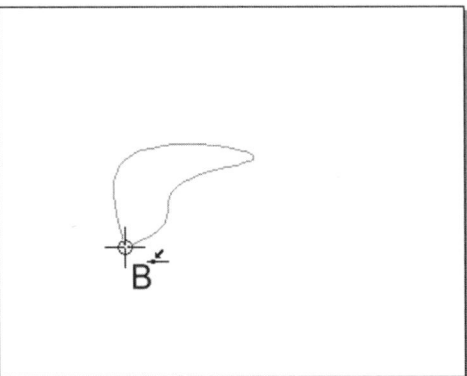

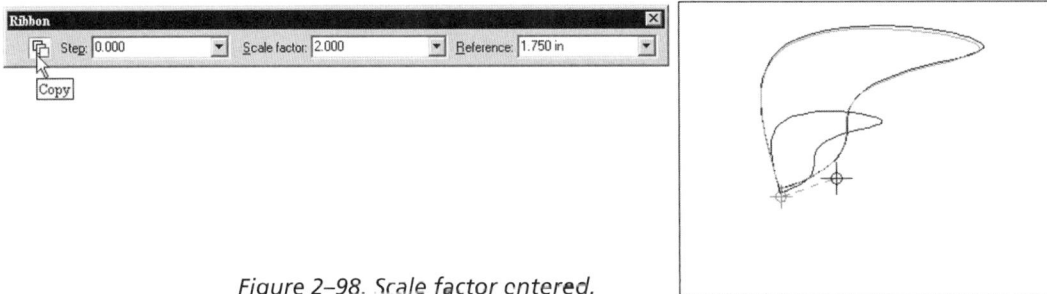

Figure 2–98. Scale factor entered.

Perform the following steps to align two drawing elements.

1 Activate Sheet 10.

2 Click on the Align icon on the Change toolbar.

3 Select A and B indicated in Figure 2–99.

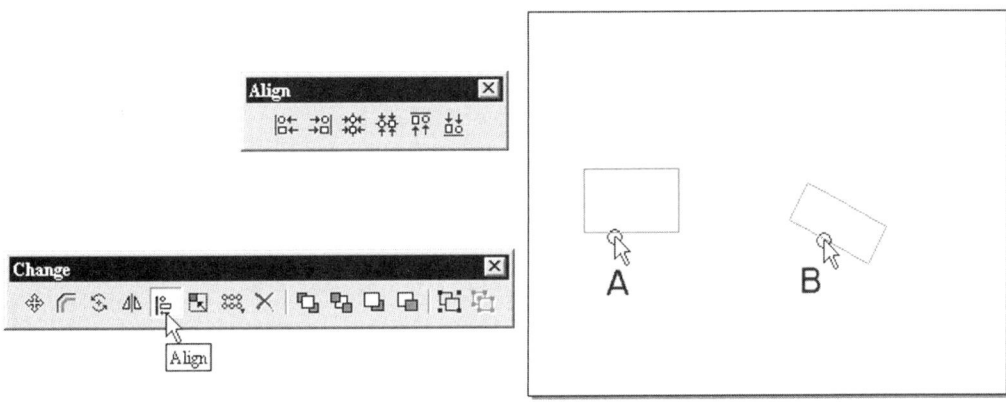

Figure 2–99. Drawing objects being aligned.

Figure 2–100. Rectangles being aligned about vertical.

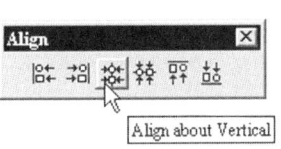

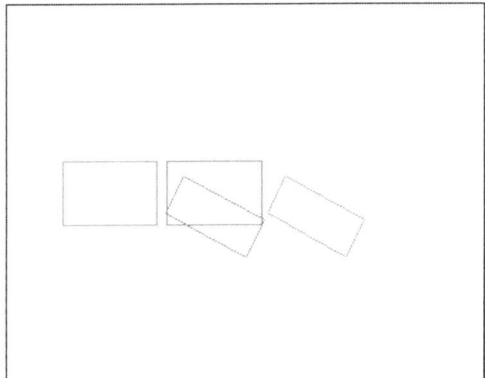

4 There are five ways to align two objects. Move the cursor over the options available from the associated ribbon to find out what they are.

5 Click on the Align about Vertical icon on the ribbon. (See Figure 2–100.) Save your file.

Offsetting and Repeating Drawing Elements

Perform the following steps to offset a drawing element.

1   Activate Sheet 5.

2   Click on the Offset icon on the Change toolbar.

3   On the associated ribbon, click on the Select Chain icon (if it is not already clicked) and specify a cumulative offset value of 0.5 inch. Select Chain means that all drawing elements connected in the form of a chain will be selected when an individual of the chain is selected.

4   Select drawing element A indicated in Figure 2–101.

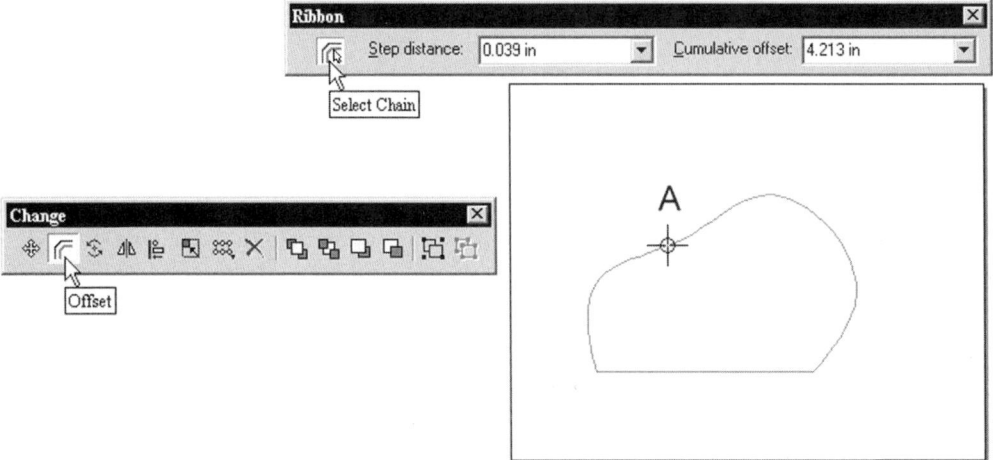

Figure 2–101. Drawing element selected and offset value specified.

5   Move the cursor to location B indicated in Figure 2–102 and click on the graphics area. A set of offset drawing elements is constructed.

6   Right click to exit the command.

Perform the following steps to construct a rectangular pattern of drawing elements.

1   Activate Sheet 6.

2   Select drawing element A indicated in Figure 2–103.

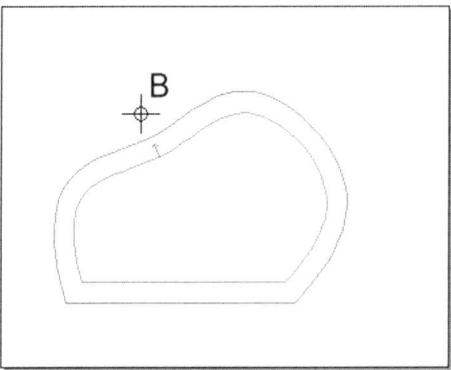

Figure 2–102. Offset location specified.

3 Click on the Rectangular Pattern icon on the Change toolbar. Note that the Rectangular Pattern and the Circular Pattern icon share the same button location on the Change toolbar.

4 In the associated ribbon, set X count to 4, Y count to 2, X offset to 1.5 inches, Y offset to 1.5 inches, and Angle to 20 degrees indicated in Figure 2-104.

5 Click on the Finish button. A rectangular pattern is constructed.

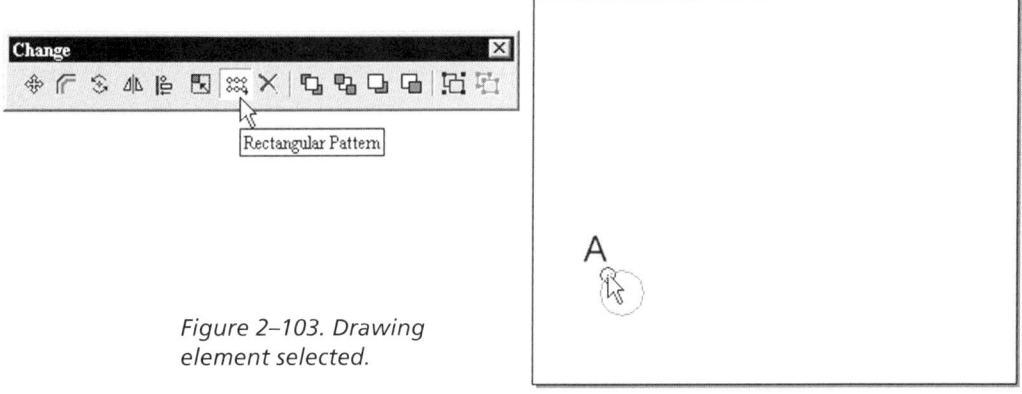

Figure 2–103. Drawing element selected.

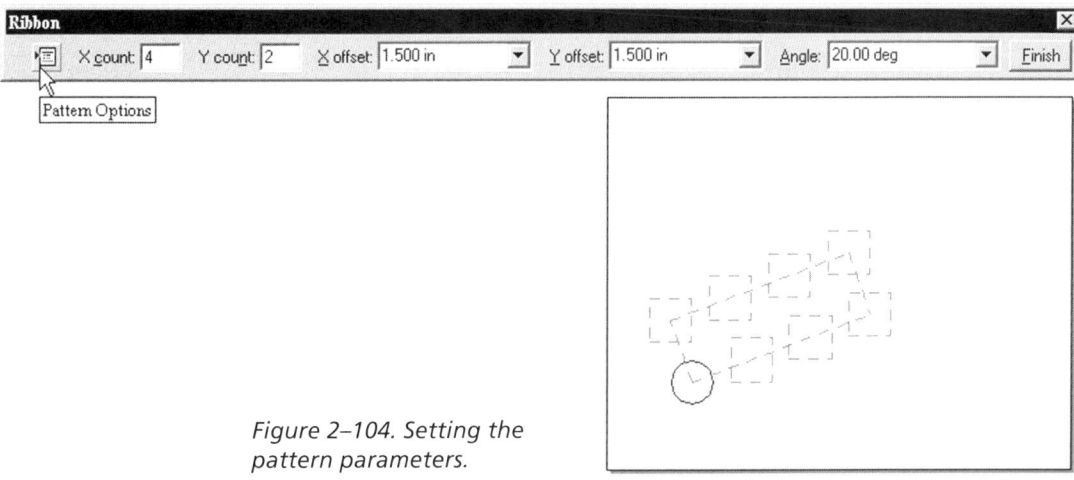

Figure 2–104. Setting the pattern parameters.

Perform the following steps to construct a circular pattern of drawing elements.

1   Activate Sheet 7.

2   Select drawing element A indicated in Figure 2–105.

3   Click on the Circular Pattern icon on the Change toolbar.

4   Refer to Figure 2–106. Click on the Pattern Options icon on the associated ribbon.

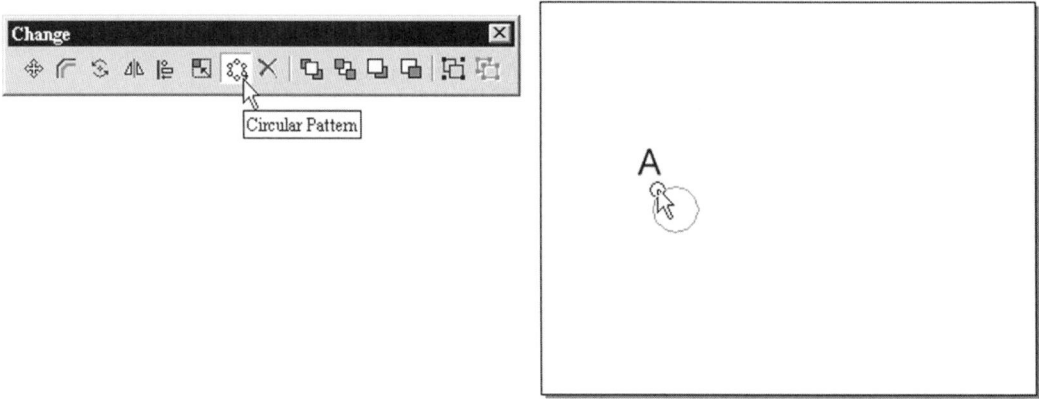

Figure 2–105. Drawing element selected.

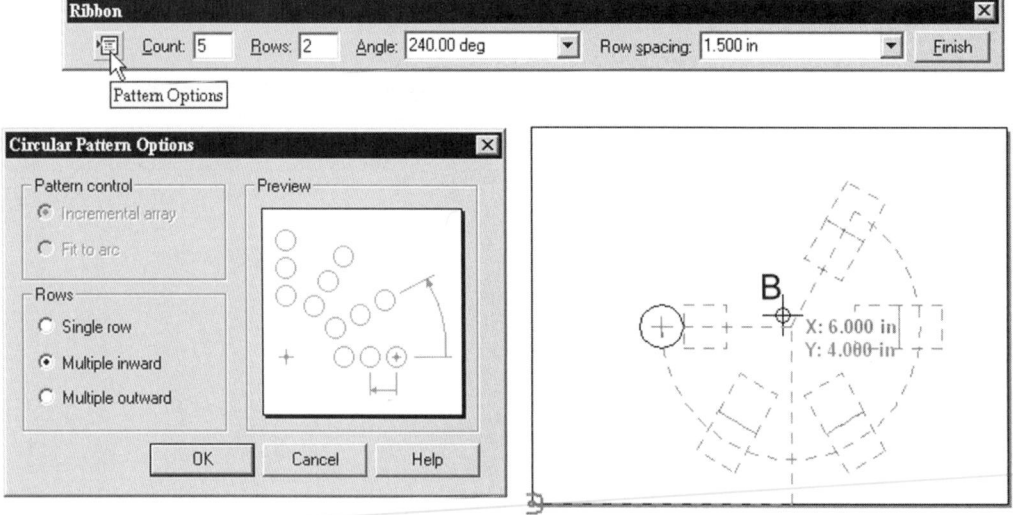

Figure 2–106. Pattern parameters specified and center of pattern selected.

5 In the Circular Pattern Options dialog box, select *Multiple inward* and click on the OK button.

6 In the ribbon, set the count to 5, the row to 2, the angle to 240 degrees, and the row spacing to 1.5 inches.

7 Click on the Finish button. A multiple-row circular pattern is constructed.

Changing Display Order

Perform the following steps to manipulate the display order of drawing elements.

1 Activate Sheet 8.

2 Select drawing element A indicated in Figure 2–107.

3 Click on the Bring to Front icon on the Change toolbar. The drawing element's display order is changed and it is now placed on top of other drawing elements.

4 Select drawing element A indicated in Figure 2–108.

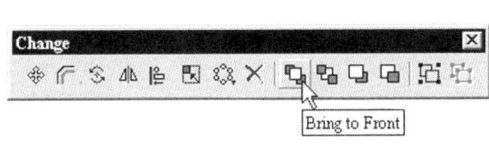

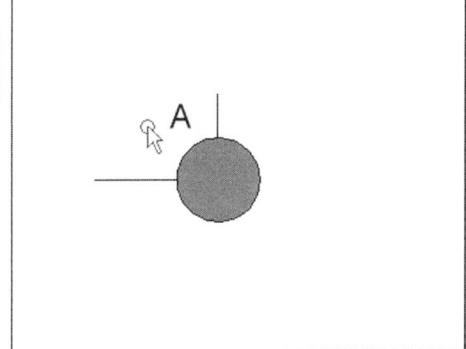

Figure 2–107. Bringing to top.

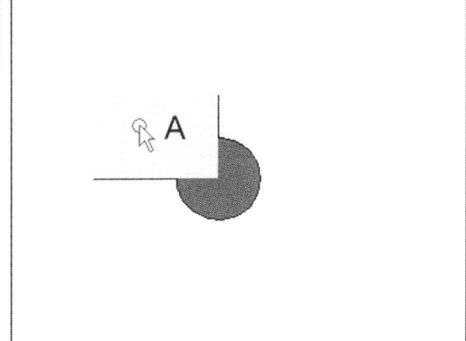

Figure 2–108. Sending to back.

5 Click on the Send to Back icon on the Change toolbar. The selected element is now placed at the back of all other drawing elements.

Grouping and Ungrouping

Perform the following steps to group and ungroup drawing elements.

1 Activate Sheet 9.

2 Hold down the Ctrl key and select drawing elements A, B, and C indicated in Figure 2–109.

3 Click on the Group icon on the Change toolbar. The selected elements are grouped. When you later select any one of them for manipulation, the entire group will be selected.

4 Select A indicated in Figure 2–110. The group is selected.

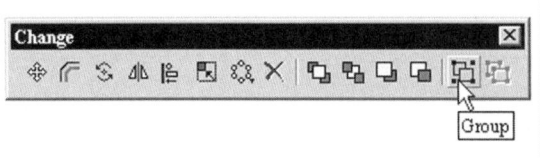

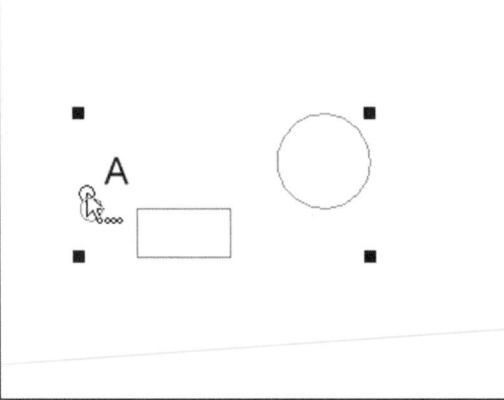

Figure 2–109. Grouping drawing elements.

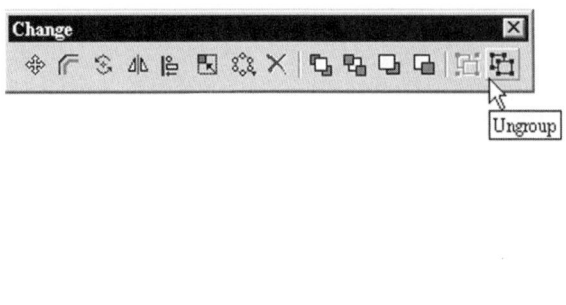

Figure 2–110. Ungrouping.

5 Click on the Ungroup icon on the Change toolbar. The drawing elements are ungrouped. They will be treated as individual drawing elements.

6 Save and close your file.

■ ■ ■ ■ Summary

Constructing engineering drawings and sketches in the computer differs from manual drafting in several respects. There is no pen and paper involved in computer drafting. Instead, you select a command, select a location, specify the object's parameter, and then let the computer construct the drawing element for you. Because drawing elements are constructed as electronic data in the database and displayed as images in the graphics area, it is advisable that you should construct the drawing elements in full size. To obtain an appropriate display of the drawing element in the graphics area, you zoom in, zoom out, or pan around. The concept of scale should only be applied to outputting the drawing on a piece of paper.

Using unique templates provided by SmartSketch, you can deploy SmartSketch as a computer-aided drafting tool to construct various types of 2D engineering drawings, sketches, and diagrams. A SmartSketch data file is called a document, and there are two types of drawing sheets: background sheets and working sheets. To construct a SmartSketch document, you start with selecting a template and then proceed with setting up background sheets and working sheets. If you are going to use the same types of settings for a number of SmartSketch documents, you should save the settings in a document file in the SmartSketch's *Templates* folder and use the template for constructing other similar documents.

Construction of drawing elements in a document concerns mainly drawing and modifying through the use of the commands available from the Draw and Change toolbars. The Draw toolbar can be categorized into five groups of commands: selection tools, drawing element construction tools, tools for constructing new drawing elements while modifying existing drawing elements, tools for modifying existing drawing elements, and symbol creation tools. While working, you need to refer to the prompts provided at the status bar because it tells you what you should do next. For many commands, there is an associated ribbon for you to input parameters for that particular command.

Relationships between drawing elements in a document can be set while constructing the elements or after the elements are constructed. Relationships can be set as temporary or persistent. To manipulate persistent relationships, you display relationship handles. To help specify precise locations while drafting, you use the PinPoint option and the Alignment Indicator, as well as inputting precise values in the ribbon associated with the commands.

The Change toolbar has four groups of commands. The first group concerns transforming drawing elements. While transforming, you may choose to keep the original elements. The second group of commands offsets and repeats existing drawing elements. The third group of commands affects the display order of overlapping drawing elements. The last group of commands helps organize drawing elements into groups for ease of manipulation.

■ ■ ■ ■   # Review Questions

1 State the key differences between manual drafting and computer-aided drafting.

2 What are the basic steps to start a SmartSketch document?

3 List the commands available from the Draw toolbar and the Change toolbar.

4 How can relationships be manipulated?

5 What tools are used to help specify precise locations in a document?

Chapter 3

Drafting II

■ ■ ■ ■ Objectives

This chapter continues to build on the material presented in Chapter 2. It explains the concept of layers, formatting objects, adding dimensions and annotations, and setting system options. After studying this chapter, you should be able to:

- ❑ Manipulate layers
- ❑ Format drawing elements
- ❑ Manipulate dimensions and annotations
- ❑ Set system options

Overview

A way to help manage drawing elements in a document is to organize them into layers. In addition, you can format them in terms of color, line width, and line type to illustrate various constituents of a drawing. Although drawing elements are the main constituents of a document, dimensions and annotations are used to help elaborate a drawing. Finally, this chapter also addresses system option settings and customization of toolbars.

■ ■ ■ ■ Managing Layers

In manual drafting, drawing elements are constructed on sheets of fixed size. You cannot zoom in and zoom out as you do when drafting via computer. As a result, individual drawing elements in a complex drawing may be difficult to see and manipulate. To overcome this difficulty, manual drafting uses physical transparent layers as a means of organizing drawing elements. By categorizing drawing elements, constructing categorized drawing elements on different transparent layers, and subsequently taking out one or more unwanted transparent layers, complex drawings become simpler and selected drawing elements can be better seen in the drawing.

The concept of layers in computer-aided drafting systems is inherited from manual drafting. Although layers are not physical layers but conceptual layers, the operation is similar. You can have a number of transparent layers in a SmartSketch document and construct drawing elements in different layers. By turning off unwanted layers, individual drawing elements can be manipulated more easily.

Three commands from the Tools pull-down menu concern layer manipulation. They are Layers, Layer Groups, and Display Manager.

Hiding Layers, Displaying Layers, and Changing Drawing Elements' Layer Assignments

In a drawing sheet having a number of layers and drawing elements residing on different layers, you may turn off a layer or a number of layers to cause the drawing elements residing on these layers to be invisible. Contrary to turning off, you can turn on a deactivated layer or a number of deactivated layers to cause the related drawing elements in the layers to be visible again.

Among the layers in a drawing sheet, one of them has to be set as the current layer (with the analogy that the layer is put on top of all other layers). As a result of setting a layer as the current layer, any drawing elements you construct thereafter will reside on this layer. However, you may still change a drawing element's layer assignment by moving it from one layer to another layer. Perform the following steps.

1   Open the file *Chapter3Layer.igr* from the *Chapter 3* folder of the companion CD-ROM.

2   Activate Sheet 1, if it is not already activated. This drawing sheet has a number of drawing elements residing on different layers. In Sheet 1, you will find two circles and a rectangle residing on three different layers.

3   Turn on the Layer toolbar by selecting Tools > Layers or by selecting View > Toolbars, checking the Layer box from the Toolbars list from the Toolbar dialog box, and clicking on the OK button. Alternatively, click on the Layer button on the Main toolbar.

Figure 3–1. Layer toolbar.

Figure 3–2. Setting layer circle to be the current layer.

Figure 3–1 shows the Layer toolbar, which contains the Layer pull-down list box showing the current layer in its normal state and listing all layers in the drawing sheet when the small triangle at the right side is selected, and two buttons, Change Layer and Layer Status (one for changing the selected drawing elements' layer assignments and the other for changing the layers' display status).

4 Select the small triangle at the right-hand side of the Layer list box and click on *circle* to make this layer the current layer. (See Figure 3–2.)

5 Click on the Layer Status button on the Layer toolbar. The Layer Display dialog box is displayed. (See Figure 3–3.)

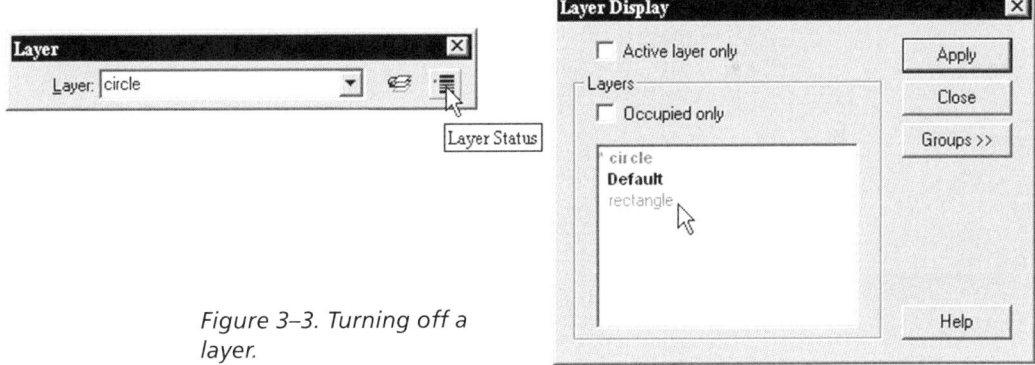

Figure 3–3. Turning off a layer.

The Layer Display dialog box contains an *Active layer only* check box, a Layers list box, and the buttons Apply, Close, Groups, and Help. As we have said, one of the layers in a drawing sheet is set as the current layer. If the *Active layer only* check box is clicked on, all layers except the current layer will be turned off.

The Layers list box lists the layers in a drawing sheet. In the list, the current layer's name is displayed in red, whereas the other layer names are in black. Clicking on a layer's name (other than the current layer) will turn its color from black to gray, denoting that the layer is turned off. To turn on a deactivated layer, click on the layer's name in the Layers list box again, changing its color from gray back to black. Above the list of layers is an *Occupied only* check box. If you click on this check box, only layers with drawing elements will be displayed. In other words, empty layers will not be displayed in the list.

As for the buttons, naturally the Apply button applies the settings made in this dialog box to the drawing sheet, and the Close button closes the dialog box. As for the Group button, it expands the dialog box, enabling you to turn groups of layers on and off. (Layer grouping is discussed later in the book.) Continue with the following steps.

6   Click on the layer *rectangle* in the Layers list box.

7   Click on the Apply button. The *rectangle* layer is turned off and the rectangle constructed on this layer is not visible. (See Figure 3–4.)

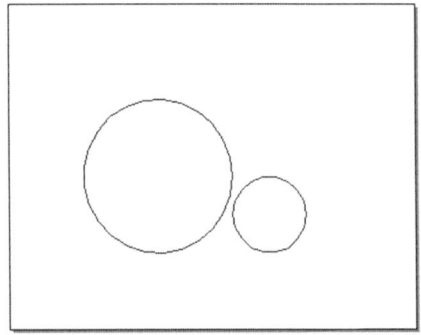

Figure 3–4. Layer rectangle *turned off.*

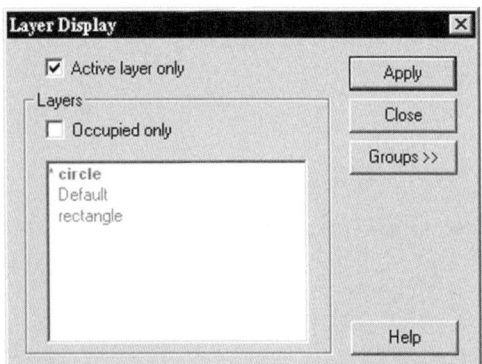

Figure 3–5. Turning on only the active layer.

8   Click on the layer *rectangle* in the Layers list box, changing the color of the layer's name from gray to black.

9   Click on the Apply button. The layer *rectangle* is turned on and the rectangle on this layer becomes visible again.

10   Click on the *Active layer only* button, as shown in Figure 3–5.

11   Click on the Apply button and then on the Close button. All layers except the current layer *circle* are turned off.

12   Click on layer *rectangle* on the Layer toolbar to set it as the current layer. Because only the current layer is displayed, layer *rectangle* is turned on and all other layers are turned off. (See Figure 3–6.)

13   With reference to Figure 3–7, select rectangle A and click on the Change Layer button on the Layer toolbar.

14   In the Change Layer dialog box shown in Figure 3–8, the Current Layers box lists the layers on which the selected drawing elements reside, and the *Change all to* box lists the layers you may change the selected drawing elements to. Click on layer *circle* in the *Change all to* box.

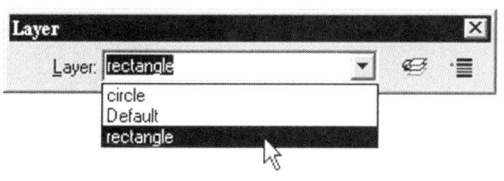

Figure 3–6. Layer rectangle *set as the current layer.*

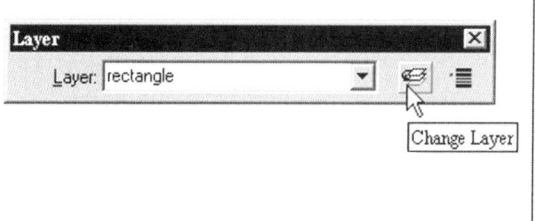

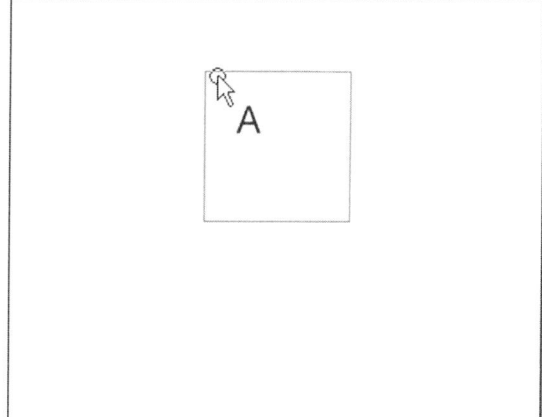

Figure 3–7. Drawing element selected.

Figure 3–8. Change Layer dialog box.

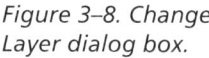

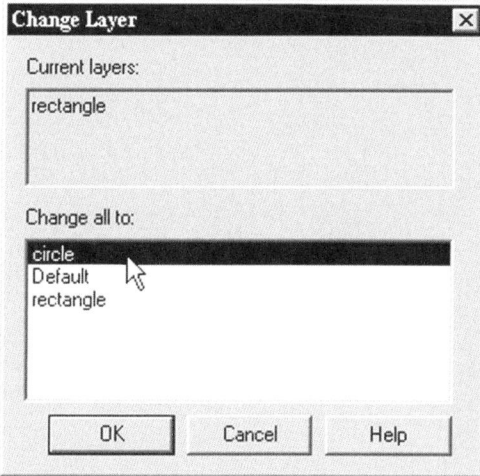

15 Click on the OK button.

Upon exiting the command and clicking on the graphics area, the selected rectangle disappears because it is moved to layer *circle* and the layer is turned off.

16 Click on layer *circle* on the Layer toolbar to set it as the current layer. The current layer is displayed. (See Figure 3–9.)

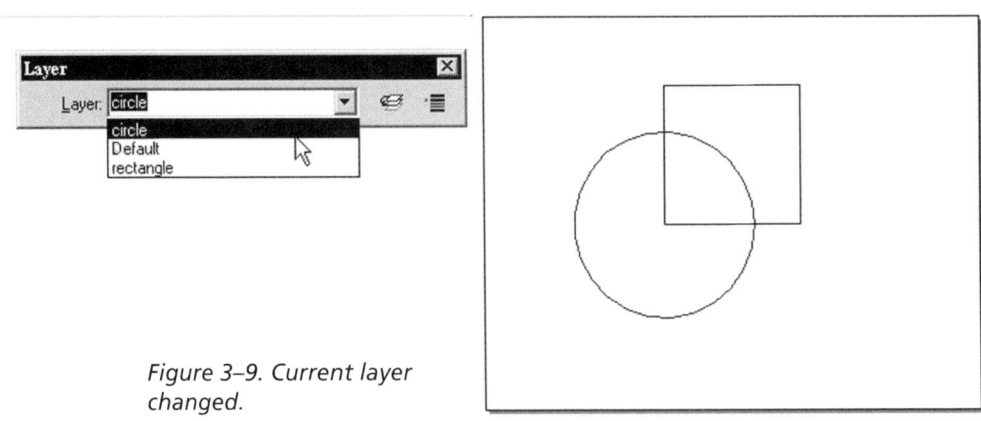

Figure 3–9. Current layer changed.

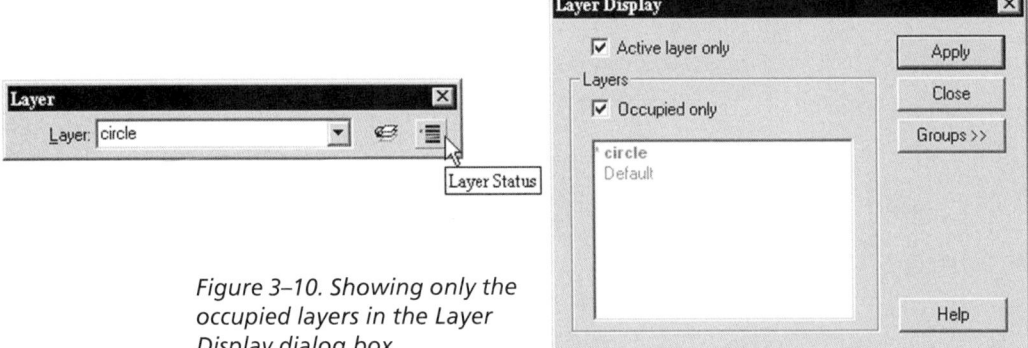

Figure 3–10. Showing only the occupied layers in the Layer Display dialog box.

17 Click on the Layer Status icon on the Layer toolbar.

18 In the Layer Display dialog box, click on *Occupied only*. Because layer *rectangle* becomes empty, it is not displayed in the list box.

Constructing, Modifying, and Deleting Layers and Layer Groups

Other than turning on and off layers and changing drawing elements' layer assignment, you can construct new layers and modify and delete existing layers. To handle two or more layers collectively, you can put them into groups. Contrary to grouping, you can ungroup grouped layers to revert them to individual layers.

Because layers are unique to individual drawing sheets of a document, you can have different sets of layers in different drawing sheets (working or background). Perform the following steps.

1 Open the file *Chapter3Layer.igr*, if you already closed it.

2 Activate Sheet 2.

3 Select Tools > Layers to display the Layer dialog box, if it is already closed. As can be seen in Figure 3–11, there is only one layer in this drawing sheet despite having three layers in Sheet 1 of this document.

Figure 3–11. Layer dialog box showing only one layer in the drawing sheet.

4 Select Tools > Layer Groups. The Layer Groups dialog box is displayed. The dialog box has two boxes: Layers and Groups. (See Figure 3–12.)

5 In the Layers box of the Layer Groups dialog box, type the text string *Layer 1* over the layer name *Default*, thus changing the name of the default layer. Note that you can also change a layer's name by typing over the existing layer name in the layer name list of the Layer dialog box.

6 In the blank area at the upper left-hand corner of the Layers box of the Layer Groups dialog box, type the text string *Layer 2* and then press the Enter key. A layer named *Layer 2* is constructed. (See Figure 3–13.)

7 Repeat step 6 to construct another layer named *Layer 3*. (See Figure 3–14.)

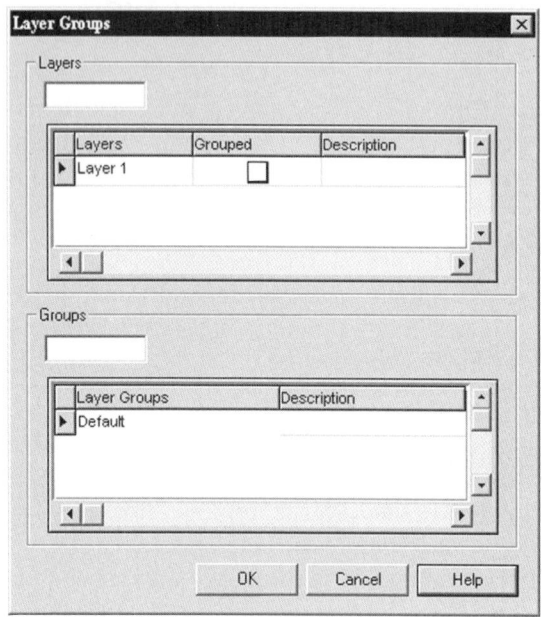

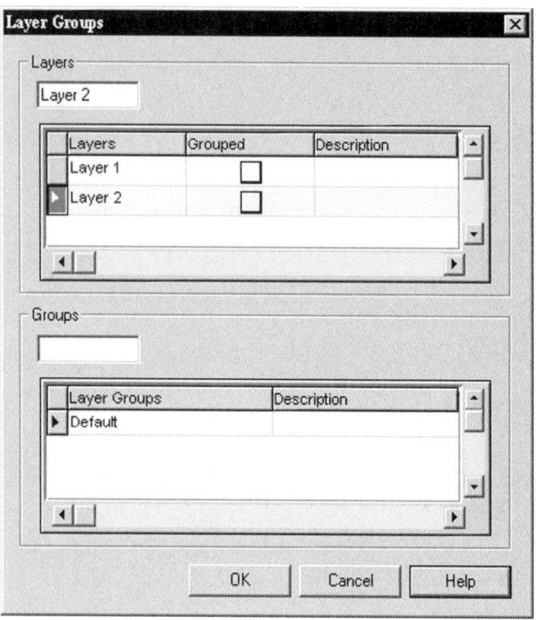

Figure 3–12. Default layer's name changed in the Layers box of the Layer Groups dialog box

Figure 3–13. New layer constructed in the Layer Groups dialog box.

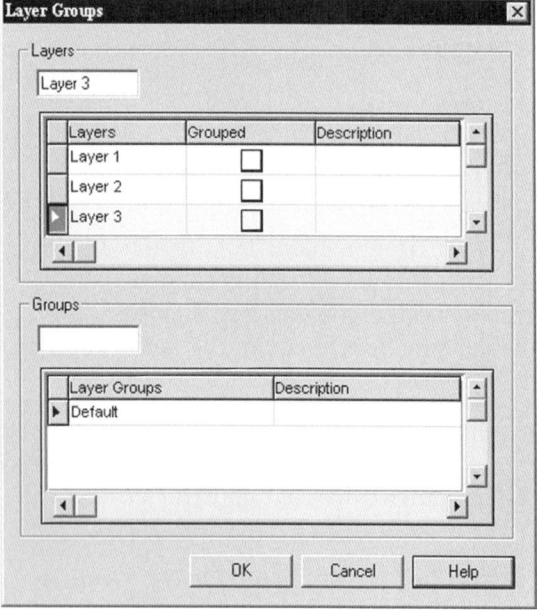

Figure 3–14. Second new layer constructed.

8 In the Groups box of the Layer Groups dialog box, type the text string *Group 1* over the existing default group name to change the group's name.

9 Click on *group 1* to activate the group and then click on the Grouped column of *Layer 1* and *Layer 2* from the Layers box. These two layers are now grouped with the group name *Group 1*. (See Figure 3–15.)

10 Type the text string *Group 2* in the box near the top left-hand corner of the Groups box of the Layer Groups dialog box. A group is constructed.

11 Click on *Group 2* to activate it (See Figure 3-16).

12 Click on *Layer 2* and *Layer 3* check boxes in the Grouped column of the Layers box. These two layers are grouped under the group name *Group 2*.

13 Click on the OK button to close the Layer Groups dialog box. Two groups are formed.

14 Click on the Layer Status icon on the Layer toolbar.

15 In the Layer Display dialog box, click on the Groups button to expand the dialog box, if it is not already expanded.

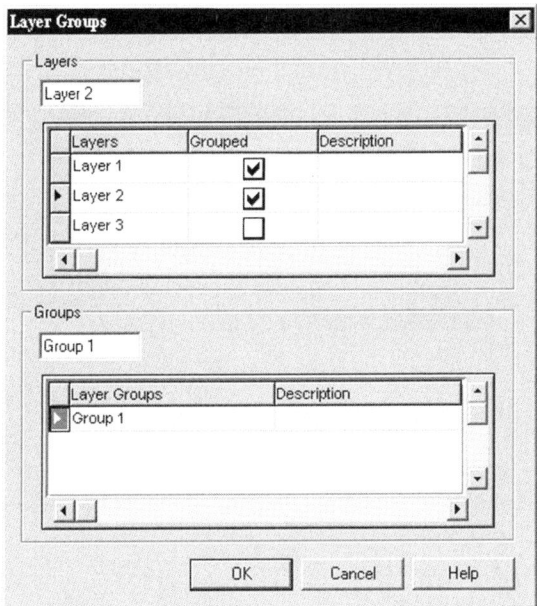

Figure 3–15. A group of layers formed.

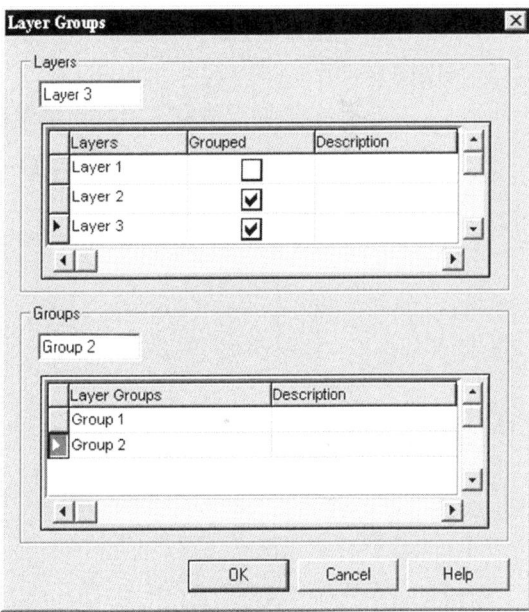

Figure 3–16. Second group of layers formed.

16 With reference to Figure 3–17, click on *Group 1* and then click on the Off button. *Layer 1* and *Layer 2*, belonging to *Group 1*, are turned off.

17 Click on *Group 1* and then click on the On button. *Layer 1* and *Layer 2* are turned on again.

18 Click on *Group 2* and then click on the Off button. This time, only *Layer 2* of *Group 2* is turned off because Layer 1, though it belongs to *Group 2*, is the current layer. (See Figure 3–18.)

19 Click on the Close button to close the dialog box.

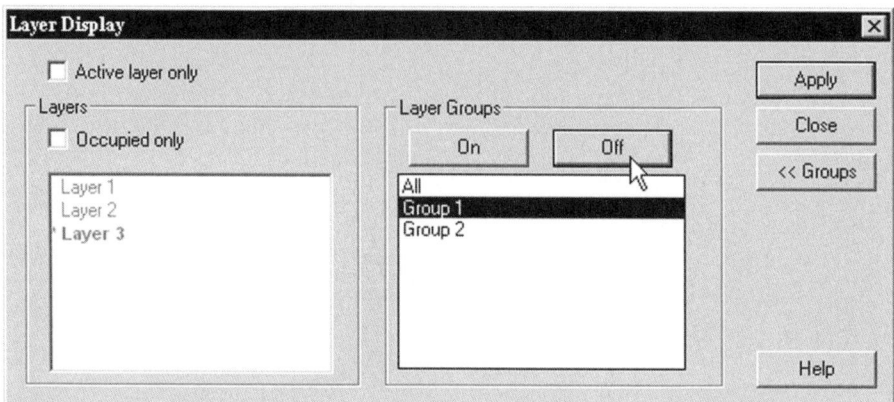

*Figure 3–17.
Turning off
a layer
group.*

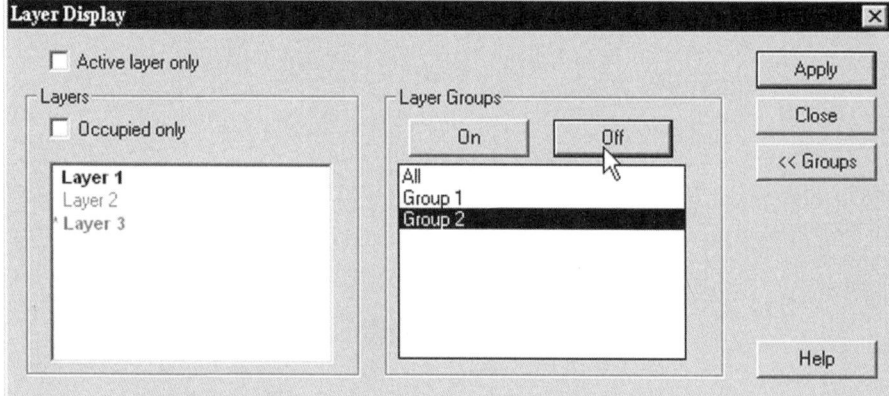

*Figure 3–18.
A layer of
Group 2
turned off.*

Using the Display Manager

The Display Manager controls the way drawing elements are displayed.
Perform the following steps.

1 Select Tools > Display Manager. The Display Manager dialog box
has two tabs: Sheets and Layers. (See Figure 3–19.) Note that the
Sheets tab determines the type of drawing sheets displayed, working sheets or background sheets.

Both the Sheets and Layers tabs have six columns. The Sheets tab controls the way drawing elements appear on a drawing sheet, and the Layers tab controls the way layers or layer groups appear. Common to them
are five columns: Display, Lock, Color, Line Type, and Width. The Display column turns on or off the display of layers. However, current layers
are not allowed to be turned off. The Lock column locks layers so that
drawing elements residing on the layers are locked and not allowed to

be manipulated. Again, current layers are not allowed to be locked. The Color, Line Type, and Width columns override the color, line type, and line width assignment to drawing elements.

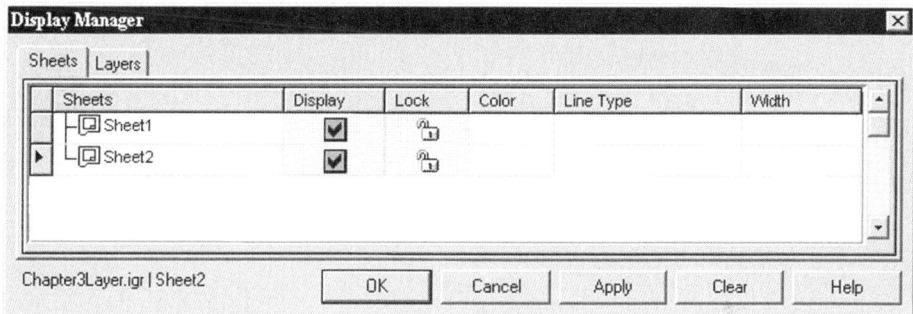

Figure 3–19. Display Manager dialog box.

2 In the Sheets tab of the Display Manager dialog box, click on Sheet 1.

3 Click on the Color column and select the color red from the pop-up color selection dialog box. (See Figure 3–20.)

4 Click on the OK button and activate Sheet 1. You will find that the color of the drawing elements in Sheet 1 are all changed to red.

5 Select Tools > Display Manager and select the Layers tab.

6 Click on layer *circle*.

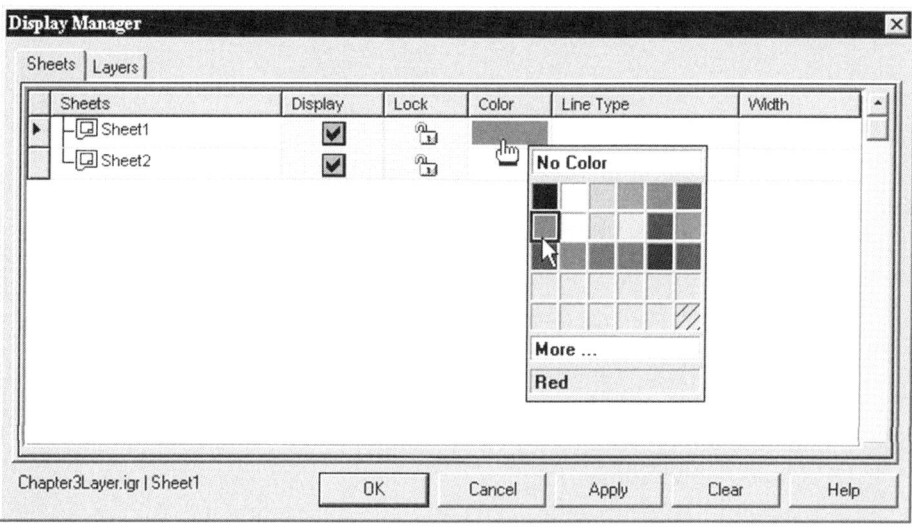

Figure 3–20. Changing the color assignment of drawing elements in all layers of Sheet 1.

7 Click on the color column and change the color to blue. (See Figure 3–21.)

8 Click on the OK button. You will find that drawing elements on layer *circle* are all changed to the color blue.

9 Select Tools > Display Manager.

10 Click on the Line Type column of the *rectangle* layer.

11 Click on the dashed line from the pop-up dialog box. (See Figure 3–22.) Line type on this layer is overridden.

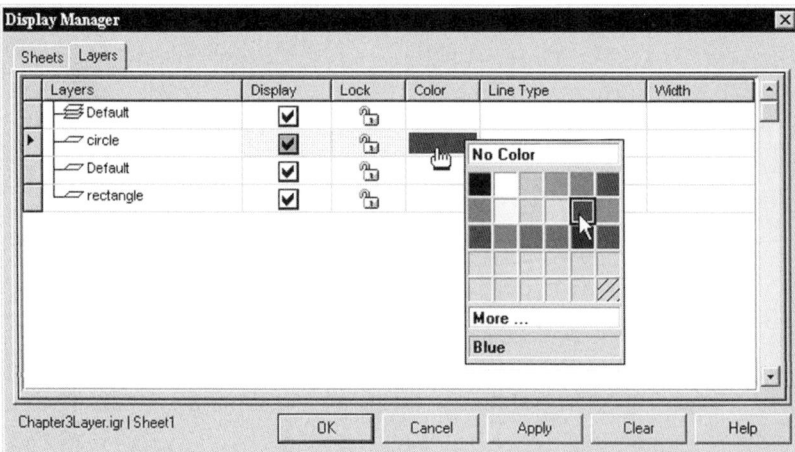

Figure 3–21. Color assignment of a layer in a sheet being overridden.

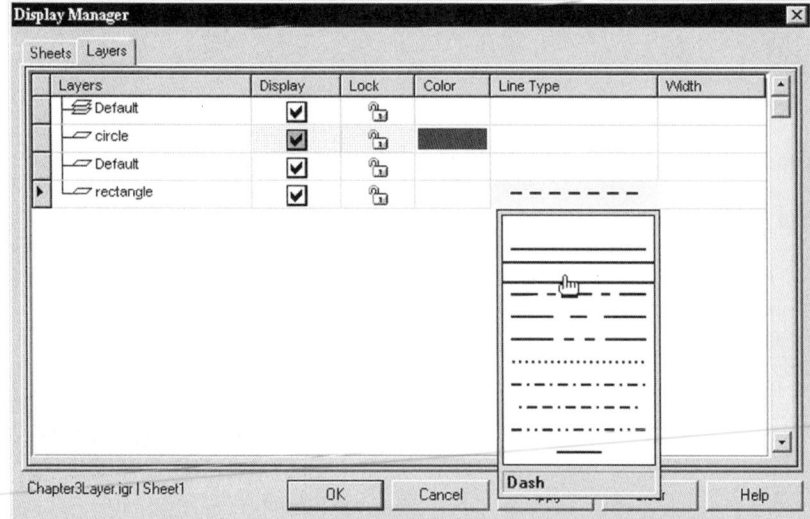

Figure 3–22. Line type being overridden.

12 Click on the OK button.

13 Save the file in the computer and close the file.

▪ ▪ ▪ ▪ Formatting Lines, Text, and Dimensions

The way drawing elements appear in a document depends on the active line style, text box style, or dimension style. In a document, you can maintain a set of styles and apply one of the styles to the drawing elements. Besides using the current style in drafting, you may override the settings in the current style.

Manipulating Line Style

Line style is determined by color, line width, and line type of line elements in a document.

Managing Line Style

In Chapter 2 and previous tutorials in this chapter you have been using the default line style in the construction of drawing elements. Perform the following steps to manipulate line styles in a document.

1 Start a new document by using the template file *Technical Drawing (Imperial.igr)* from the *Drawing* subfolder of the *templates* folder.

2 Select Format > Style. (See Figure 3–23.)

Figure 3–23. Style dialog box.

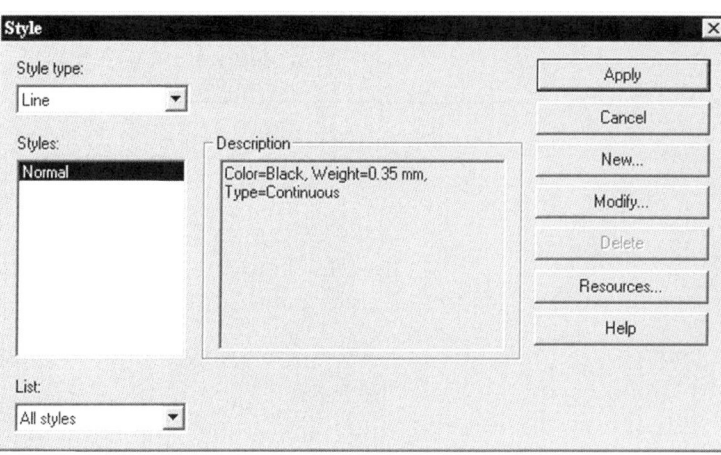

3   Select Line in the *Style type* pull-down list box of the Style dialog box.

The Style dialog box can be used to manipulate dimension, line, and text box styles. Common to three styles, you can select a style from the style list box at the left and then click on the Apply button to apply the style to the document. You can click on the New button to construct a new style, click on the Modify button to modify an existing style, click on the Delete button to remove a style, and click on the Resources button to include a resource document.

4   Click on the Resources button. This is particularly useful if you already set up styles in another document.

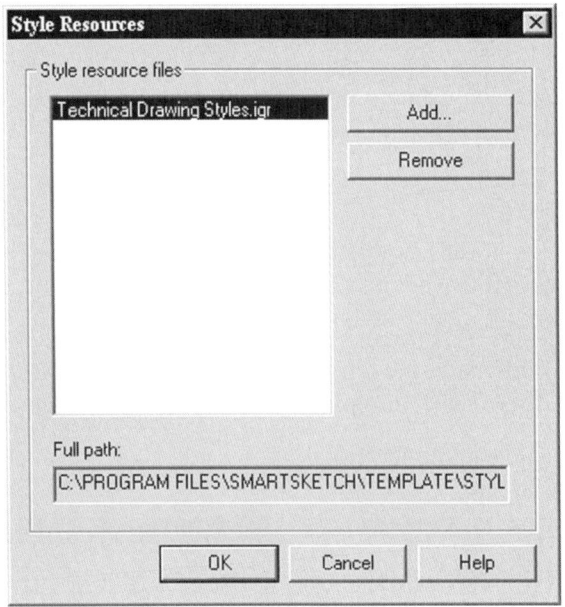

In the Style Resources dialog box shown in Figure 3–24, there are a list box, a number of buttons, and a path list. The list box shows the styles available. To include the styles from another document, click on the Add button and select a document. To remove a resource reference from the list box, select the resource file and click on the Remove button.

5   Click on the Cancel button to exit without adding or removing any resource files.

6   Click on the Modify button.

7   In the Modify Line Style dialog box shown in Figure 3–25, select the General tab, change the line width to 0.5 mm, and click on the OK button.

Figure 3–24. Style Resources dialog box.

8   Return to the Style dialog box shown in Figure 3–23 and click on the New button.

9   In the New Line Style dialog box shown in Figure 3–26, specify the line style name *Line01* in the Name box of the Name tab.

10   Select the General tab of the New Line Style dialog box. (See Figure 3–27.)

11   Set the color to blue, line width to 0.25 mm, and line type to a dashed line.

12   Click on the OK button.

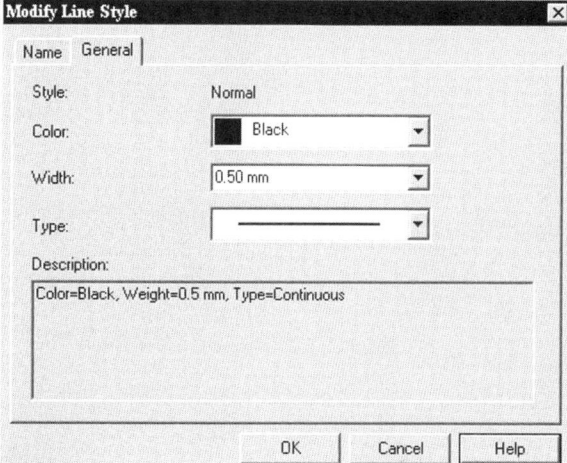

Figure 3–25. Modify Line Style dialog box.

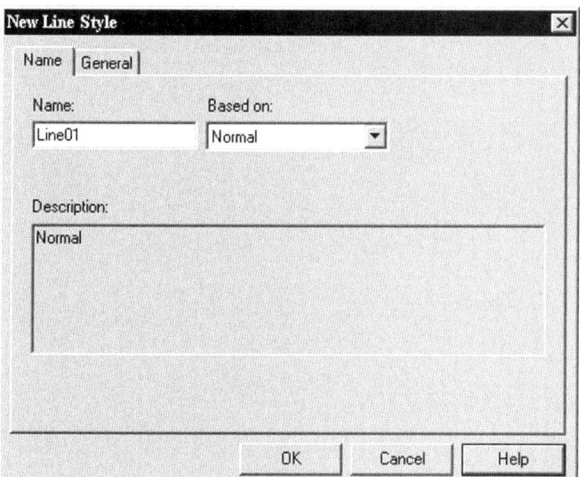

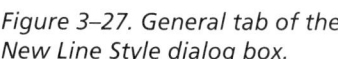

Figure 3–26. Line style name specified.

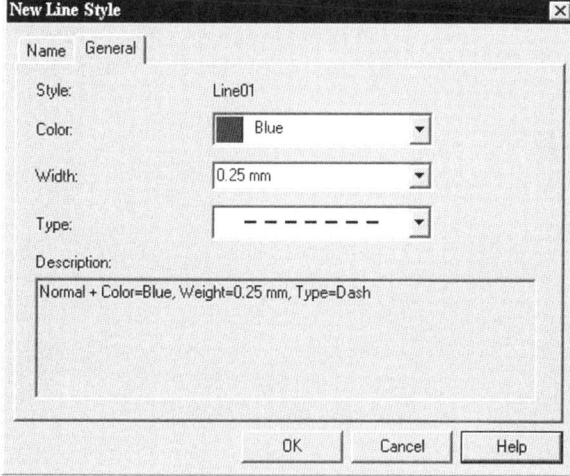

Figure 3–27. General tab of the New Line Style dialog box.

13 Return to the Style dialog box. You will find two line styles listed in the list box. Select the Normal style. Then click on the Apply and Close buttons. Now you have two line styles. The Normal style is the active line style.

Perform the following steps to construct a line with Normal line style.

1 Select Tools > PinPoint, if PinPoint is not already activated.

2 Click on the Line/Arc Continuous icon on the Draw toolbar.

3 Select location A (Figure 3–28, coordinates X:2/Y:2) and then location B (Figure 3–28, coordinates X:4/Y:5) to construct a line segment.

4 Right click to terminate. A line segment taking on Normal line style is constructed.

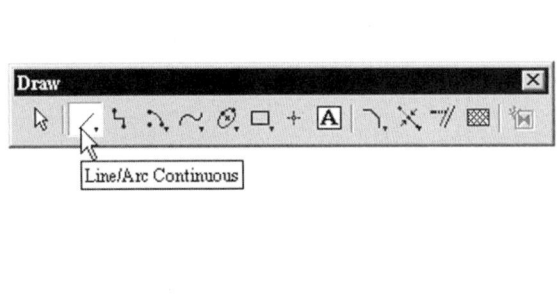
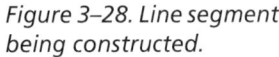

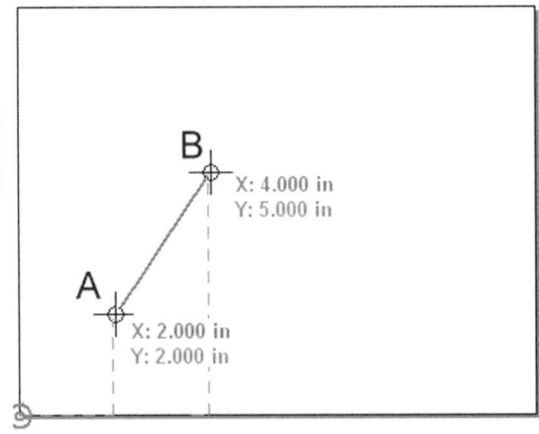

Figure 3–28. Line segment being constructed.

Perform the following steps to construct another line with *Line01* line style.

1 Select Format > Style.

2 Select Line from the *Style type* pull-down list box of the Style dialog box.

3 Select the *Line01* line style and click on the Apply button.

4 Close the dialog box by clicking on the Cancel button.

5 Click on the Line/Arc Continuous icon on the Draw toolbar.

6 Select location A (Figure 3–29, coordinates X:6/Y:2) and then location B (Figure 3–29, coordinates X:8/Y:5) to construct a line segment.

7 Right click to terminate. A line segment taking on *Line01* line style is constructed.

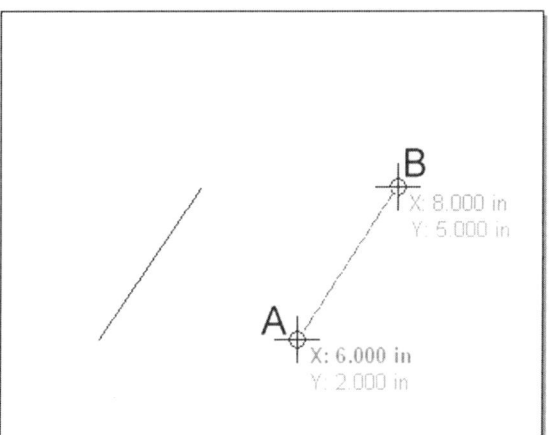

Figure 3–29. Second line segment, with a different line style, constructed.

Using the Line Formatting Tool

Other than using the current line style in construction of drawing elements, you can override the settings by performing the following steps.

1 Select Format > Line.

2 Because *Line01* is the current line style, the initial settings shown in the Format Line dialog box should be blue in color, 0.25 mm in width, and dash in line type.

3 Click on the arrow at the right-hand side of the Color pull-down list box and select the Red option.

4 Click on the arrow at the right-hand side of the Width pull-down list box and click on 2.0 mm.

5 Click on the OK button. As can be seen in the description shown in Figure 3–30, the line format is now red in color, 2.00 mm in width, and continuous in line type. In other words, any drawing elements you construct will now take on this format.

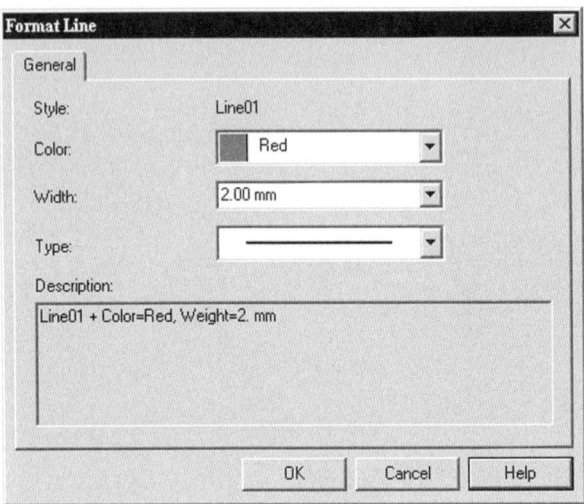

Figure 3–30. Format Line dialog box.

Perform the following steps to construct a circle with the current line format.

1 Click on the Line/Arc Continuous icon on the Draw toolbar.

2 Click on location A (Figure 3–31, coordinates X:3/Y:2) and then location B (Figure 3–31, coordinates X:6/Y:6). A line is constructed.

3 While the Line/Arc Continuous command is active, click on the Line Color button on the ribbon and change the color to purple. (See Figure 3–32.)

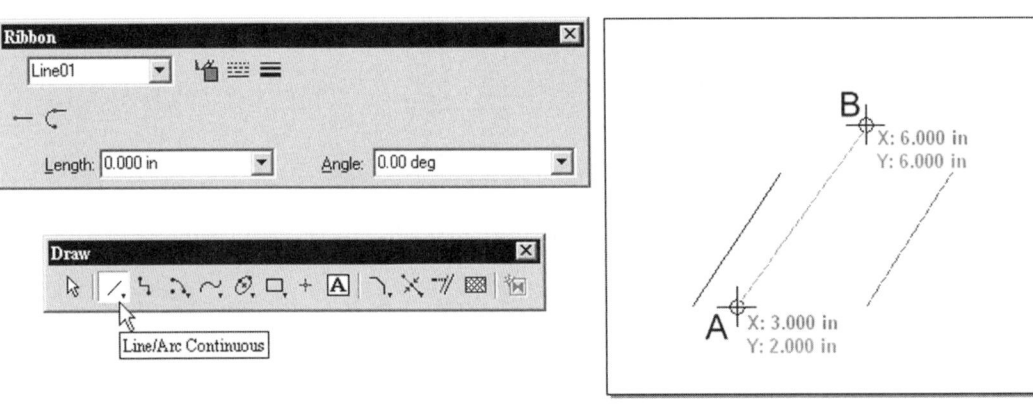

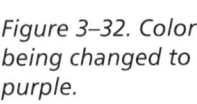

Figure 3–31. Line with a different style constructed.

Figure 3–32. Color being changed to purple.

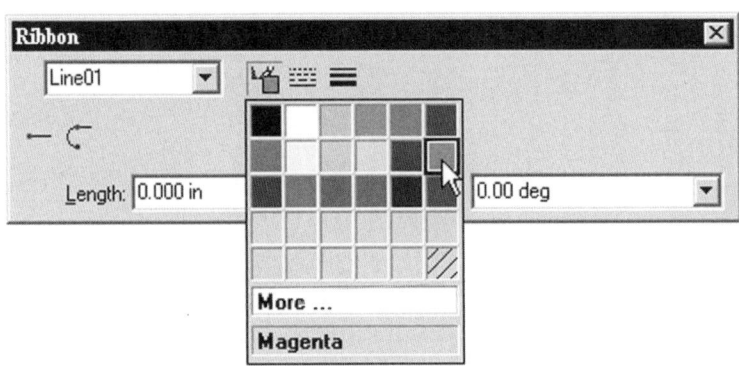

4 Click on the Line Type button and select the line style shown in Figure 3–33.

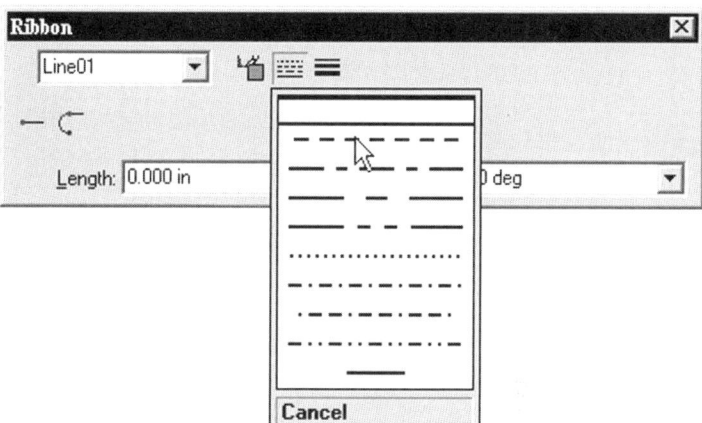

Figure 3–33. Line type being changed.

5 Click on the Line Width button and change the line width to 0.5 mm. (See Figure 3–34.)

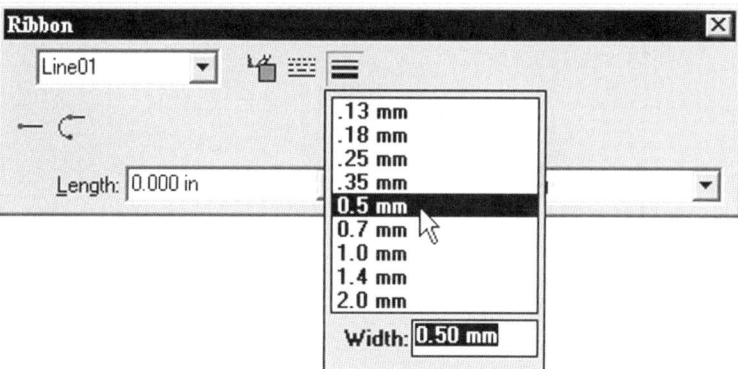

Figure 3–34. Line width being changed to 0.5 mm.

6 With the Line/Arc Continuous command active, click on location A (Figure 3–35, coordinates X:4/Y:1) and then location B (Figure 3–35, coordinates X:8/Y:7). A line with a different line style is constructed. Note that any drawing elements you construct will now take on this line format.

7 Save the file as *LineStyle.igr* and then close the file.

If you wish to set the line style to one of the saved line styles, select Format > Style, select a line style, and click on the Apply button.

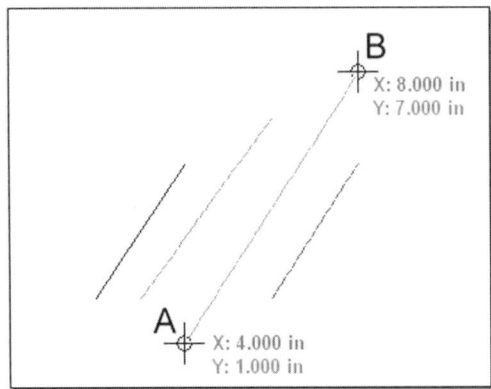

Figure 3–35. Line with a different line format being constructed.

Manipulating Text Box Style

Text box style is the display appearance of a text box in a document.

Setting Text Box Style

Prior to constructing text objects in your document you should spend some time to set a text box style or to override the settings of the current text box style. Perform the following steps.

1   Start a new document by using the template file *Technical Drawing (Imperial.igr)* from the *Drawing* subfolder of the *templates* folder.

2   Select Format > Style.

3   In the Style dialog box, select Text from the *Style type* pull-down list box at the upper left-hand corner. (See Figure 3–36.)

Figure 3–36. Text box style selected in the Style dialog box.

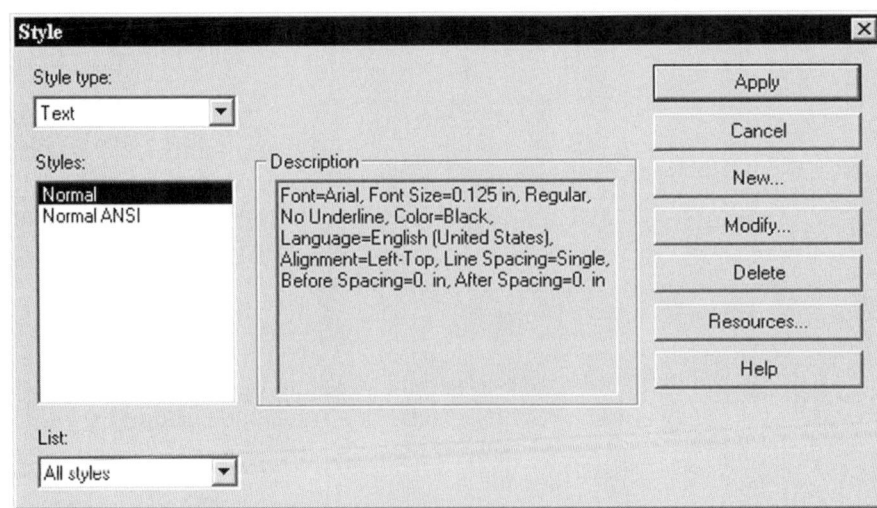

4 Click on the New button to construct a new text box style.

5 In the Name tab of the New Text Box Style dialog box, specify *Text01* in the Name field. (See Figure 3–37.)

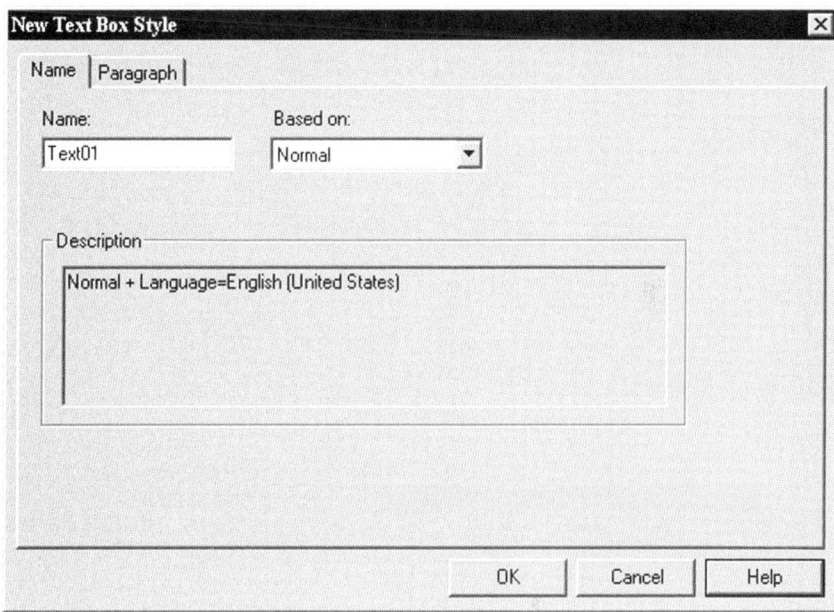

Figure 3–37. Specifying text box style name.

6 Select the Paragraph tab.

7 In the Paragraph tab, shown in Figure 3–38, set the font size to 0.5 inch, set the font style to bold, and click on the OK button. A text box style is constructed.

8 Return to the Style dialog box and click on the Apply and Close buttons.

9 Select Tools > PinPoint, if PinPoint is not already activated.

10 Click on the Text Box icon on the Draw toolbar.

11 Click on location A indicated in Figure 3–39 (coordinates X:2/Y:5).

12 Type the text string *THIS IS A TEXT STRING*.

13 Right click. A text string is constructed.

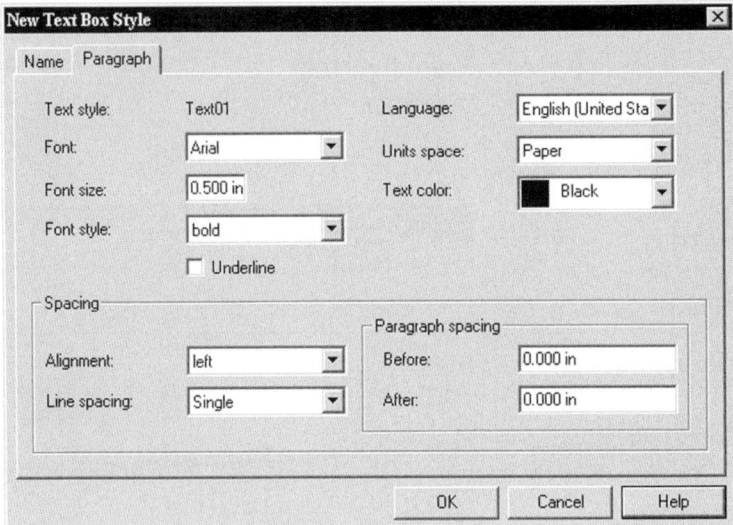

Figure 3–38. Paragraph parameters being set.

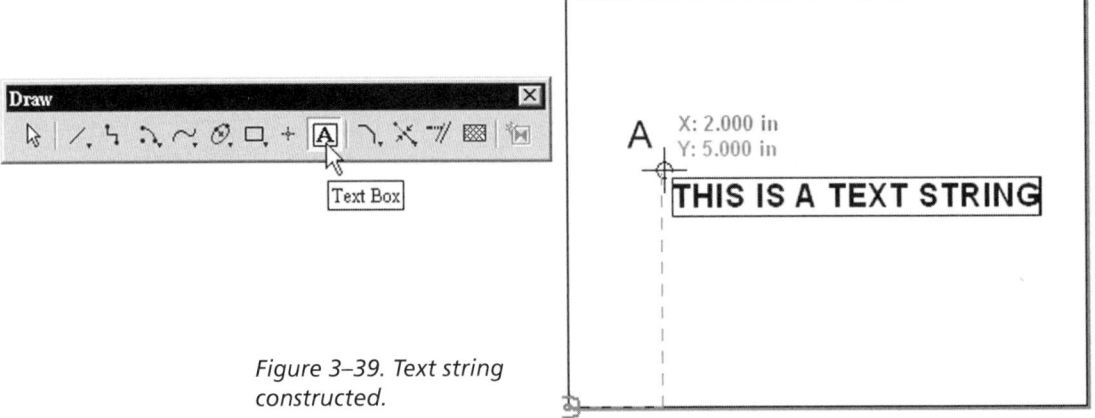

Figure 3–39. Text string constructed.

Using the Text Formatting Tool

Like working with line styles, you can override settings concerning the appearance of a text box without setting up a new style. Perform the following steps.

1 Select Format > Text Box.

2 In the Format Text Box dialog box, shown in Figure 3–40, set the font size to 1 inch, font style to italic, and text color to dark blue.

3 Click on the OK button. The current text box style is overridden.

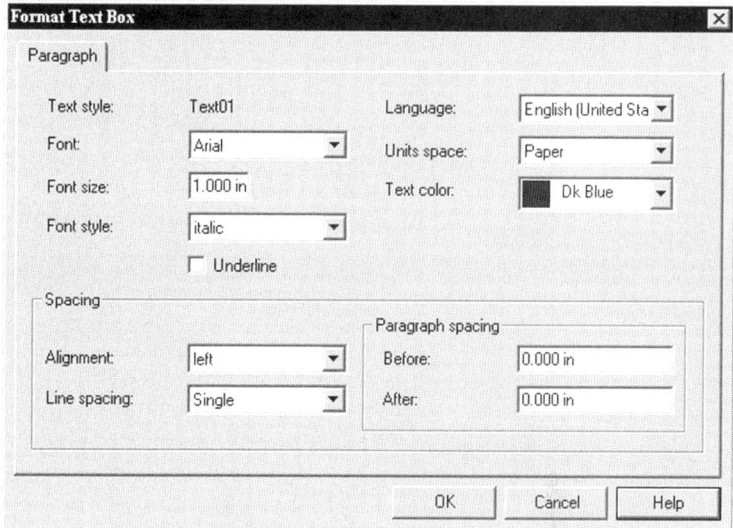

Figure 3–40. Format Text Box dialog box.

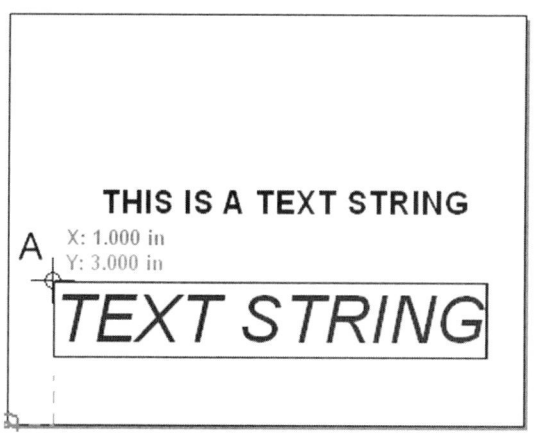

Figure 3–41. Text string constructed.

4 Click on the Text Box icon on the Draw toolbar. Click on location A indicated in Figure 3–41 (coordinates X:1/Y:3) and type *TEXT STRING*.

5 Right click.

The ribbon associated with the Text Box command has most of the settings related to text box formatting. (See Figure 3–42.)

This ribbon has a number of options. They are, from left to right, Text Style Name, Font Type, Font Size, Font Color, Bold, Italic, Underline, Paragraph Alignment, Border, and More (options). The Text Style box enables you to select a text style, and the other options enable you to override the settings of the selected text box style.

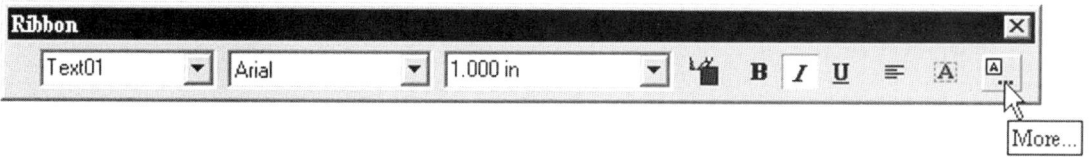

Figure 3–42. Ribbon associated with the Text Box command.

The extended ribbon has seven additional boxes/options. These options are the Height, Width and Angle drop-down boxes and the Horizontal Text Orientation, Vertical Text Orientation, Left to Right, and Right to Left icons. Perform the following steps.

1   Click on the More button on the ribbon. (See Figure 3–43.)

Figure 3–43. Text ribbon extension.

2   While the Text Box command is active, click on the B button (B stands for bold) on the ribbon, clear the I button (I stands for italic), and set the color to red. (See Figure 3–44.)

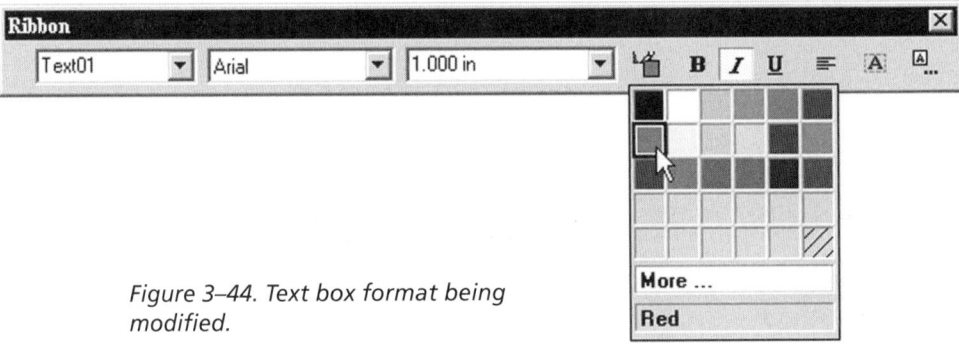

Figure 3–44. Text box format being modified.

3   Click on location A indicated in Figure 3–45 (coordinates X:1/Y:7) and type *TEXT STRING.*

4   Right click.

5   Save the file as *TextBoxStyle.igr* and then close the file.

Understanding Dimension Style

Dimension style concerns the display appearance of dimensions in a document. Similar to line style and text box style management, you can construct a number of dimension styles in a document and make one of them the current style. Dimensions you construct thereafter will take on

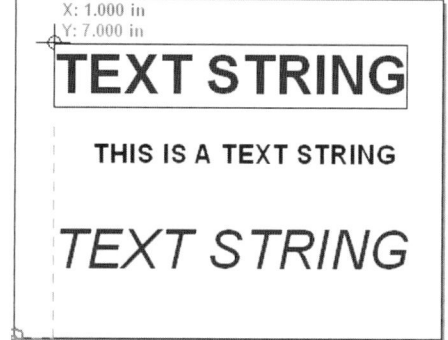

Figure 3–45. Text string constructed.

this style. Also similar to line style and text box style, you can override the current dimension style without constructing a new dimension style. Perform the following steps to familiarize yourself with the settings for dimension styles.

1 Start a new document by using the template file *Technical Drawing (Imperial.igr)* from the *Drawing* subfolder of the *templates* folder.

2 Select Format > Style.

3 Select Dimension from the Style list and then click on the New button.

4 The New Dimension Style dialog box has a number of tabs. Specify a style name *Dimension01* in the Name tab, shown in Figure 3–46.

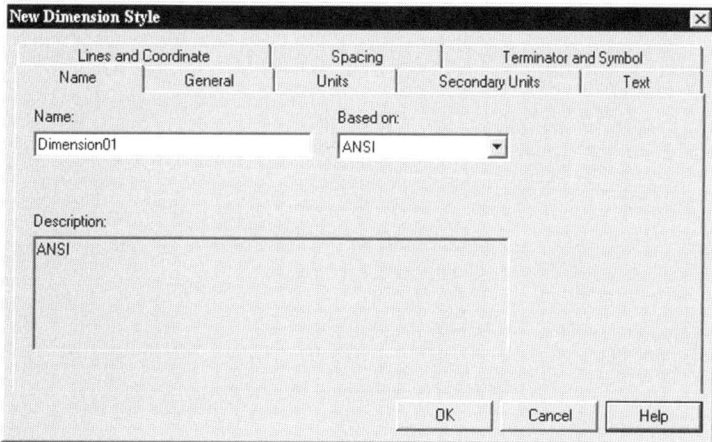

Figure 3–46. Name tab of the New Dimension Style dialog box.

5 Select the General tab. (See Figure 3–47.) This tab sets the color and scale.

There are two types of dimensions: driving and driven. They will be explained in more detail later. Error dimension concerns dimensions that cannot be solved after drawing elements are modified. Because each drawing sheet has its own scale, you can select Automatic scale to let the system determine the actual dimension value or you can input a scale value for the dimensions.

Figure 3–47.
General tab.

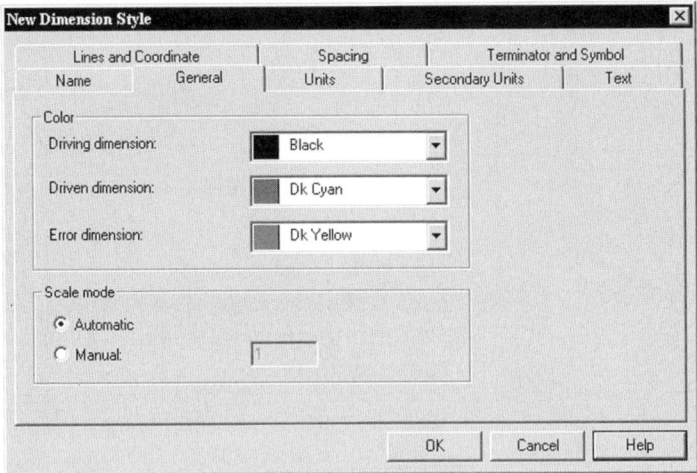

6   Select the Units tab. (See Figure 3–48.) This tab establishes the units of dimensions.

7   Select the Secondary Units tab. (See Figure 3–49.) Secondary units are units complementary to the basic units. For example, you can state metric units as secondary units in conjunction with primary units.

Figure 3–48.
Units tab.

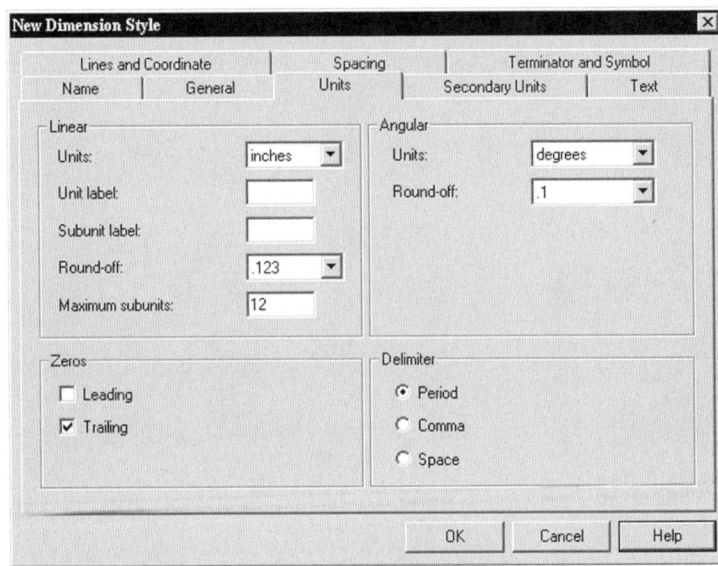

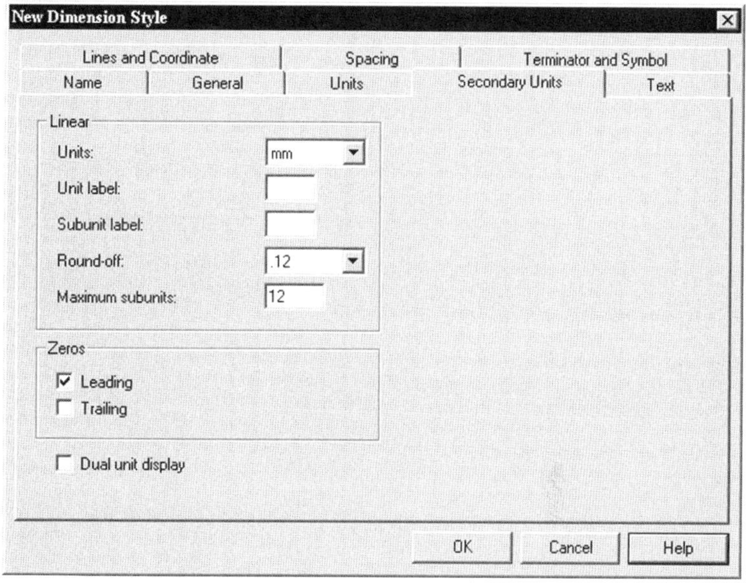

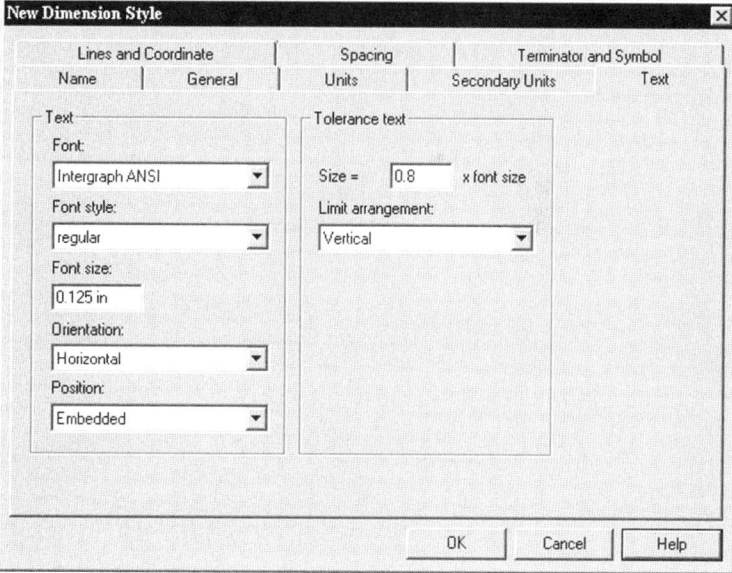

Figure 3–49.
Secondary
Units tab.

Figure 3–50.
Text tab.

8 Select the Text tab. (See Figure 3–50.) The Text tab establishes dimensions' text values.

9 Select the Lines and Coordinate tab. (See Figure 3–51.) This tab establishes the dimension lines, projection lines, and coordinates of a dimension.

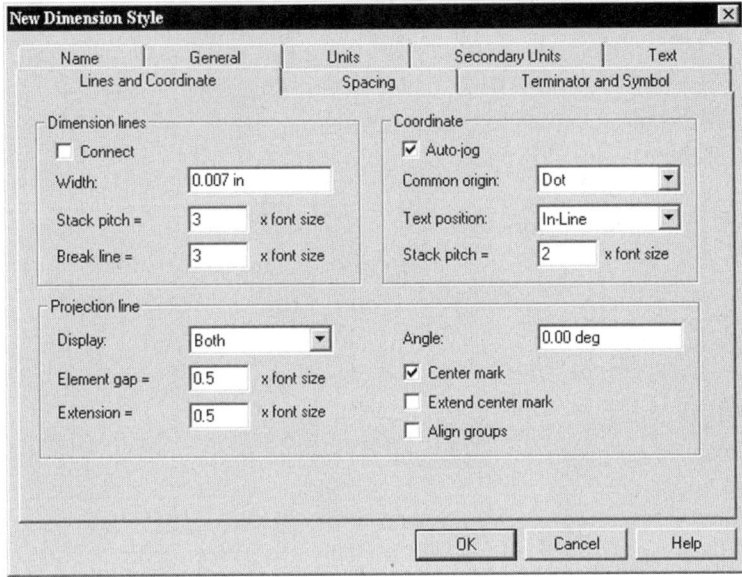

*Figure 3–51.
Lines and
Coordinate
tab.*

10 Select the Spacing tab. (See Figure 3–52.) This tab establishes the spacing of lines and text of a dimension.

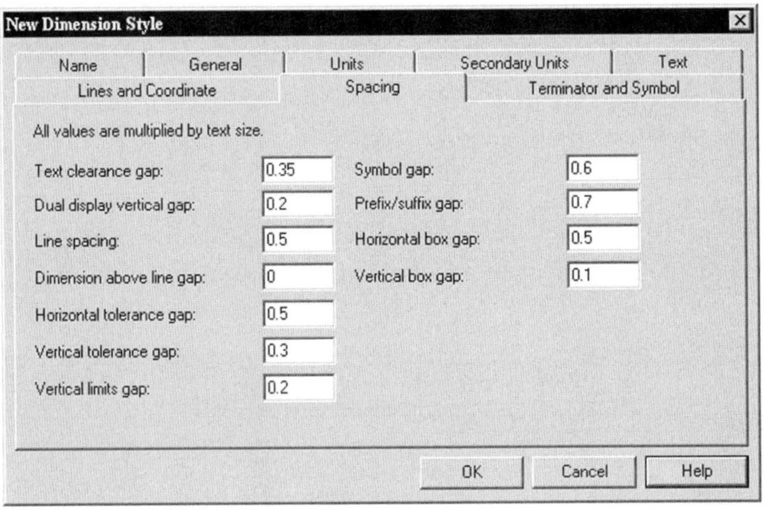

*Figure 3–52.
Spacing tab.*

11 Select the Terminator and Symbol tab. (See Figure 3–53.) This tab establishes the type of terminator (arrowhead type) used in a dimension symbol, if any, to be included in the dimension.

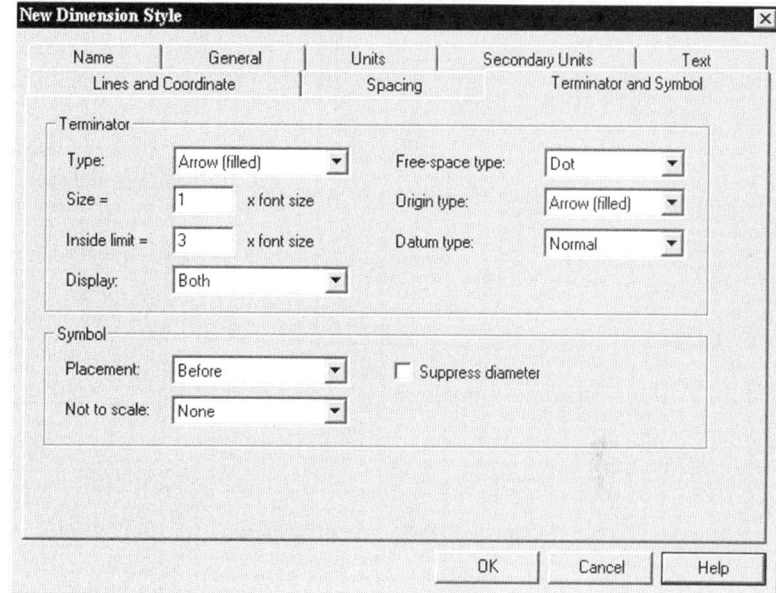

Figure 3–53. Terminator and Symbol tab.

12 Click on the OK button.

13 Close the file without saving.

■ ■ ■ ■ Working on Drawing Element Properties

Previously you have worked with setting up and managing layers and manipulating line styles, text box styles, and dimension styles. One thing all of these tools have in common is that they all affect the drawing elements that are constructed after the settings are made. If you already constructed drawing elements and wanted to change their layer assignment or their display style, you would use the Properties tool from the Edit pull-down menu. Perform the following steps.

1 Open the file *Chapter3Edit.igr* from the companion CD-ROM and save it to the working folder of your computer.

2 Click on the Select Tool icon on the Draw toolbar.

3 Select A indicated in Figure 3–54.

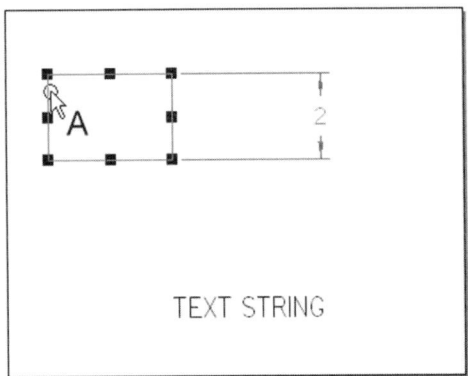

Figure 3–54. Rectangle selected.

4 Select Edit > Properties. This command is context sensitive. If the selected object is a line object, the Element Properties dialog box is displayed. If the selected object is a dimension, the Dimension Properties dialog box is displayed. If the selected object is a text object, the Text Box Properties dialog box is displayed.

The Element Properties dialog box has three tabs: Info, Format, and User. The Info tab concerns the parameters of the drawing element, the Format tab concerns the line style of the drawing element, and the User tab concerns attributes to be discussed later in this chapter.

5 Select the Info tab, if it is not already selected.

6 Change the height to 3 inches. (See Figure 3–55.)

Figure 3–55.
Info tab.

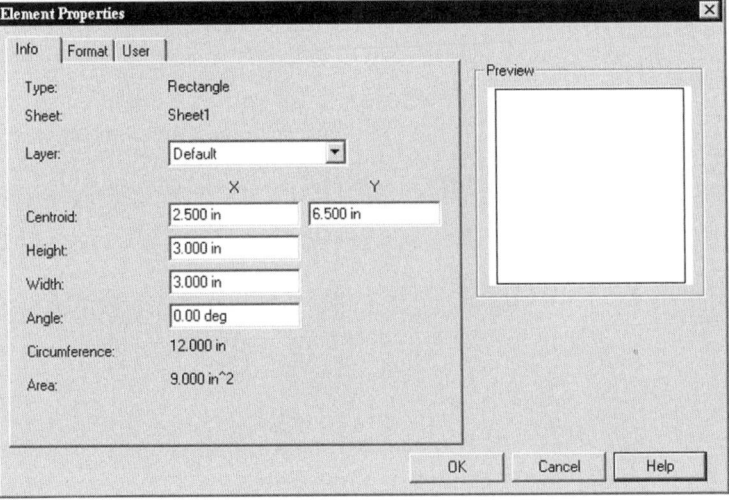

7 Select the Format tab.

8 Change line color to blue and line width to 1 mm. (See Figure 3–56.)

9 Click on the OK button. Note that not only the rectangle's height, color, and line width are changed but the dimension value is changed. We will explain more about driving and driven dimensions later. (See Figure 3–57.)

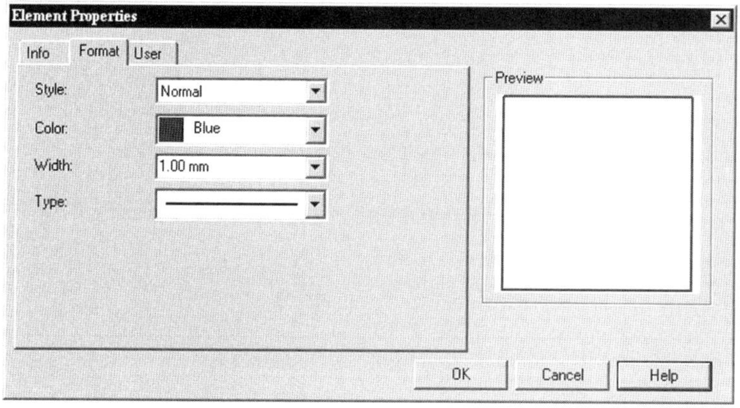

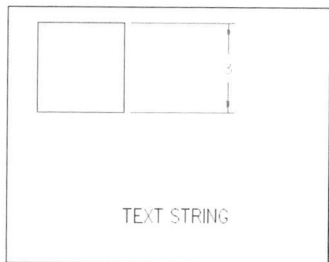

Figure 3–56. Format tab.

Figure 3–57. Rectangle edited.

10 Click on the Select Tool icon on the Draw toolbar and select text box A indicated in Figure 3–58.

11 Select Edit > Properties.

The Text Box Properties dialog box has four tabs: Info, Paragraph, Border and Fill, and User.

12 Select the Info tab, if it is not already selected.

13 Change the angle to 15 degrees. (See Figure 3–59.)

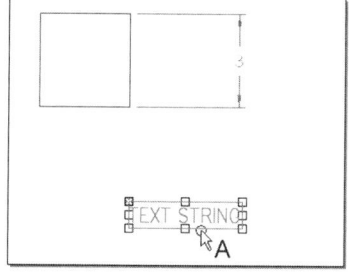

Figure 3–58. Text box selected.

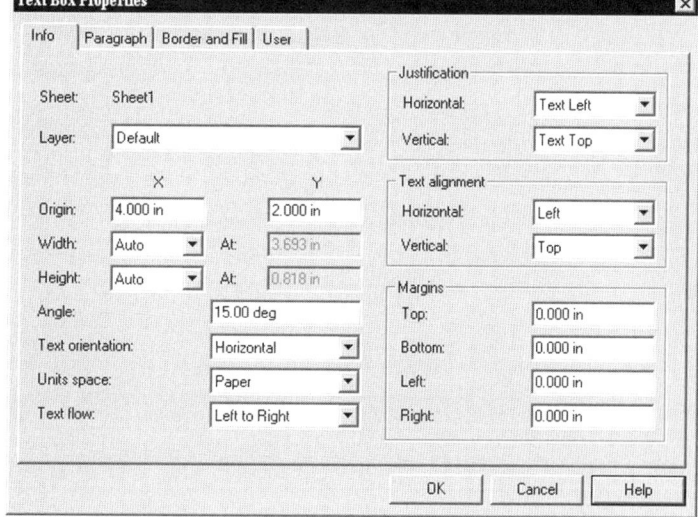

Figure 3–59. Info tab.

14 Select the Paragraph tab.

15 Change the font size to 0.7 inch.

16 Change the font color to red. (See Figure 3–60.)

17 Select the Border and Fill tab.

18 Check the *Show border* and Shadow boxes.

19 Change the color of text to blue.

20 Set the fill color to light gray. (See Figure 3–61.)

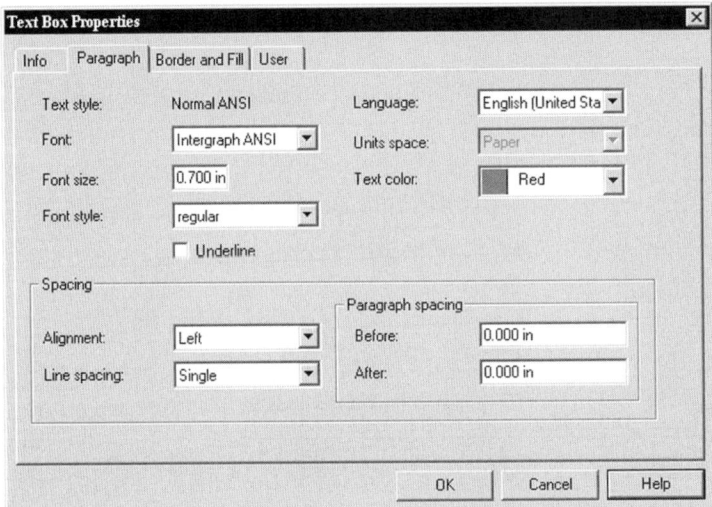

Figure 3–60.
Paragraph tab.

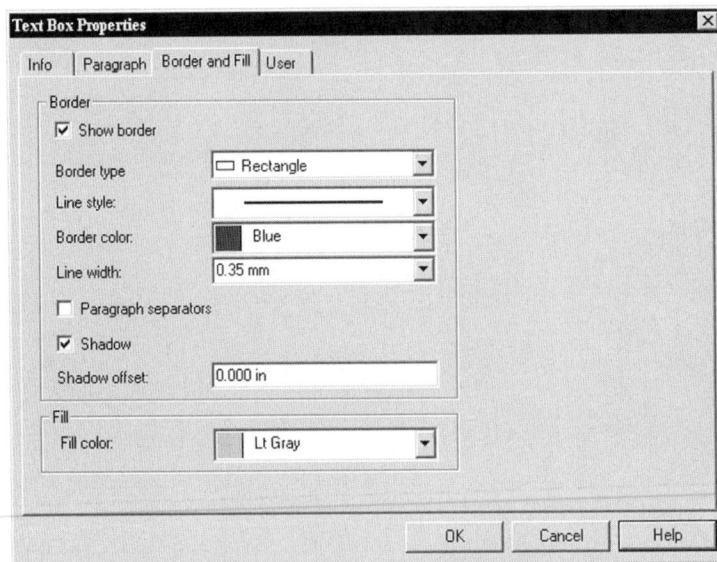

Figure 3–61.
Border and
Fill tab.

21 Click on the OK button. The text box is edited. (See Figure 3–62.)

22 Select dimension A indicated in Figure 3–63.

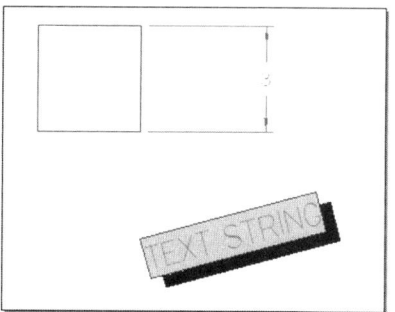

Figure 3–62. Text box edited.

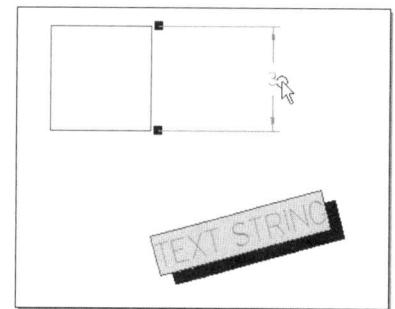

Figure 3–63. Dimension selected.

23 Select Edit > Properties.

24 Select the Units tab.

25 Change Round-off to 0.1. (See Figure 3–64.)

Figure 3–64.
Dimension's round-
off value changed.

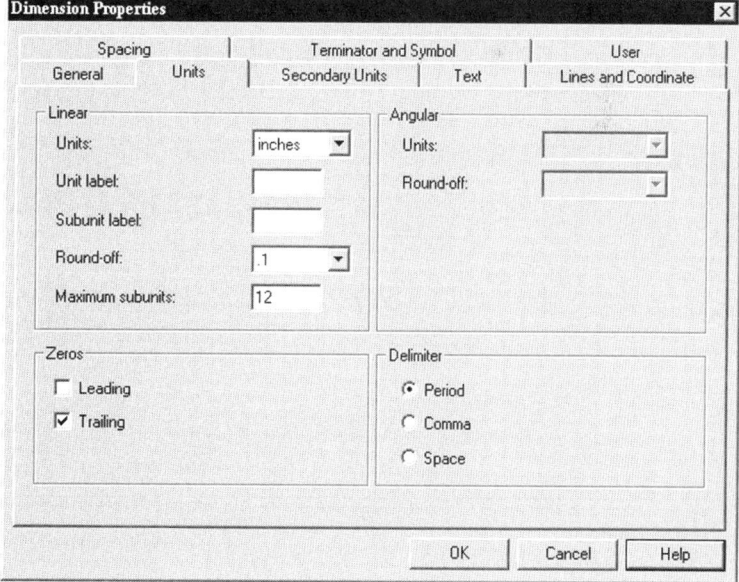

26 Click on the OK button. The round-off value is changed. (See Figure 3–65.)

27 Save the file in a folder of your computer and close the file.

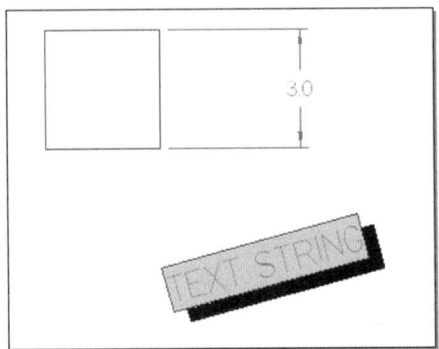

Figure 3–65.
Round-off value
changed.

■ ■ ■ ■ Dimensioning and Annotating a Document

Although drawing elements are constructed as full size and display scales are set in a document, it may be necessary to further illustrate a drawing element by placing dimensions and adding annotations.

Using the Dimension Toolbar

To dimension an engineering drawing, add leaders and balloons, and make measurements, you use the Dimension toolbar, shown in Figure 3–66. There are two ways to open the Dimension toolbar. In addition to using the Toolbar dialog box accessible from the View > Toolbar menu, you can click on the Dimension button on the Main toolbar.

Figure 3–66.
Dimension toolbar.

Options available from the Dimension toolbar are delineated in Table 3–1.

Table 3–1 Dimension Toolbar Options and Their Functions

Option	Function
Smart Dimension	Constructs a dimension for a line, circle, arc, ellipse, or curve
Distance Between	Constructs a dimension that measures the distance between two drawing elements or key points

Option	Function
Angle Between	Constructs a dimension that measures the angle between two drawing elements or key points
Axis	Sets the dimension axis for a drawing
Coordinate Dimension	Constructs coordinate dimensions
Symmetric Diameter	Constructs a dimension that measures the distance between a center line and another element or a key point
Measure Distance	Measures the distance between two selected drawing elements
Measure Area	Measures the area of selected elements
Character Map	Displays the Character Map dialog box
Leader	Constructs a leader
Balloon	Constructs a balloon
Dimension Text	Overrides a driven dimensional value with a text string

Commands on the Dimension toolbar can be divided into three groups. The first group governs dimensioning. It consists of the Smart Dimension, Distance Between, Angle Between, Axis, Coordinate Dimension, and Symmetric Diameter commands.

The second group, including the Measure Distance and Measure Area commands, is used for taking measurements. Finally, the Character Map command concerns insertion of special characters, the Leader command inserts a leader, and the Balloon command inserts a balloon and the dimension's text.

Understanding Driving and Driven Dimensions

Dimensions in a document may serve two purposes: to report the size and to control the size. In SmartSketch terminology, there are two types of dimensions: driving dimensions and driven dimensions.

Driving dimensions control the size, and driven dimensions simply report the size. If dimensions are set as driving, changing their values will modify the size of the associated drawing elements. For a driving dimension to be effective you need to click on Maintain Relationship from the Tools pull-down menu. Perform the following steps.

1   Open the file *Chapter2Dimension.igr* from the companion CD-ROM and save it to the working folder of your computer.

2 Activate Sheet 1, if it is not already the current working sheet.

3 Select Tools > Maintain Relationships.

Note that it is very important to click on Maintain Relationship if you are going to place a driving dimension.

4 If a ribbon is not displayed, select View > Toolbars, select Ribbon in the Toolbar list, and click on the OK button.

5 Click on the Smart Dimension icon on the Draw toolbar.

A ribbon common to all other dimension commands except the Axis command will display after selecting a dimension command. (See Figure 3–67.) Options on the ribbon are context sensitive, meaning that their availability depends on the types of commands and the types of drawing elements selected. Table 3–2 details these options.

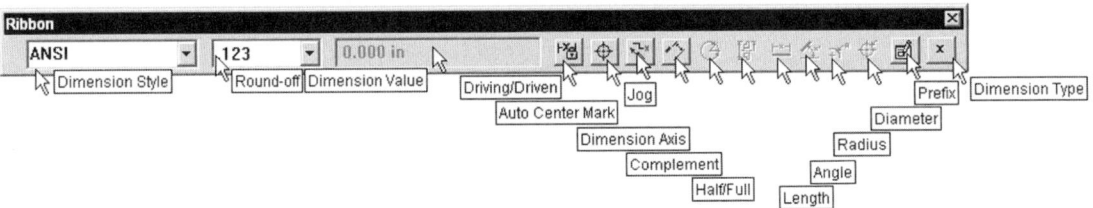

Figure 3–67. Dimension ribbon.

Table 3–2 Dimension Ribbon Options and Their Functions

Option	Function
Dimension Style	Lists the dimension styles available. (Dimension style will be discussed later.)
Round-off	Sets the round-off for the dimension value.
Dimension Value	Displays or sets the dimension value.
Driving/Driven	Toggles between setting the dimension as a driving dimension or a driven dimension. To set dimension to be driving, you must set Maintain Relationship from the Tools pull-down menu.
Auto Center Mark	Places a center-line crosshair at the center of the radius of curvature measured by the dimension.
Jog	Offsets the project lines of a coordinate or radial dimension.

Option	Function
Dimension Axis	Sets the orientation of dimensions constructed by the Distance Between or Coordinate Dimension command.
Complement	Places an angular dimension at the 180-degree complement.
Half/Full	Toggles between half and full for the Symmetric Diameter command.
Length	Places a length dimension for a line or arc.
Angle	Places an angle dimension for a line or arc.
Radius	Places a radial dimension for a circle, arc, or curve.
Diameter	Places a diameter dimension for a circle or an arc.
Prefix	Adds prefix, suffix, superfix, and subfix for a dimension.
Dimension Type	Sets dimension types and their related tolerances.

Because Maintain Relationships is already active, the Driving/Driven button on the ribbon is checked, denoting that the dimension you are going to place is a driving dimension. If Maintain Relationship is not clicked on, the Driving/Driven button will be grayed out, denoting that only driven dimensions can be created.

6 Select line A indicated in Figure 3–68.

7 Click on location B indicated in Figure 3–68. A driving dimension is constructed.

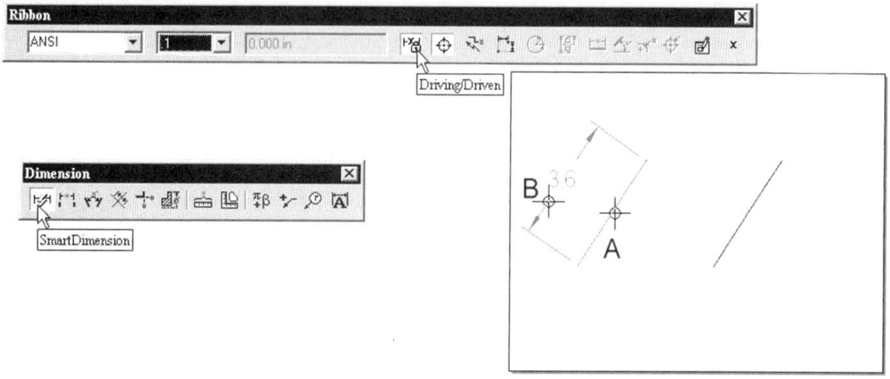

Figure 3–68. Driving dimension constructed.

8 Uncheck the Driving/Driven button on the ribbon.

9 Select line A indicated in Figure 3–69.

10 Click on location B indicated in Figure 3–69. A driven dimension is constructed.

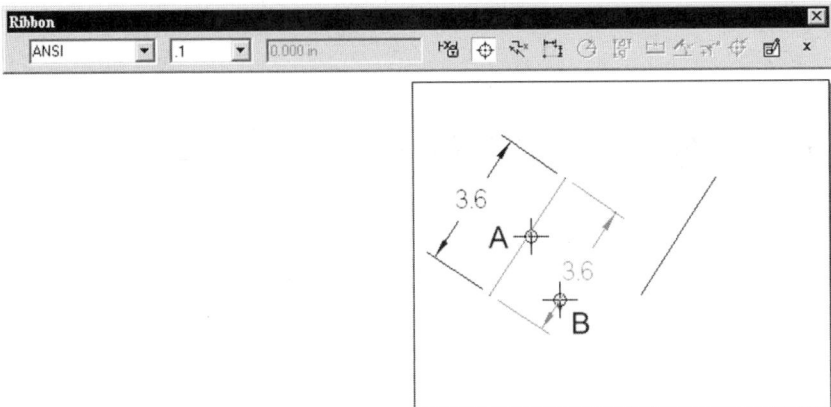

Figure 3–69. Driving/Driven button unchecked.

11 Select line A indicated in Figure 3–70.

12 Click on location B indicated in Figure 3–70. Another driven dimension is constructed.

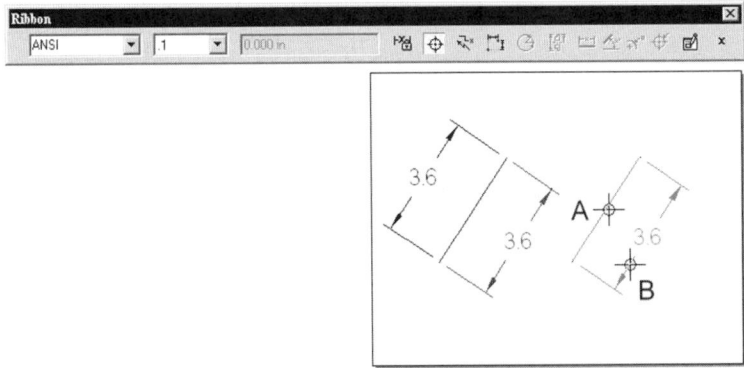

Figure 3–70. Another driven dimension constructed.

To appreciate how driving and driven dimensions differ, perform the following steps.

1 Click on the Select Tool icon on the Draw toolbar.

2 Select driving dimension A indicated in Figure 3–71.

3 In the ribbon, change the dimension's value to 4 inches. Note that the line changes in length and the driven dimension changes its value to reflect the change in the length of the line.

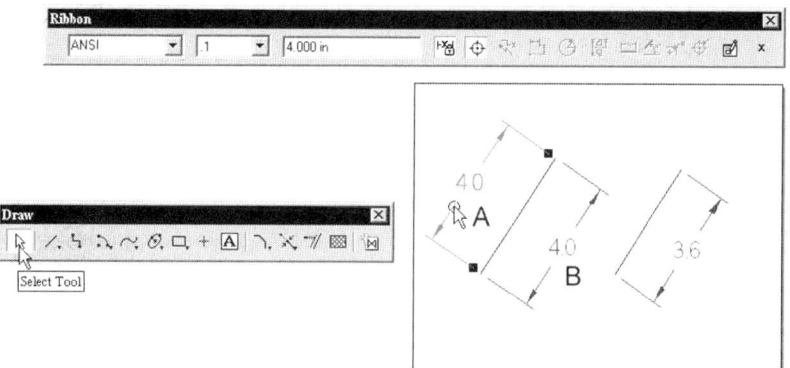

Figure 3–71.
Driving
dimension's value
changed.

4 Select dimension B indicated in Figure 3–72.

5 In the ribbon, change the dimension's value to 5 inches. Because this is a driven dimension, only the dimension's value is changed (the line does not change).

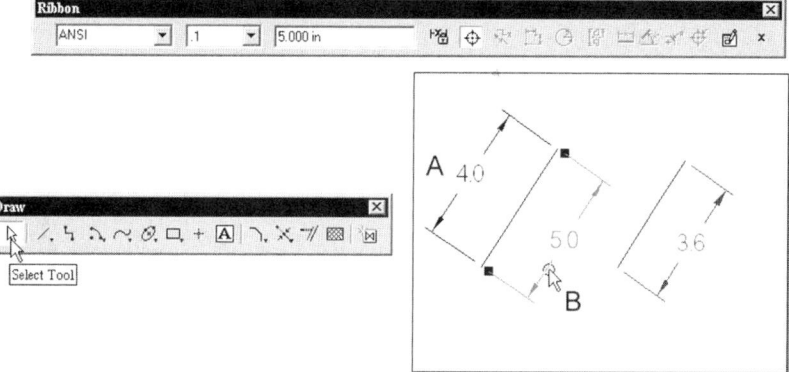

Figure 3–72.
Driven dimension
changed.

6 Select driven dimension C indicated in Figure 3–73.

7 Change its value to 4 inches. Again, only the dimension's value is changed.

8 Select line X indicated in Figure 3–74.

9 In the ribbon, change the length of the line to 5 inches. Although the line's length is changed, the dimension is not changed because the driven dimension has been modified.

Figure 3–73.
Driven dimension
changed.

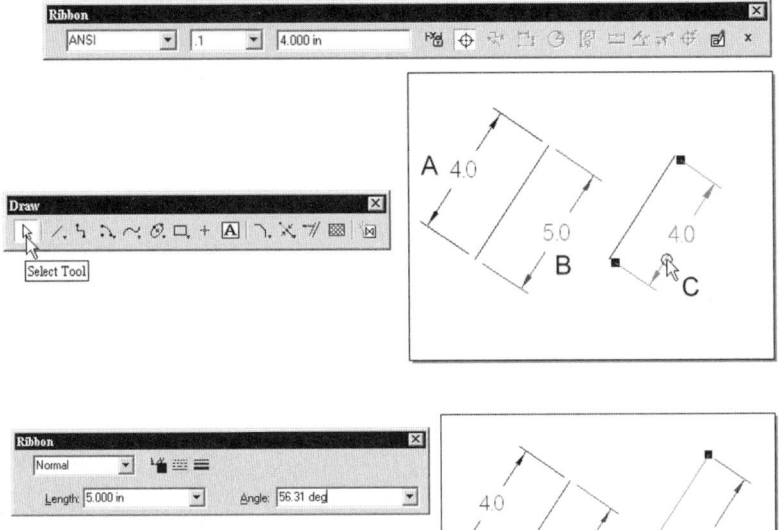

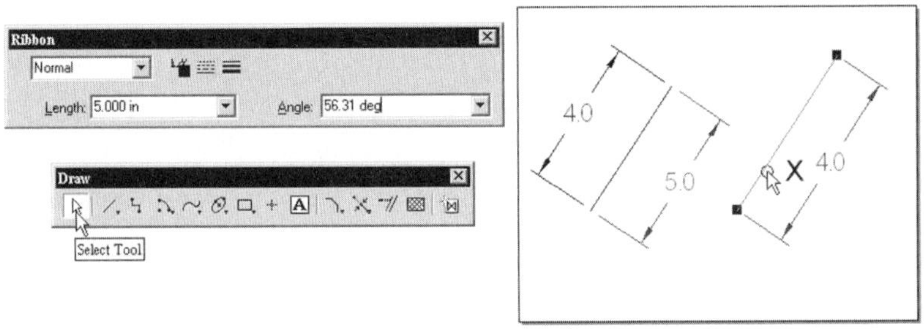

Figure 3–74.
Line's length
changed but the
overridden driven
dimension does
not follow the
change.

10 Select line Y indicated in Figure 3–75.

11 Change the line's length to 5 inches. However, it does not change. Instead, a warning message displays, telling you that the change conflicts with existing relationships, the relationship maintained by the driving dimension.

If you really want to change the length of the line, you can either modify or delete the driving dimension.

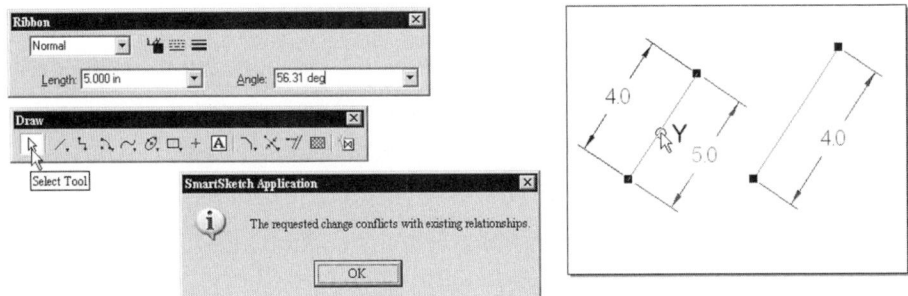

Figure 3–75. Line's length cannot be changed.

12 Select the driving dimension and press the Delete key. The driving dimension is deleted.

13 Select line Y indicated in Figure 3–76.

14 In the ribbon, change the line's length to 5 inches.

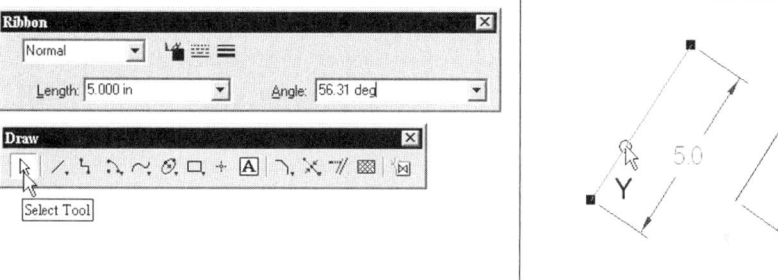

Figure 3–76. Driving dimension deleted and the line's length changed.

Perform the following steps to modify dimension text.

1 Click on the Dimension Text icon on the Dimension toolbar.

2 Select dimension A indicated in Figure 3–77.

3 In the ribbon, type the text string *text*. The dimension text is changed.

*Figure 3–77.
Endpoints
(dimension A)
selected.*

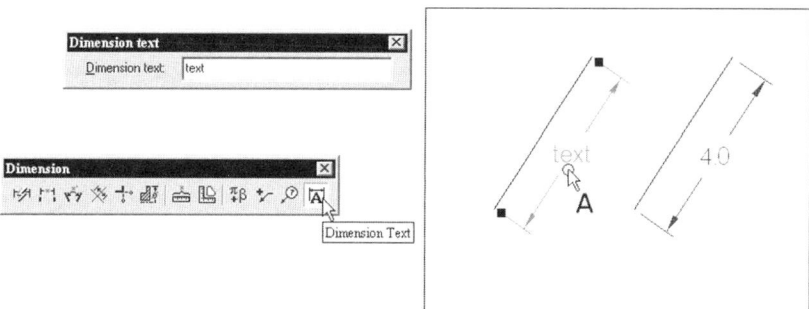

Adding Dimensions

To understand how to apply dimensions and annotations to a document, perform the following steps.

1   Activate Sheet 2.

2   Click on the Distance Between icon on the Dimension toolbar.

3   Click on endpoint A and then endpoint B indicated in Figure 3–78.

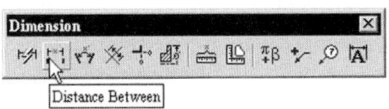

Figure 3–78. Endpoints selected.

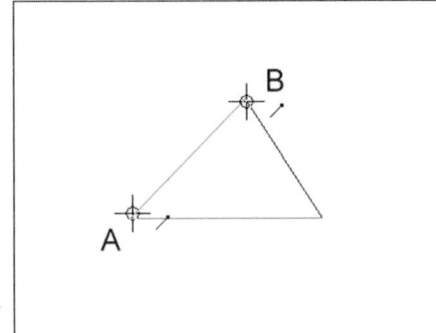

4   Move the cursor to location C indicated in Figure 3–79 but do not click on the graphics area. You will find a horizontal dimension displayed at the cursor.

5   Move the cursor to location D indicated in Figure 3–80.

6   Click on the graphics area. A vertical dimension is constructed.

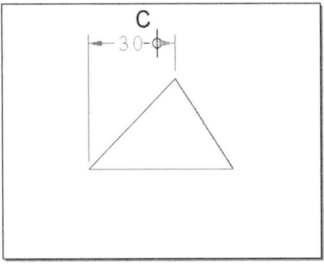

Figure 3–79. Horizontal dimension displayed at the cursor.

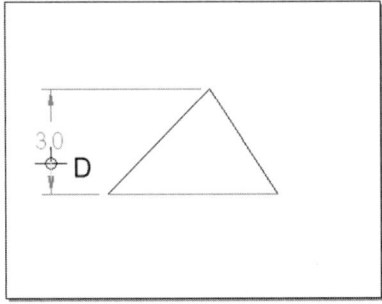

Figure 3–80. Vertical dimension being constructed.

7 Click on the Angle Between icon on the Dimension toolbar.

8 Select lines A and B and then click on location C indicated in Figure 3–81. An angular dimension is constructed.

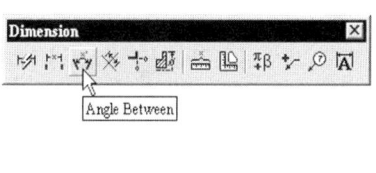

Figure 3–81.
Angular
dimension
constructed.

9 Click on the Axis icon on the Dimension toolbar.

10 Select line A indicated in Figure 3–82. The orientation of the Distance Between command is changed.

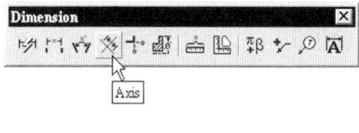
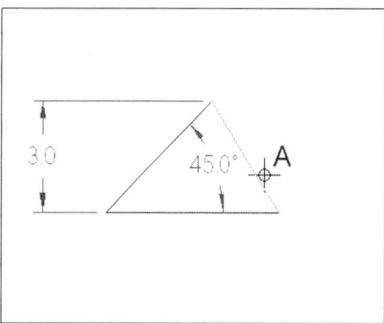

Figure 3–82.
Axis command
applied.

11 Click on the Distance Between icon on the Dimension toolbar.

12 Select endpoints A and B indicated in Figure 3–83.

13 Click on location C indicated in Figure 3–84. A dimension is constructed. Compare the result with Figure 3–81.

14 To reset the axis, click on the Axis icon on the Dimension toolbar and select line A indicated in Figure 3–85.

15 Activate Sheet 3.

16 Click on the Coordinate Dimension icon on the Dimension toolbar.

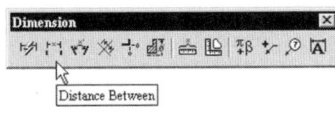

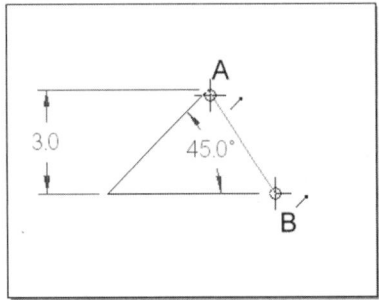

Figure 3–83. Endpoints selected.

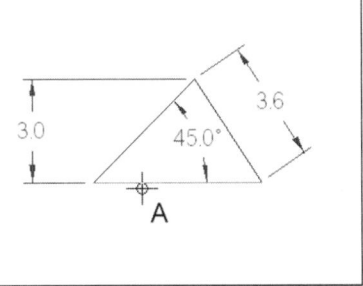

Figure 3–84. Dimension's orientation affected by the Axis command.

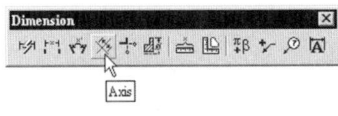

Figure 3–85. Axis reset.

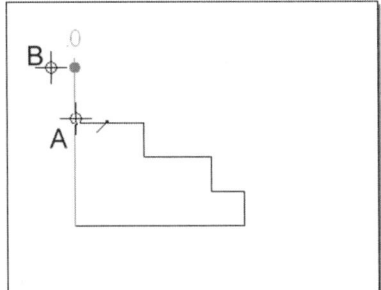

17 Select endpoint A and then location B indicated in Figure 3–86. A coordinate dimension is constructed.

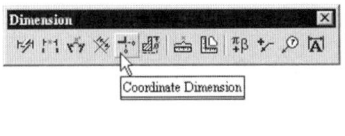

Figure 3–86. A coordinate dimension constructed.

18 Select endpoint C and then location D indicated in Figure 3–87. Another coordinate dimension is constructed.

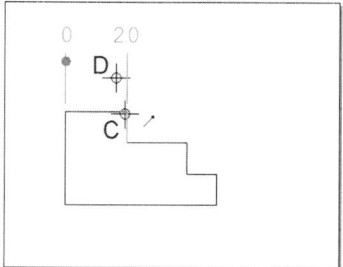
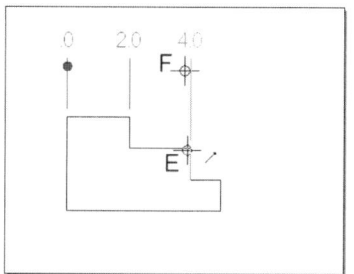
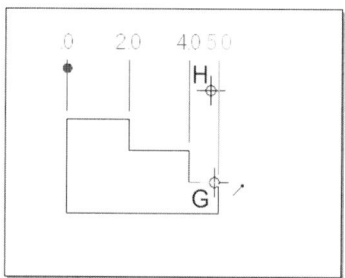

Figure 3–87. Second coordinate dimension constructed.

Figure 3–88. Third coordinate dimension constructed.

Figure 3–89. Fourth coordinate dimension constructed.

19 Select endpoint E and then click on location F indicated in Figure 3–88. A third coordinate dimension is constructed.

20 Select endpoint G and then click on location H indicated in Figure 3–89. A fourth coordinate dimension is constructed.

21 Right click.

22 Activate Sheet 4.

23 Click on the Symmetric Diameter icon on the Dimension toolbar.

24 Select endpoints A and B and click on location C indicated in Figure 3–90. A symmetric diameter dimension is constructed.

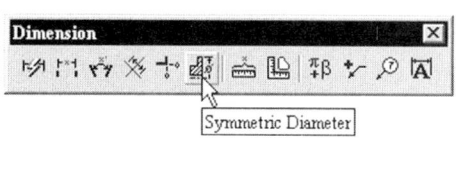
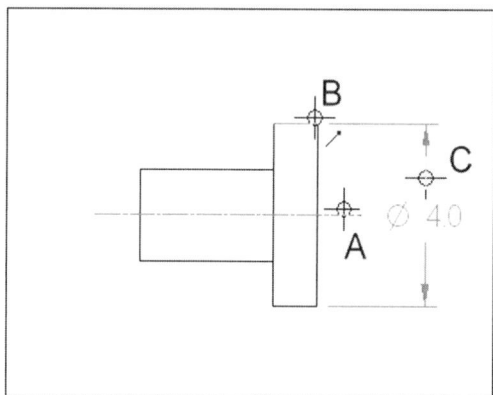

Figure 3–90. Symmetric diameter dimension being constructed.

Taking Measurements

Other than adding dimensions, you can take distance and area measurements. Perform the following steps.

1   Click on the Measure Distance icon on the Dimension toolbar or
    select Tools > Measure Distance.

2   Select endpoint A and then endpoint B indicated in Figure 3–91.
    The distance between the two endpoints is displayed at the cursor.

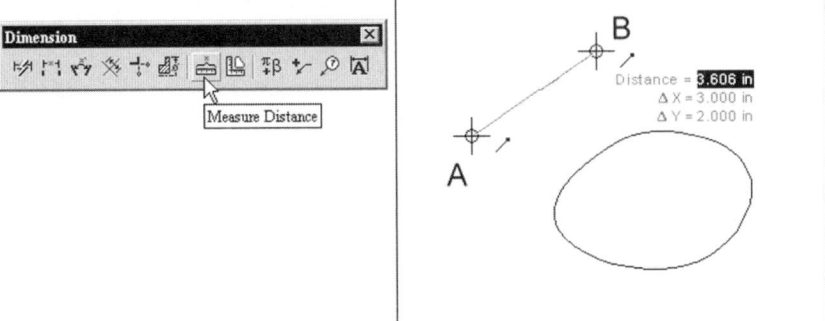

Figure 3–91.
Distance measured.

3   Click on the Measure Area icon on the Dimension toolbar or select
    Tools > Measure Area.

4   Select curve A indicated Figure 3–92. The area of the closed loop is
    displayed at the cursor.

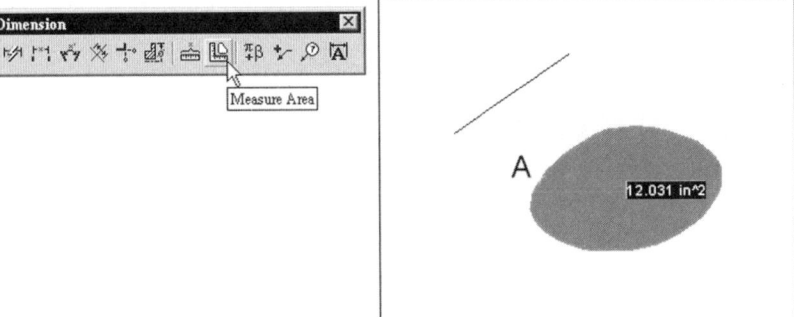

Figure 3–92.
Area measured.

Adding Text and Annotations

In addition to using the Text Box command to construct a text box, you
can use the Character Map dialog box to insert special characters and
add leaders and balloons. Perform the following steps.

1   Activate Sheet 6.

2   Click on the Character Map icon on the Dimension toolbar or
    select Tool > Character Map.

3 In the Character Map dialog box, shown in Figure 3–93, click on a sign from the Arial font and then click on the Select and Copy button.

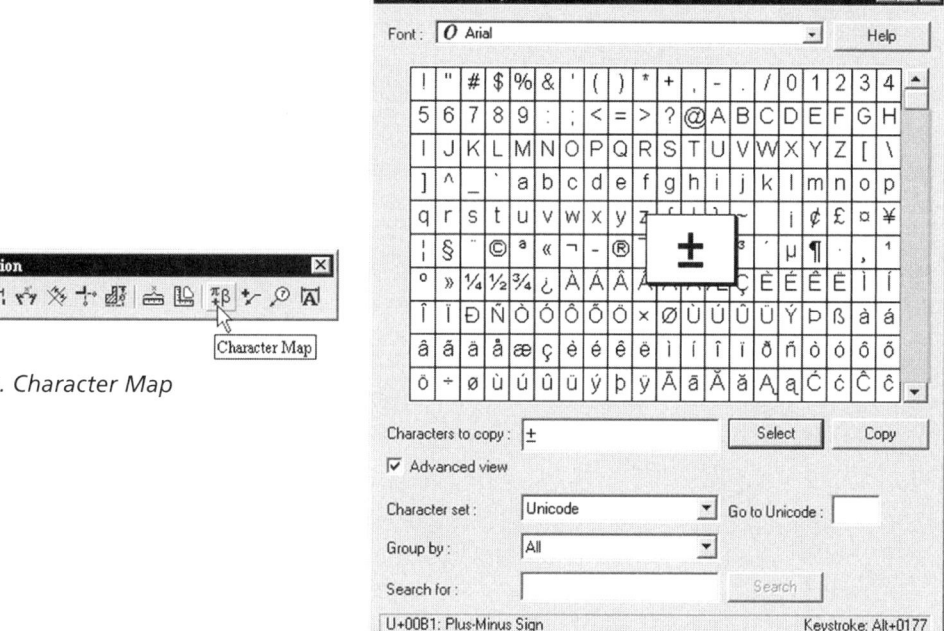

Figure 3–93. Character Map dialog box.

4 Click on the Text Box icon on the Dimension toolbar.

5 Click on location A (coordinate X:1/Y:2) indicated in Figure 3–94.

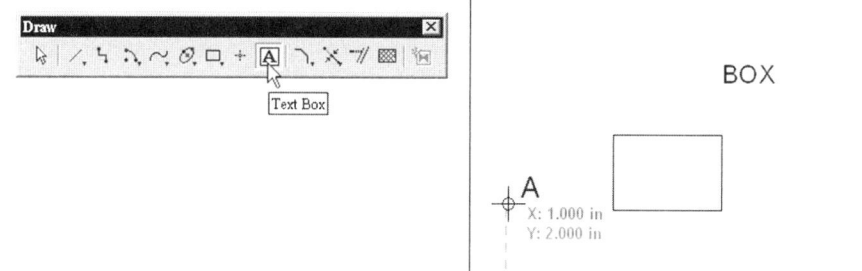

Figure 3–94. Pasting a character.

6 Select Edit > Paste. The character is pasted.

7 Click on the Select Tool icon on the Draw toolbar.

8   Select the pasted character A indicated in Figure 3–95.

9   In the ribbon, select Normal ANSI to change the text box style.

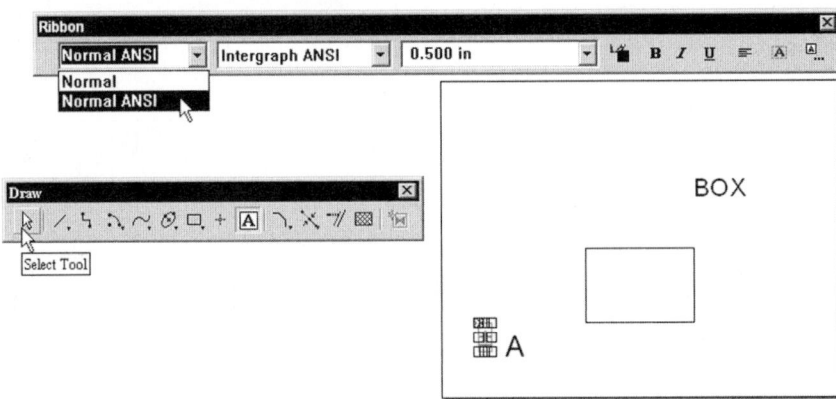

Figure 3–95.
Text box
style being
changed.

10   Click on the Leader icon on the Dimension toolbar.

11   Select midpoint A indicated in Figure 3–96.

12   Click on text box B indicated in Figure 3–96. A leader is constructed.

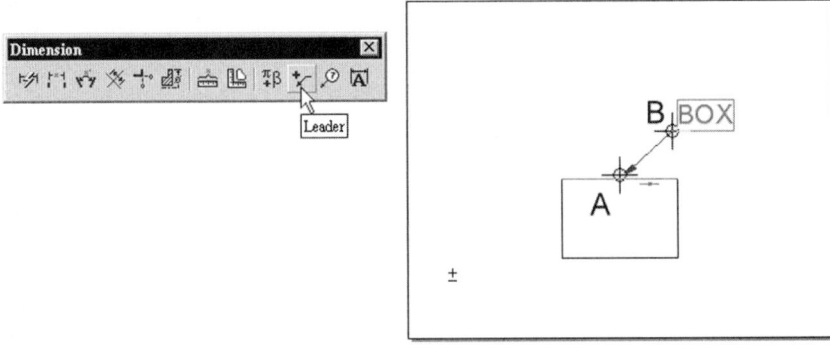

Figure 3–96.
Leader being
constructed.

13   Click on the Balloon icon on the Dimension toolbar.

14   Type *1* in the text box of the ribbon.

15   Select endpoint A and then location B indicated in Figure 3–97.

16   A balloon is constructed. (See Figure 3–98.)

17 Save the file in a folder of your computer and close the file.

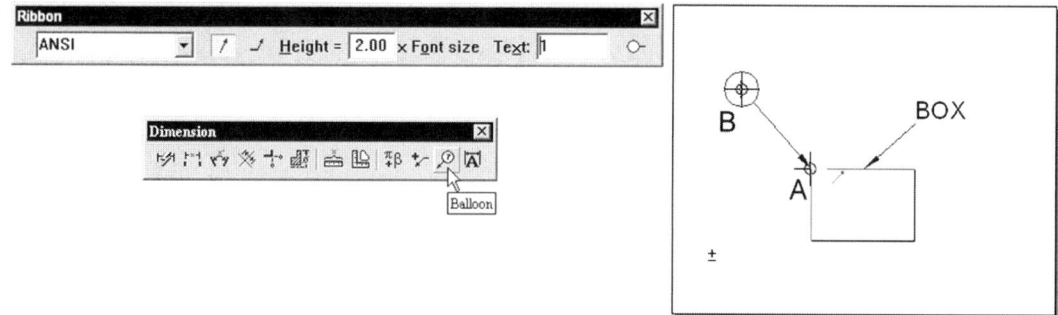

Figure 3–97. Balloon being constructed.

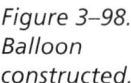

Figure 3–98. Balloon constructed.

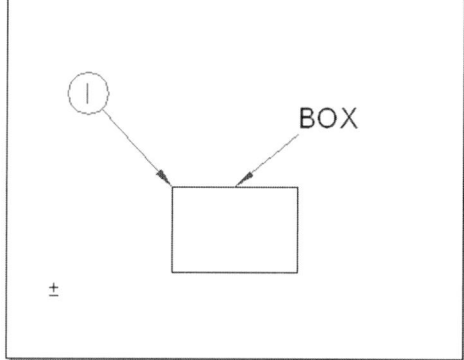

Understanding System Options

Now you should take some time understanding system options that affect the way the application behaves. Perform the following steps.

1 Select Tools > Options.

The Options dialog box is displayed. It has seven tabs: Colors, General, File Locations, View, Symbols, Reference Files, and Foreign Data.

Color Tab

The color tab concerns the color of objects. (See Figure 3–99.)

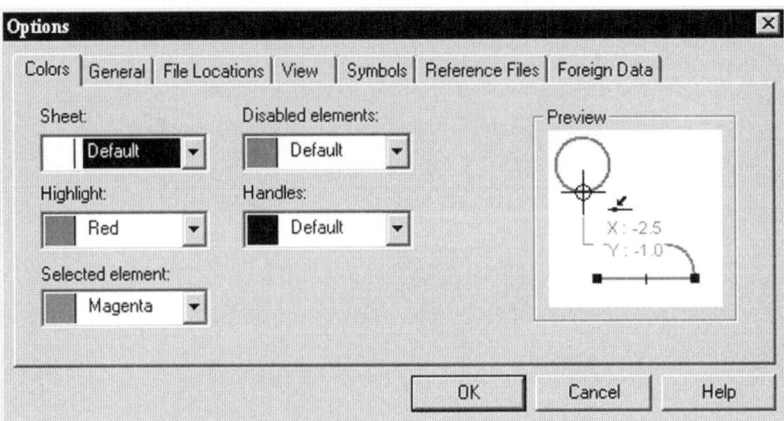

Figure 3–99.
Color tab.

General Tab

The General tab enables you to update links automatically when a document is opened, set the number of recently used files in the list, dimension key-in values automatically, display the unit of measurement labels, and set the number of undo steps. (See Figure 3–100.)

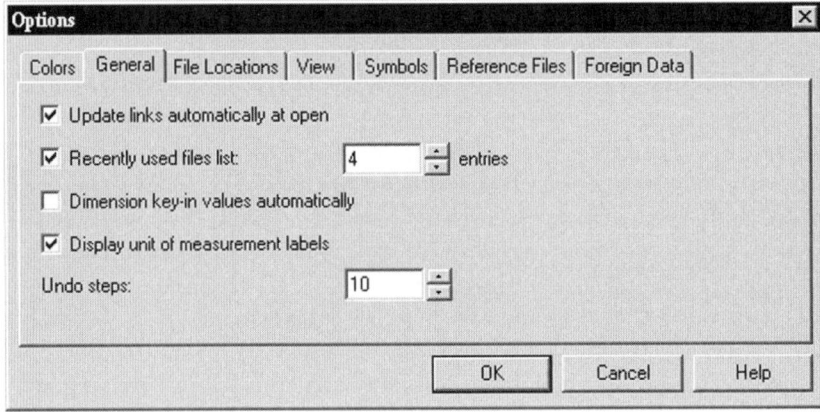

Figure 3–100.
General tab.

File Locations

The File Locations tab indicates file locations and enables you to edit these locations. (See Figure 3–101.)

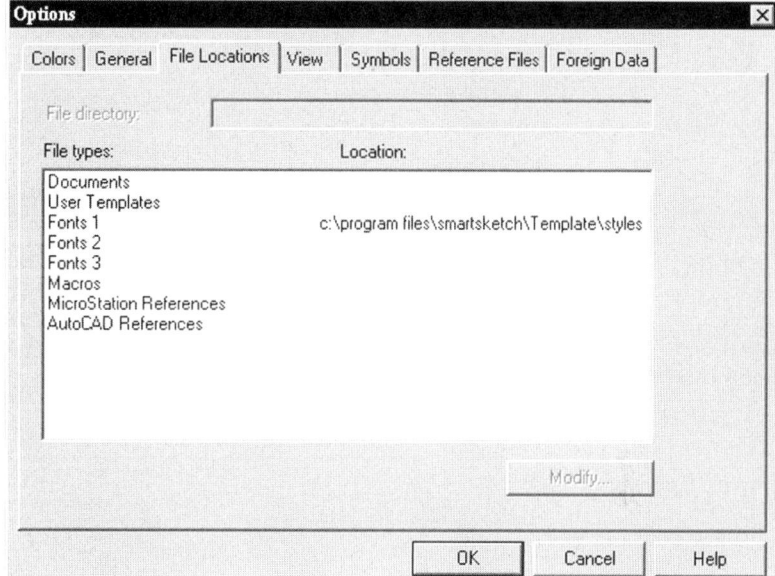

Figure 3–101.
File Locations
tab.

View Tab

The View tab concerns display of the document. You can set the display of the document in its printed form. You can set the display of the window's scroll bars, status bar, and sheet tabs. You set grid and grid snap. You can also set the display of the sheet's outline. (See Figure 3–102.)

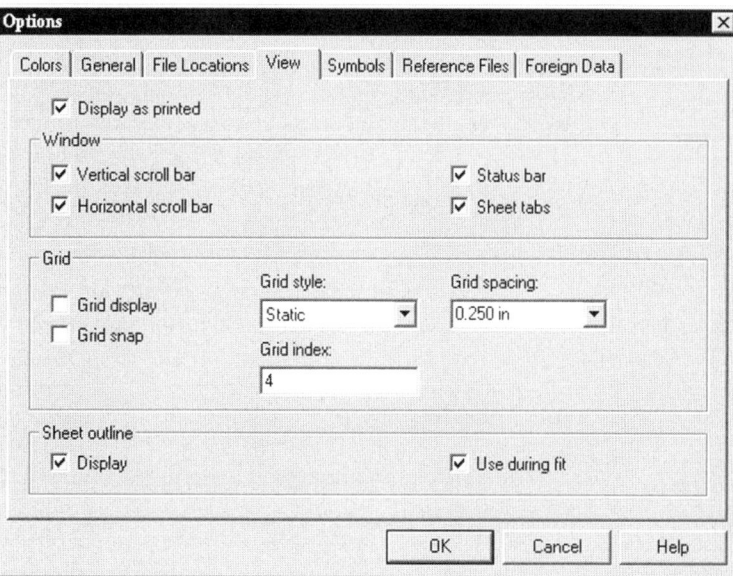

Figure 3–102.
View tab.

Symbols

The Symbols tab decides whether the drag-and-drop symbols are embedded in or linked to the document. (See Figure 3–103.)

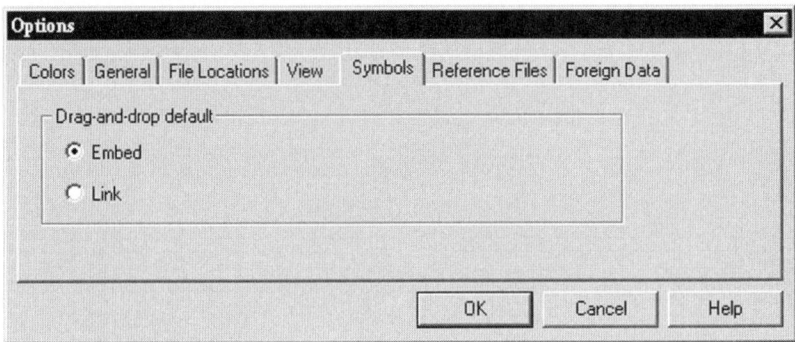

Figure 3–103.
Symbols tab.

Reference Files

The Reference Files tab sets the scale of reference files. (See Figure 3–104.)

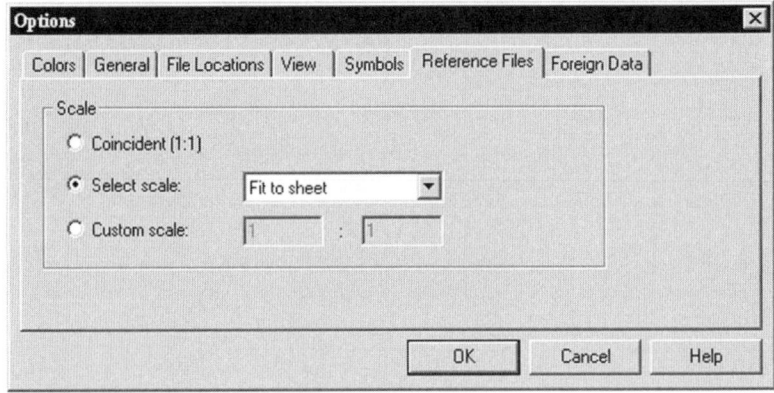

Figure 3–104.
Reference
Files tab.

Foreign Data

The Foreign Data tab controls the format of import and export data. (See Figure 3–105.)

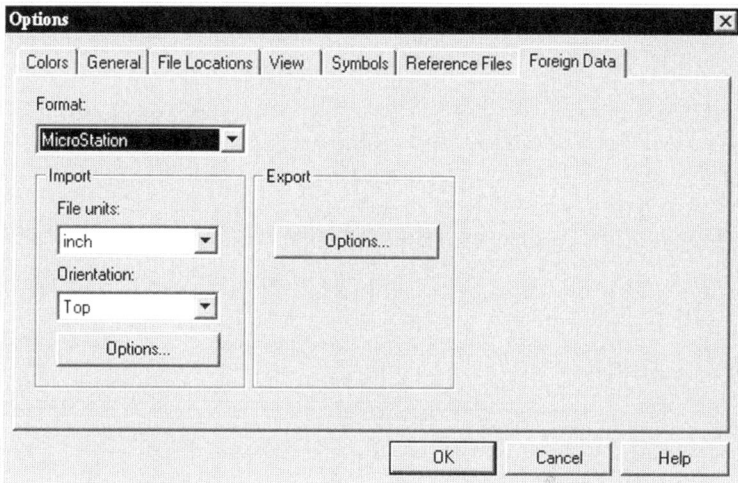

Figure 3–105.
Foreign Data tab.

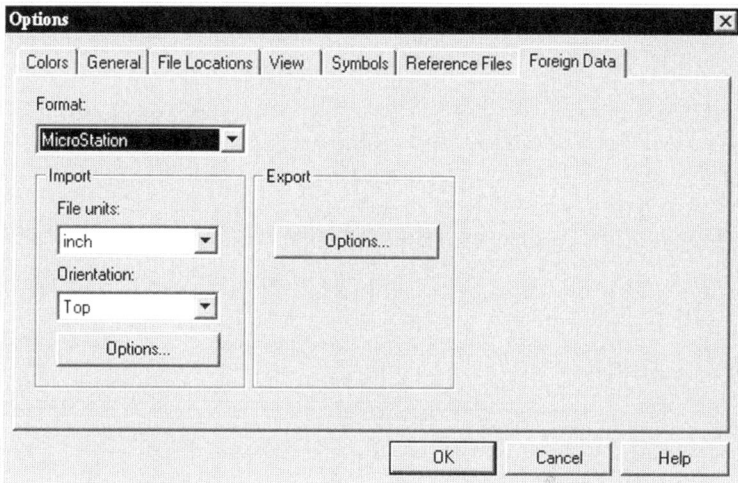

■ ■ ■ ■ **Summary**

Analogous to physical layers used in manual drafting, layers in a Smart-Sketch document are conceptual layers used to help organize drawing elements. You can set a layer to be invisible to hide drawing elements residing on that layer. You can also lock a layer so that drawing elements residing on it can be seen but cannot be modified.

Three types of styles affect the appearance of objects in a document: line style, text box style, and dimension style. Line style concerns the color, line type, and line width of drawing elements. Text box style concerns the font style, paragraph setting, and spacing of text. Dimension style concerns the appearance of dimensions.

You can maintain a number of styles in a document and set one of the line styles as the default line style. In addition, you can override the settings of the default style by formatting or using the ribbon associated with the drawing element construction command. Apart from setting the line styles to be constructed, you can modify the style of drawing elements already constructed.

There are two types of dimensions in a document: driving dimensions and driven dimensions. By activating Maintain Relationship, you can use driving dimensions to control the size of a drawing element. To simply specify the size of a drawing element you use driven dimensions.

■ ■ ■ ■   # Review Questions

1 Briefly explain the commands related to manipulation of layers in a document.

2 How is a line style set, and how are the settings in the default line style overridden?

3 What parameters of a text box are controlled by the Text Box style?

4 Explain how driving dimensions can be placed in a document.

CHAPTER 4

Symbols

■ ■ ■ ■ Objectives

The goals of this chapter are to explore the use of Symbol Explorer in symbol insertion and the use of symbol authoring tools in defining various properties of a symbol. After studying this chapter, you should be able to:

❐ Use the Symbol Explorer

❐ Use symbol authoring tools to construct custom symbols

❐ Use the Attribute Viewer

Overview

A symbol is a collection of drawing elements intended to be used repeatedly in many drawing sessions. In addition to using the symbols provided by the system, you can construct your own symbols. Insertion of symbols can be made through the use of the Symbol Explorer. Construction of custom symbols can be done by using the set of symbol authoring tools. To manipulate the attributes of an element or symbol, you use the Attribute Viewer.

■ ■ ■ ■ Using Symbols

Symbols are used in a drawing session to enhance document productivity. You may use library symbols provided by the system or use custom-built symbols.

Using the Symbol Explorer

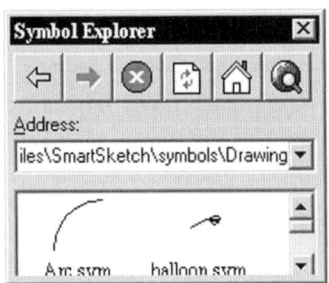

Figure 4–1. Symbol Explorer.

A symbol is saved in the computer with the extension of .*sym*. By default, files are saved in subdirectories under the *Symbols* folder of the SmartSketch installation folder. However, you may save them in any specific folders as may be convenient. To help visualize the effect of a symbol before inserting it into a document, you use the Symbol Explorer. (See Figure 4–1.)

The Symbol Explorer, as its name implies, is an explorer enabling you to explore the symbols available. To display the Symbol Explorer, if it is not already displayed, click on Symbol Explorer on the Main toolbar or select Views > Toolbars and click on Symbol Explorer from the Toolbars list of the Toolbars dialog box. By default, Symbol Explorer is docked at the right-hand side of the SmartSketch application window.

The Symbol Explorer has six buttons, an address bar, and a preview window. Symbol Explorer behaves in a way very similar to Windows Explorer. Not only exploring the World Wide Web, the Symbol Explorer enables you to explore symbols that can be used for insertion into a document. The six buttons at the top of the Symbol Explorer are: Back, Forward, Stop, Refresh, Home, and Explore Elsewhere. The Back button takes you backward to the previous location. The Forward button takes you forward to a location. The Stop button halts the loading of a selected location. The Refresh button reloads the current selection. The Home button takes you to the preset directory. The Explore Elsewhere button enables you to explore elsewhere.

Setting Symbol Explorer's Home Directory

Perform the following steps to set the home directory of the Symbol Explorer.

1   Start a new document by selecting File > New.

2   In the New dialog box, select *Technical Drawing (Imperial).igr* from the template list and then click on the OK button.

3   Click on Symbol Explorer on the Main toolbar, if the Symbol Explorer is not already displayed.

4   Click on the Home button on the Symbol Explorer.

The default home location of the explorer should be the symbols directory associated with the template the template is based on. For example, if SmartSketch is installed in *C:\Program Files\SmartSketch*, the default home address should be *C:\Program Files\SmartSketch\symbols\Drawing*. Continue with the following to change the home address.

5 Select File > Properties.

6 Select the Browser tab. (See Figure 4–2.)

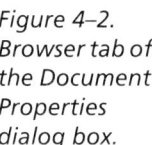

Figure 4–2. Browser tab of the Document Properties dialog box.

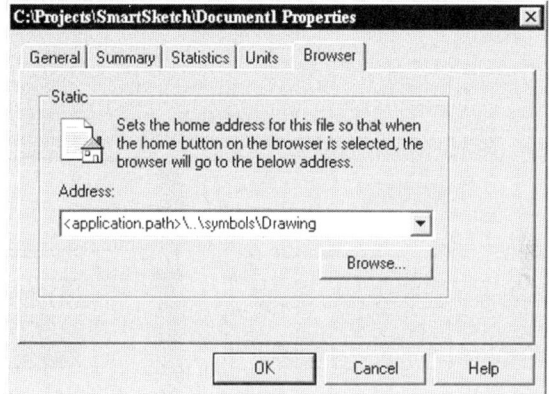

7 Click on the Browse button.

8 In the Modify Location dialog box, select a directory and then click on the OK button. (See Figure 4–3.) It is assumed that you already have constructed a folder *C:\Projects\SmartSketch* in your computer.

9 Click on the OK button on the Document Properties dialog box.

Figure 4–3. Modify Location dialog box.

Exploring Symbols

You will find that the address bar now displays the directory you just selected. Perform the following steps to learn more about the Symbol Explorer.

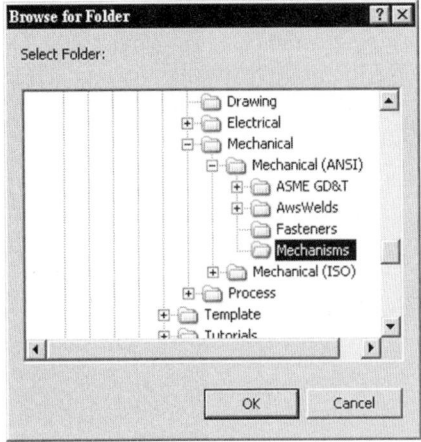

Figure 4–4. Browse for Folder dialog box.

1 Click on the Back button. This brings you back to the previous address, the default address.

2 Click on the Forward button. This brings you to the address where you are backed from.

3 Click on the Back button again. Now your address bar should show the default address of the template file your document is based on.

4 Click on the Explore Elsewhere button. The Browse for Folder dialog box is displayed. (See Figure 4–4.)

5 Select the directory *Symbol\Mechanical\ Mechanical (ANSI)\Mechanisms.*

6 Click on the OK button.

Upon exiting from the Browse for Folder dialog box the address bar changes to the *Mechanisms* folder of the *Symbols* directory and the viewer at the bottom of the Symbol Explorer shows the symbols available in this folder.

7 Select *Ground sym* (A of Figure 4–5) from the viewer of the Symbol Explorer, hold down the left mouse button, drag to location B of Figure 4–5, and release the mouse button. The symbol is inserted.

Figure 4–5. Mechanisms folder displayed in the Symbol Explorer.

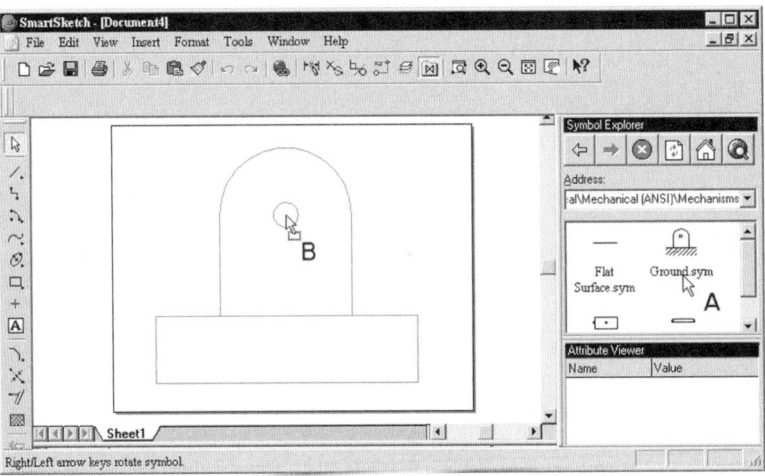

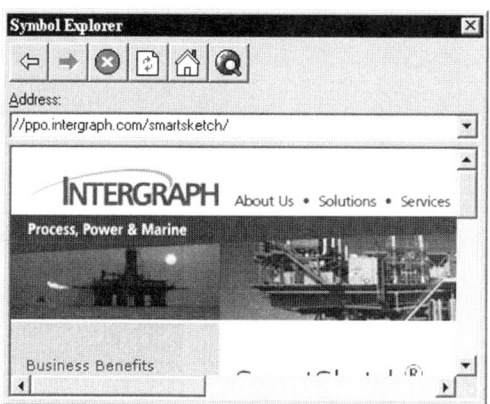

Figure 4–6. Exploring www.smartsketch.com *via the Symbol Explorer.*

Browsing the World Wide Web

As we have said, the Symbol Explorer's functions are quite similar to those of Windows Explorer. As such, you can type an Internet address in the Address box of the Symbol Explorer. Figure 4–6 shows *www. martsketch.com* being explored through the Symbol Explorer. Note that the web site is redirected after you type the address in the address box.

Embedded Symbols and Linked Symbols

Depending on the setting in the Symbols tab of the Options dialog box (accessible from the Tools > Options menu), an inserted symbol may either be embedded in or linked to the current document. (See Figure 4–7.) With an embedded symbol, a copy of the origin symbol is saved in the document. Changing an instance of the symbol in the document will modify all instances. However, changes in the original symbol will have no effect on the instances of the document. On the other hand, a linked symbol is always related to the source symbol. Any change in the original symbol will be reflected in the instances of the document.

Figure 4–7. Setting the Drag-and-drop default option.

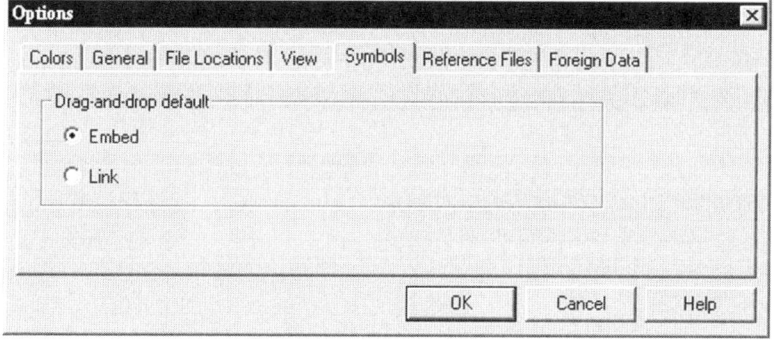

Manipulating a Symbol's Handles

After you release the mouse button following a drag-and-drop operation, the inserted symbol will be selected automatically and a number of handles will be highlighted. By default, a symbol has a number of handles, exhibiting three behaviors: rotation, alignment, and association.

❐ *Rotation:* It can be rotated by selecting and dragging the rotate handle.

❐ *Alignment:* It will be aligned automatically with a target element or object when the symbol is placed in a document.

❐ *Association:* It will automatically associate with a target element or object. If the associated element or object moves, it follows the movement.

If you click in a blank space of the graphics area, the symbol will be deselected. Perform the following steps.

1 Select the inserted symbol, if it is already unselected.

2 As shown in Figure 4–8, there are three handles (A, B, and C) enabling you to move, rotate, and mirror the inserted symbol.

3 Select handle A, drag it to location D, and release the mouse button. (See Figure 4–9.) The inserted symbol is moved.

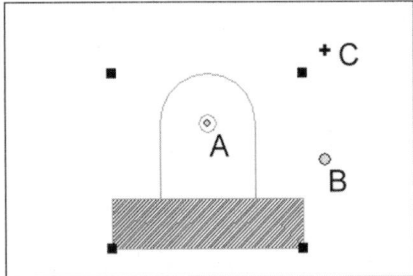

Figure 4–8. Handles displayed after the symbol is selected.

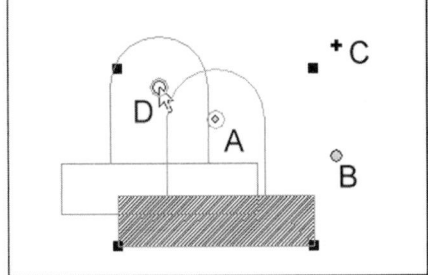

Figure 4–9. Symbol being moved.

4 Select handle B (indicated in Figure 4–9), drag it to location E (indicated in Figure 4–10), and release the mouse button. The symbol is rotated.

Figure 4–10. Symbol being rotated.

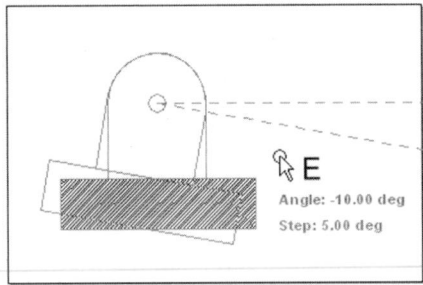

5 Select handle C (indicated in Figure 4–11), drag it to location F of Figure 4–12, and release the mouse button. The symbol is mirrored.

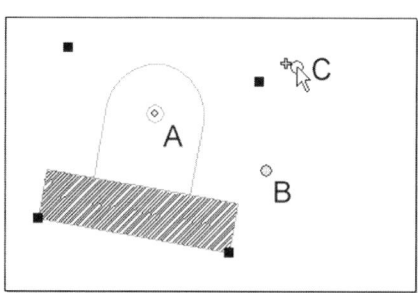

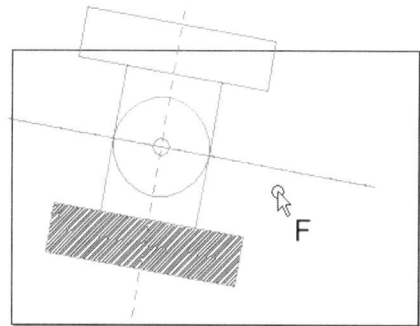

Figure 4–11. Symbol being mirrored.

Figure 4–12. Symbol mirrored.

Adding and Removing Favorites in the Symbol Explorer

To help save time in exploring symbols frequently used, you include links to these symbols by adding favorites in the Symbol Explorer. Perform the following steps.

1 With reference to Figure 4–13, select *Ground sym* from the viewer of the Symbol Explorer, right click, and select Add To Favorites. The symbol now appears as a header bar in the viewer (see A of Figure 4–14).

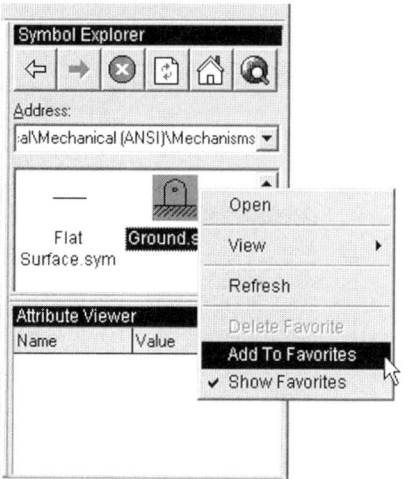

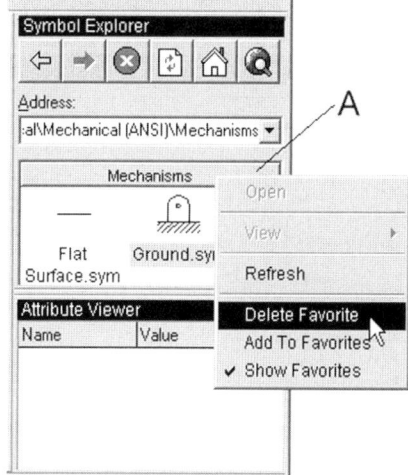

Figure 4–13. Adding a favorite in the Symbol Explorer's viewer.

Figure 4–14. Removing the favorite.

2 To remove the favorite, select A of Figure 4–14, right click, and select Delete Favorite. Note that if you merely want to hide the favorite you can select the favorite, right click, and clear the check mark prefixing Show Favorites.

3 Save the file as *Chapter4SymbolExploring.igr* in a folder of your computer.

■ ■ ■ ■ Symbol Authoring

Apart from using existing symbols available from the symbol library, you can construct your own symbols and use them in later drawing sessions. Symbols can be built from scratch by merging two or more existing symbols or by modifying an existing symbol.

As explained in Chapter 2, you can use the Create Symbol command on the Draw toolbar to construct a symbol from existing drawing elements. To further control how a symbol will behave when it is being dragged and dropped into a document, you need to perform symbol authoring, which requires prior installation of SmartSketch's Symbol Authoring module in your computer.

To use the Symbol Authoring tool, select Tools > Add-In and click on Symbol Authoring Tools in the Add-In Manager dialog box. (See Figure 4–15.)

Figure 4–15. Add-In Manager dialog box.

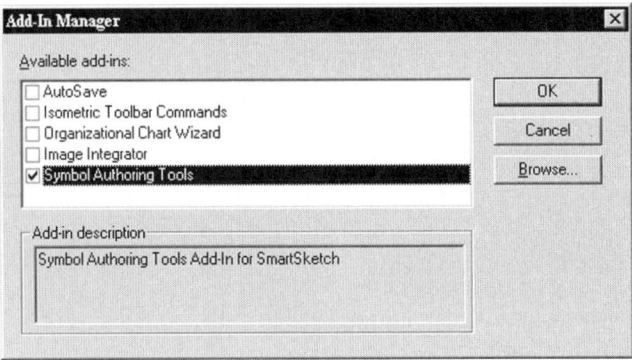

Using the Symbol Authoring Tools Toolbar

To perform symbol authoring, you use the commands available from the Symbol Authoring Tools toolbar. You can display the toolbar by selecting View > Toolbars, clicking Symbol Authoring Tools from the toolbars list, and clicking on the OK button. (See Figure 4–16.) Details of the commands are delineated in Table 4–1.

Figure 4–16.
Symbol Authoring
Tools toolbar.

Table 4–1 Symbol Authoring Tools Toolbar Options and Their Functions

Option	Function
SmartPoint Properties	Enables you to place SmartPoints (connect points, drop points, and drag points) on a symbol.
Symbol Origin	Defines the origin of a symbol.
Symbol Properties	Enables you to define properties of a symbol.
Lookup Table	Enables you to import data from a database source.
Symbol Representation	Enables you to define a different representation of the same symbol.
Edit SmartText	Enables you to place SmartText in a SmartLabel.

Constructing a Symbol from Scratch

To refresh your memory of how to construct a symbol from a collection of drawing elements, perform the following steps.

1 Open the file *Chapter4SymbolAuthoring.igr* from the *Chapter 4* folder of the companion CD-ROM.

2 With reference to Figure 4–17, click on the Select Tools icon on the Draw toolbar, select location A, and then drag the mouse cursor to location B to select the drawing elements.

3 Click on the Create Symbol icon on the Draw toolbar.

4 Select endpoint C indicated in Figure 4–18 to specify the origin point.

5 In the Save As Symbol dialog box, specify a symbol file name (*symbo01.sym*) and then click on the OK button. A symbol file is constructed.

To facilitate the reuse of the symbol you construct, create a folder in your computer for holding custom symbols and add the folder as a favorite in the Symbol Explorer.

6   Close the document file without saving.

Now you should have a symbol file identical to the document file given to you.

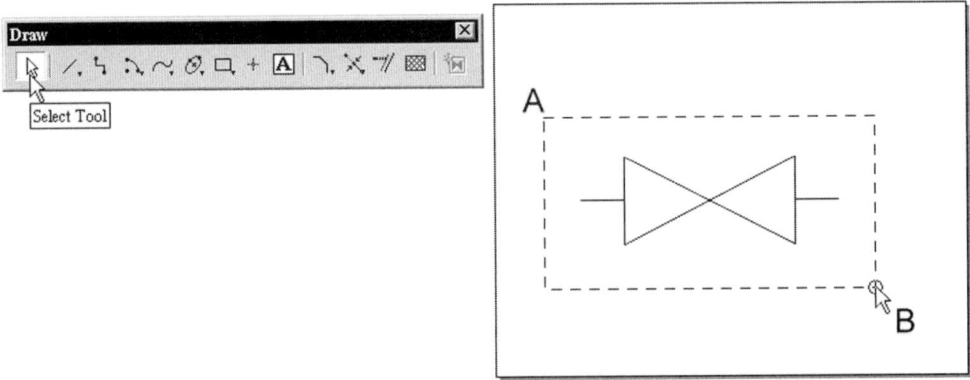

Figure 4–17. Drawing elements selected.

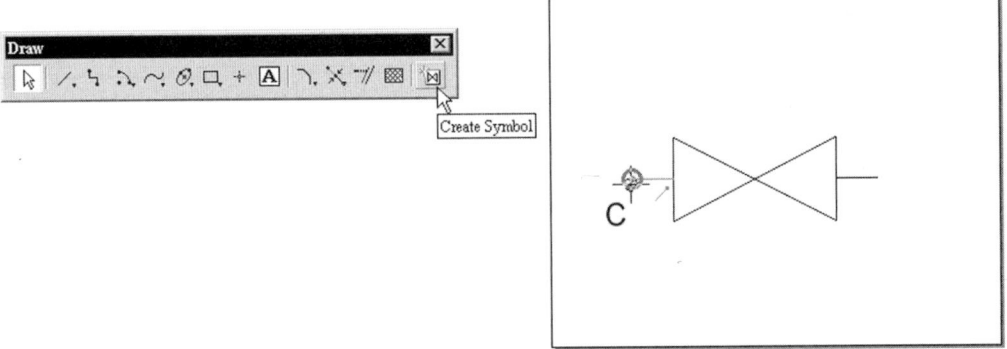

Figure 4–18. Origin point being defined.

Modifying a Symbol

To modify the drawing elements in the symbol, you do not have to recreate the symbol from scratch. Instead, you can open it as if it were a document file and use the tools explained in Chapters 2 and 3 to modify it.

Constructing and Modifying Drawing Elements

You can use the tools from the Draw toolbar to add drawing elements to a symbol, together with the PinPoint ribbon. While adding new drawing elements, relationships can be incorporated in exactly the same way as working on a document file. Perform the following steps.

1 Open the symbol file *Symbol01* you constructed previously.

2 Select File > Save As.

3 In the Save As dialog box, specify the symbol file name *Symbol02*.

Now the symbol file *Symbol01.sym* is closed and you will be working on the symbol file *Symbol02*.

4 Click on the Circle by Center Point icon on the Draw toolbar.

5 Select midpoint A indicated in Figure 4–19 to specify the center point.

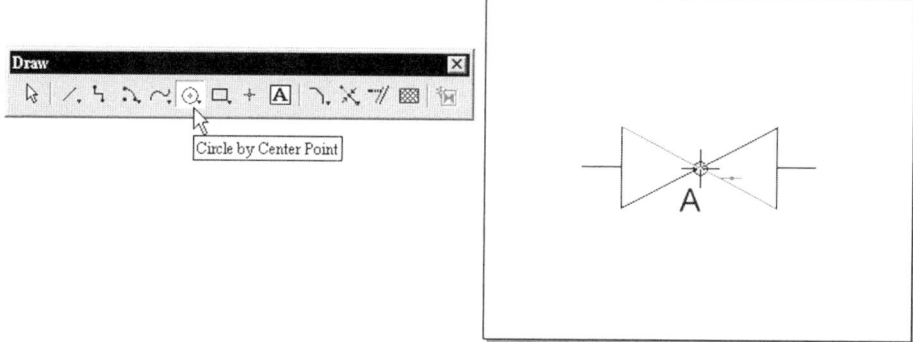

Figure 4–19. Circle being constructed.

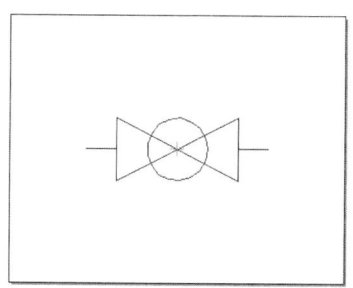

Figure 4–20. Symbol modified.

6 Specify a diameter of 2 inches in the ribbon.

7 Right click to exit the command.

8 A new symbol file is constructed by modifying an existing symbol. (See Figure 4–20.) Save your file.

Manipulating Layers, Formats, Dimensions, and Annotations

Like an ordinary document, you can organize individual drawing elements of a symbol into different layers, set formats, and add dimensions and annotations. Perform the following steps.

1 Open the symbol file *Symbol02*, if you already closed it.

2 Select File > Save As.

3   In the Save As dialog box, specify the symbol file name *Symbol03*.

Now the symbol file *Symbol02.sym* is closed and you will be working on the symbol file *Symbol03*.

4   Select Tools > Layer Options.

5   In the Groups dialog box, change the default layer's name to *Layer1*.

6   Add a new layer named *Layer2*. (See Figure 4–21.)

7   Click on the OK button to close the dialog box.

Figure 4–21. Default layer's name changed and new layer constructed.

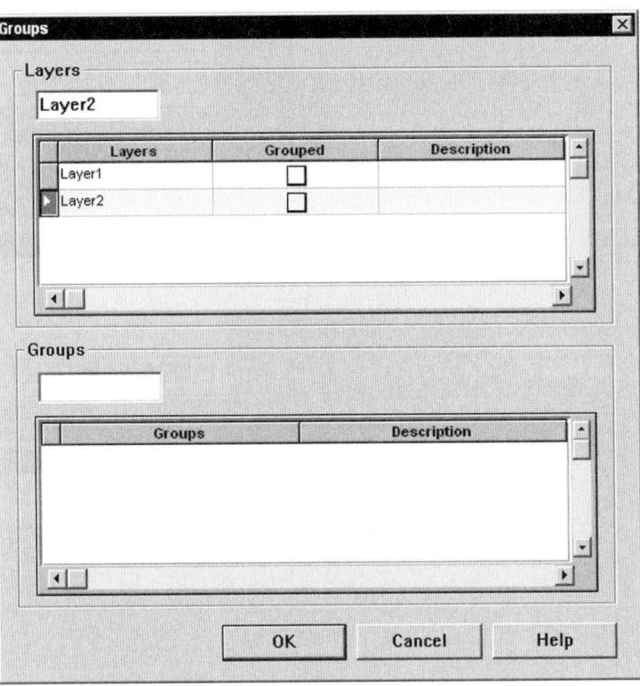

8   Select Tools > Display Manager.

9   Select the Layers tab.

10   Change *Layer1*'s color to green and *Layer2*'s color to magenta. (See Figure 4–22.)

11   Click on the OK button to close the dialog box.

12   Select circle A indicated in Figure 4–23.

13   Select Edit > Properties.

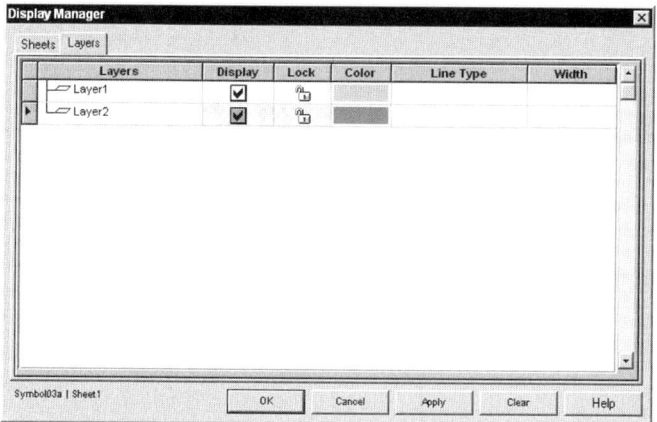

Figure 4–22. Layers' colors changed.

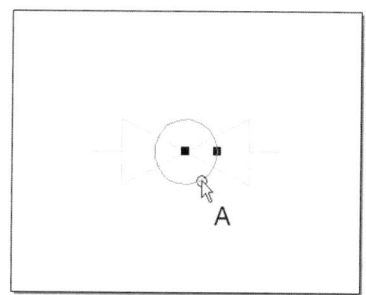

Figure 4–23. Circle selected.

14 In the Element Properties dialog box, shown in Figure 4–24, select *Layer2* from the Layers pull-down list box of the Info tab.

Figure 4–24. Layer property being changed.

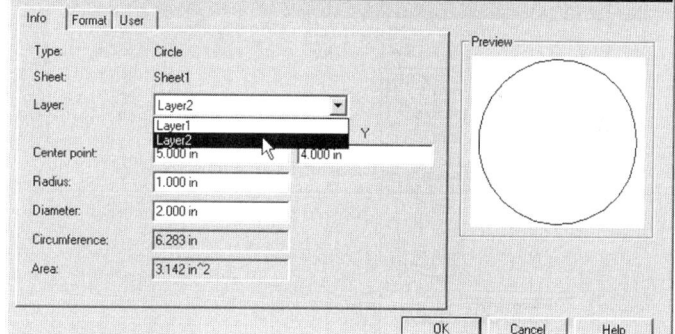

15 Click on the OK button.

16 Save and close your file.

Merging Two or More Symbols

Apart from constructing a symbol from scratch and modifying an existing symbol, you can construct a new symbol by merging two or more symbols. Perform the following steps.

1 Start a new document file. Use *Technical Drawing (Imperial).igr* as the template.

2 Select File > Sheet Setup.

3 In the Sheet Setup dialog box, select B Wide (17 inches x 11 inches) from the *Standard sheet size* box.

4 Click on the OK button.

5 Select Fit from the Main toolbar to fit the entire sheet in the graphics area.

6 Insert the symbol *Symbol01* into the document. (See Figure 4–25.)

7 Insert the symbol *Symbol03* into the document at endpoint A. (See Figure 4–26.)

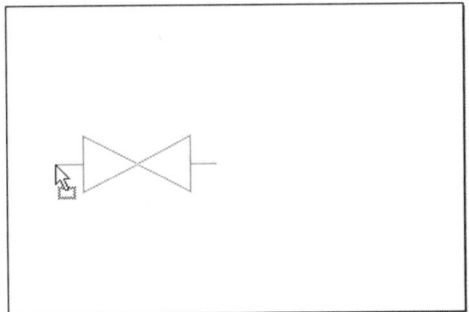

Figure 4–25. A symbol inserted in the document.

Figure 4–26. Second symbol inserted.

8 Click on the Select Tool icon on the Draw toolbar.

9 Click on A and drag the mouse to location B indicated in Figure 4–27 to select the two symbols.

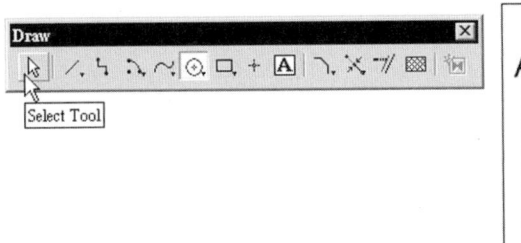

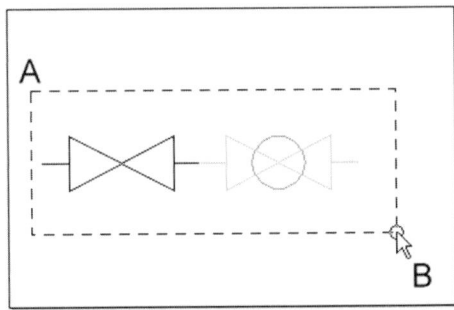

Figure 4–27. Symbols selected.

10 Click on the Create Symbol icon on the Draw toolbar.

11 Select endpoint A indicated in Figure 4–28 to specify the origin point.

12 In the Save as Symbol dialog box, specify the symbol name *Symbol04* and click on the Save button.

13 A new symbol is constructed. Close the document file without saving.

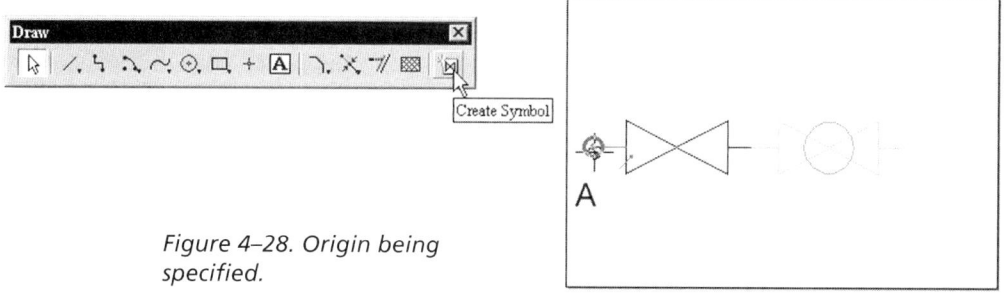

Figure 4–28. Origin being specified.

Constructing SmartPoints

To enhance the performance of a symbol when you place it in a document, you can add SmartPoints. There are three types of SmartPoints you can place on a symbol. They are connect points, drop points, and drag points.

Constructing Connect Points

A connect point is a point at which a connector constructed by using the Connector command of the Draw toolbar attaches to a symbol. Perform the following steps.

1 Select File > Open.

2 In the Open dialog box, select *Symbol (*.sym)* from the Files of Type list box and select the symbol file *Symbol01.sym* you constructed. Alternatively, you can select the *Symbol01.sym* file from the *Chapter 4* folder of the companion CD-ROM.

3 Click on the SmartPoint Properties icon on the Symbol Authoring Tools toolbar.

The SmartPoint Properties dialog box has two tabs: Behaviors and Information and Format. The Behaviors tab enables you to insert connect points, drop points, and drag points. The Information and Format tab enables you to set the points' format (style, color, and width), layer, and dimensions.

4 Select the Information and Format tab and set the color to black. (See Figure 4–29.)

5 Select the Behaviors tab. Click on the *Connect point* check box, set the connect point angle to 45 degrees, and enter the text string *Add connector* in the ToolTip box. (See Figure 4–30.)

6 Click on the Insert button.

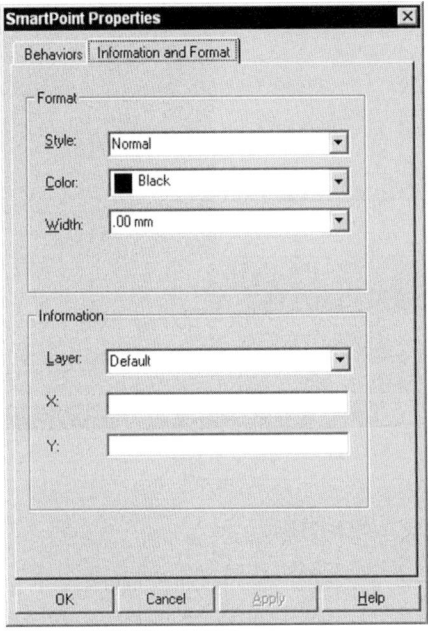

Figure 4–29. SmartPoint Properties dialog box.

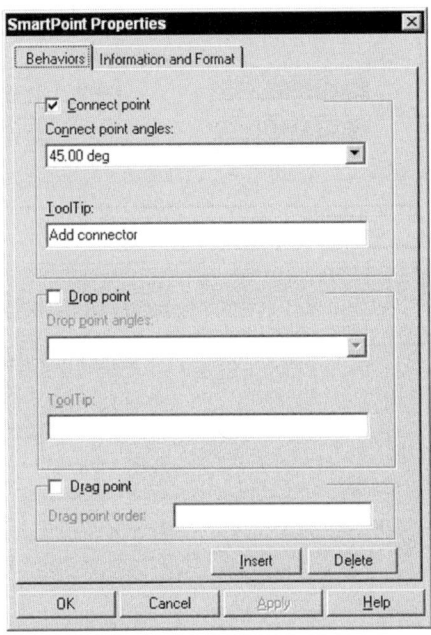

Figure 4–30. Connect point being defined in the SmartPoint Properties dialog box.

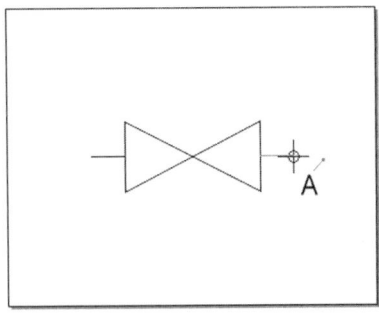

Figure 4–31. Connect SmartPoint being added to the symbol.

7 Select endpoint A indicated in Figure 4–31.

8 Click on the OK button in the SmartPoint Properties dialog box. A connect SmartPoint is defined.

9 Save your symbol file.

To appreciate how the symbol will behave when inserted, perform the following steps.

1 Start a new file. Use the template *Technical Drawing (Imperial).igr.*

2 Click on the Explore Elsewhere button on the Symbol Explorer.

3 In the Browse for Folder dialog box, select the folder where you saved the symbol you have just constructed.

4 Drag and drop the symbol *Symbol01* into the document.

5 Click on the Connector icon on the Draw toolbar and move the cursor over endpoint A of the dropped symbol. The tooltip you defined through the use of the SmartPoint Properties dialog box is displayed. (Note: Figure 4–32 shows the dropped symbol, not the symbol that is being authored.)

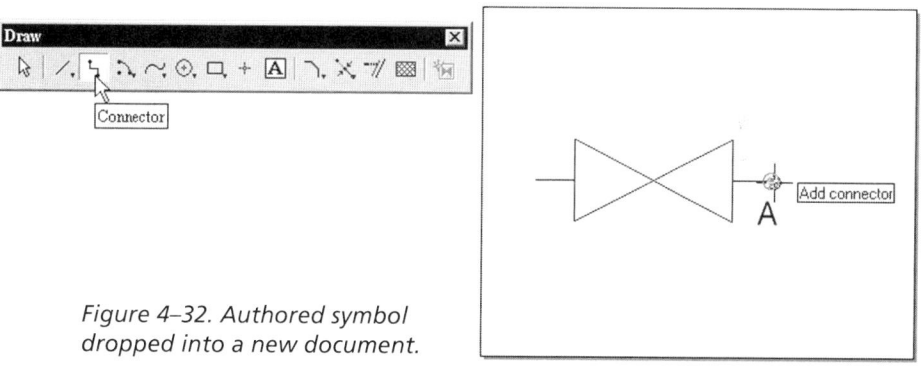

Figure 4–32. Authored symbol dropped into a new document.

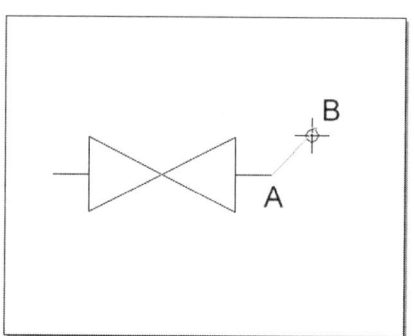

Figure 4–33. First segment of the connector being constructed at an angle.

6 Click on endpoint A and then move to location B indicated in Figure 4–33. The connector is connected at an angle of 45 degrees, defined in the SmartPoint Properties dialog box.

7 Click on location B and then right click to terminate the command.

8 Click on the Connector icon on the Draw toolbar.

9 Select endpoint C indicated in Figure 4–34.

10 Select location D indicated in Figure 4–35 and right click.

11 Close the file without saving.

As can be seen, the connector SmartPoint defined on the symbol provides a tooltip when the cursor is moved over it, and the first segment of the connector is inclined at a defined angle.

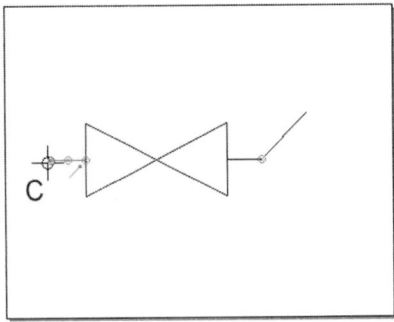

Figure 4–34. Connector being constructed again.

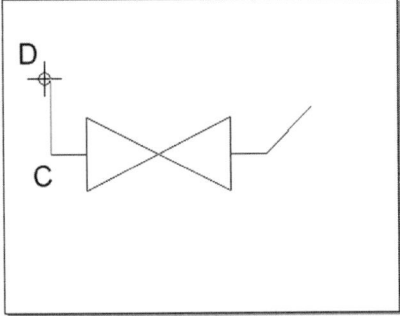

Figure 4–35. A segment of the connector being constructed.

Constructing Drop Points

The second type of SmartPoint is the drop point. A drop point facilitates the attachment of a symbol to another symbol. In constructing a drop point, you can define the angle to which the other symbol is attached, and you can add a tooltip. Perform the following steps.

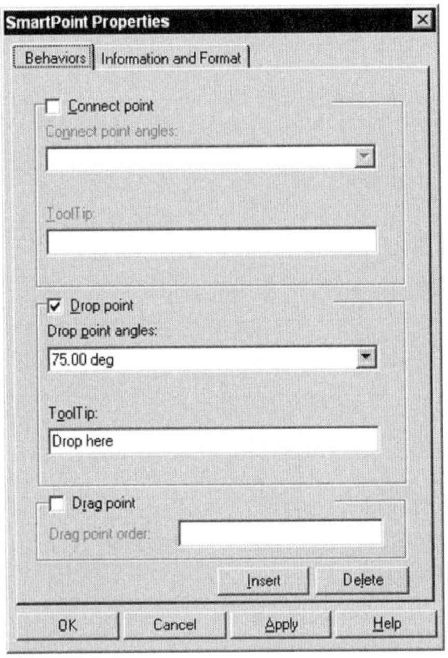

1 Open the symbol file *Symbol01*, if you already closed it.

2 Click on the SmartPoint Properties icon on the Symbol Authoring Tools toolbar.

3 Click on the *Drop point* box.

4 Set the drop point angle to 75 degrees.

5 Add the text string *Drop here* in the ToolTip box.

6 Click on the Insert button. (See Figure 4–36.)

7 Select endpoint A indicated in Figure 4–37.

8 Click on the OK button. A drop point is defined. Save your symbol file.

Figure 4–36. Drop SmartPoint being defined.

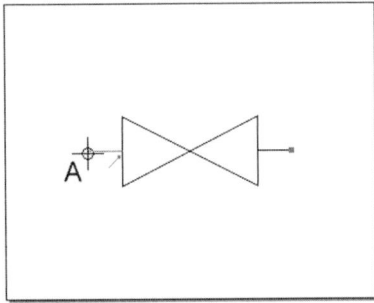

Figure 4–37. Drop SmartPoint being inserted.

To appreciate how to use the drop point, perform the following steps.

1 Start a new file. Use the template *Technical Drawing (Imperial).igr.*

2 Click on the Explore Elsewhere button on the Symbol Explorer.

3 In the Browse for Folder dialog box, select the folder where you saved the symbol you have just constructed.

4 Drag and drop the symbol file *Symbol* (constructed in Chapter 2) into the document.

5 Drag and drop the symbol file *Symbol01* (constructed earlier in this chapter) into the document.

6 Select symbol A indicated in Figure 4–38.

7 Drag symbol A to the other symbol at endpoint B, indicated in Figure 4–39. Because a drop point with a tooltip is defined, the tooltip *Drop here* is displayed.

8 Release the mouse button. The symbol is dropped at a defined angle of 75 degrees.

9 Close your file without saving.

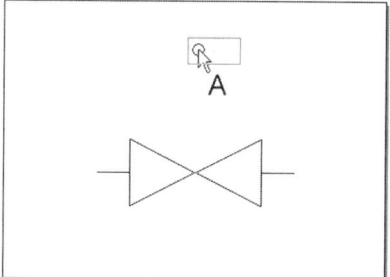

Figure 4–38. Two symbols inserted and a symbol being selected.

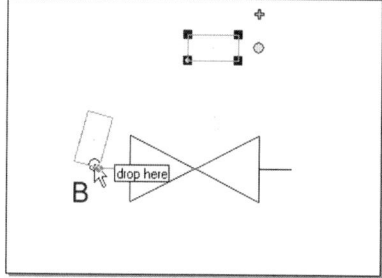

Figure 4–39. A symbol being dragged to the drop point of another symbol.

Constructing Drag Points

A drag SmartPoint is that point on the symbol you select to drag the symbol. Perform the following steps.

1 Open the symbol file *Symbol01*, if you already closed it.

2 Click on the SmartPoint Properties icon on the Symbol Authoring Tools toolbar.

3   Click the *Drag point* box in the SmartPoint Properties dialog box.

4   Type *1* in the *Drag point order* box. (See Figure 4–40.)

5   Click on the Insert button.

6   Select endpoint A indicated in Figure 4–41.

7   Click on the Apply button in the SmartPoint Properties dialog box.

8   Type *2* in the *Drag point order* box.

9   Select endpoint B indicated in Figure 4–41.

10 Click on the OK button. Two drag points are defined.

11 Save your symbol file.

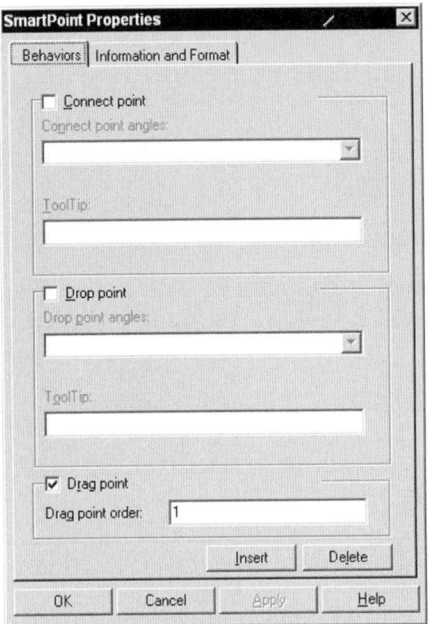

Figure 4–40. Drag point order being specified.

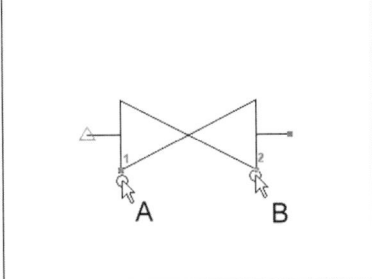

Figure 4–41. Drag points being defined.

Perform the following steps to learn how to use the drag point of the symbol.

1   Start a new file. Use the template *Technical Drawing (Imperial).igr.*

2   Click on the Explore Elsewhere button on the Symbol Explorer.

3   In the Browse for Folder dialog box, select the folder where you saved the symbol you have just constructed.

4   Drag and drop the symbol file *Symbol0* from the *Chapter 4* folder of the companion CD-ROM.

5   Drag and drop the symbol file *Symbol01* you constructed earlier in this chapter.

6   Click on symbol A indicated in Figure 4–42.

7 Select drag point B and drag the symbol to endpoint C of the other symbol. (See Figure 4–43.)

8 Close your file without saving.

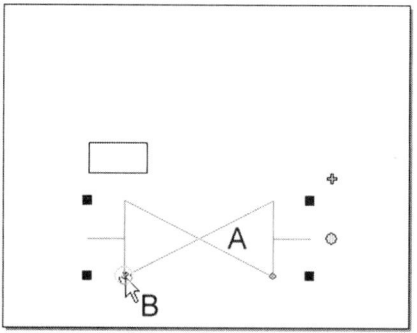

Figure 4–42. Symbol's drag point selected.

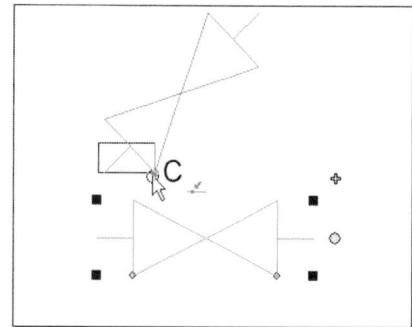

Figure 4–43. Symbol dragged.

Editing SmartPoints

SmartPoints can be edited using the same command by which you defined them. Perform the following steps.

1 Open the symbol file *Symbol01*, if you already closed it.

2 Click on the SmartPoint Properties icon on the Symbol Authoring Tools toolbar.

3 Select connect point A indicated in Figure 4–44.

Figure 4–44. Connect point selected.

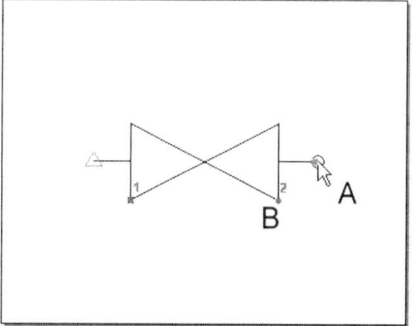

4 Change the connect point angle to 45 degrees and click on the OK button. The SmartPoint is modified. (See Figure 4–45.)

5 Select drag point B indicated in Figure 4–44.

6 Press the Delete key. The drag point is deleted.

7 Save your symbol file.

Figure 4–45.
Connect Point
being redefined.

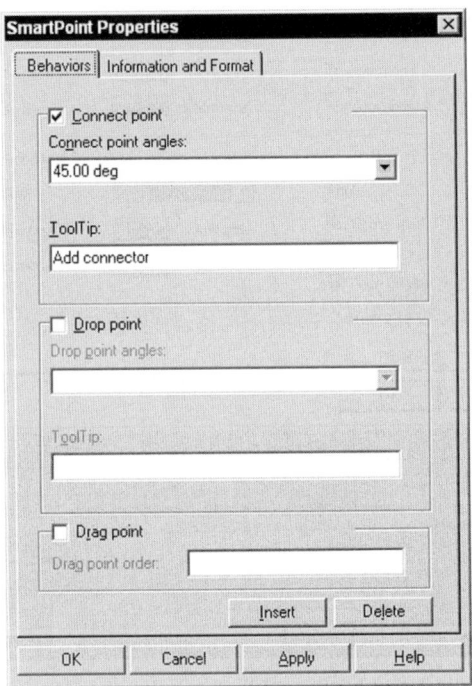

Defining a Symbol's Origin

A symbol's origin is the point the mouse cursor uses to place the symbol in a document. When you construct a symbol from a collection of drawing elements, this point has to be defined as well. Perform the following steps to redefine a symbol's origin point.

1 Open the symbol file *Symbol01*, if you already closed it.

2 Click on the Symbol Origin icon on the Symbol Authoring Tools toolbar. The current origin point of the symbol is displayed as a red circular symbol. (See A of Figure 4–46.)

3 Select endpoint B indicated in Figure 4–46. The origin is redefined.

4 Save your symbol file.

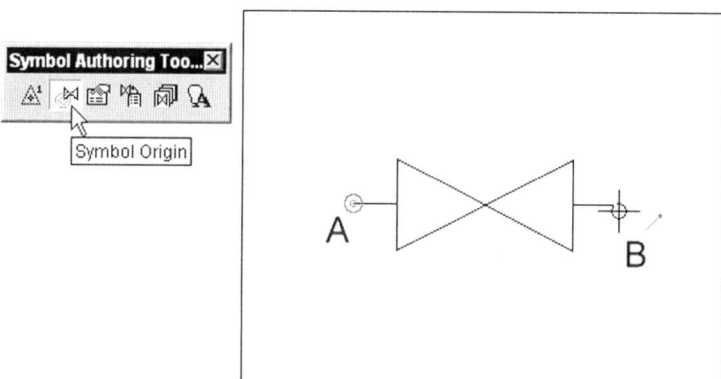

*Figure 4–46.
Symbol's origin
being redefined.*

Defining a Symbol's Properties

The Symbol Properties command on the Symbol Authoring Tools toolbar
enables you to manipulate various properties of a symbol. Perform the
following steps.

1 Open the symbol file *Symbol01*, if you already closed it.

2 Click on the Symbol Properties icon on the Symbol Authoring
Tools toolbar. (See Figure 4–47.)

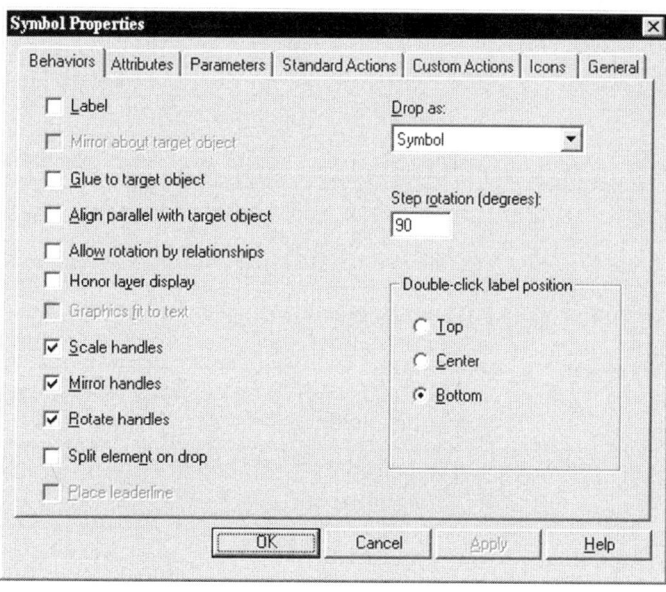

*Figure 4–47.
Symbol Properties
dialog box.*

The Symbol Properties dialog box has seven tabs: Behaviors, Attributes, Parameters, Standard Actions, Custom Actions, Icons, and General.

3 Select the Behaviors tab, if it is not already selected.

Defining Behavior Properties

The Behaviors tab of the Symbol Properties dialog box enables you to specify how a symbol behaves when it is dropped into a target document. This tab has a number of buttons. Details are delineated in Table 4–2.

Table 4–2 Behaviors Tab Options and Their Functions

Option	Function
Label	Enables the construction of Smart Label in the symbol.
Mirror about target object	Allows the symbol to mirror about the element the symbol is attached to.
Glue to target object	Sets the symbol to move with the attached object.
Align parallel with target object	Sets the symbol to align parallel to the attached object.
Allow rotation by relationships	Enables change to the orientation of the symbol in accordance with the relationship set to it.
Honor layer display	Enables turning on and off the layers of individual objects of a symbol after inserted to a document. Layers, if not already created in the target document, will be automatically created after insertion.
Graphics fit to text	Causes the symbol to stretch to fit the text height and width of a text box in the symbol.
Scale handles	Enables the scale handles of the symbol when the symbol is selected.
Mirror handles	Enables the mirror handles of the symbol when the symbol is selected.
Rotate handles	Enables the rotate handles of the symbol when the symbol is selected.
Split element on drop	Enables the symbol to split the connector and most drawing elements when the symbol is dropped onto the connector or drawing element.
Drop as	Causes the symbol to be dropped as a symbol, a group of elements, or individual elements in the target document.
Step rotation	Sets the symbol's incremental angle of rotation when it is being dropped into the target document.
Double-click label position	Sets the location of the symbol's label when being double clicked.

Setting Graphics to Fit to Text Box

The graphics of a symbol can be set to fit to a text box. Perform the following steps.

1 Open the symbol file *symbol05.sym* from the *Chapter 4* folder of the companion CD-ROM.

2 Click on the Symbol Properties icon on the Symbol Authoring Tools toolbar. (See Figure 4–48.)

3 In the Behaviors tab of the Symbol Properties dialog box, click on the Label box and then the *Graphics fit to text* box.

4 Click on the OK button.

5 Save the symbol in a folder of your computer and close the symbol file.

Figure 4–48. Symbol file opened.

To appreciate how the symbol's size fits to the size of the text box, now insert the symbol in a document.

1 Open the document file *Chapter4SymbolInsertion.igr* from the *Chapter 4* folder of the companion CD-ROM.

2 Activate Sheet 1, if it is not already activated.

3 Click on the Explore Elsewhere button on the Symbol Explorer.

4 In the Browse for Folder dialog box, select the folder for your saved symbols and click on the OK button.

You should put all symbols you constructed in a single folder and make that folder a favorite in the Symbol Explorer. This way, you can quickly browse all custom symbols you have constructed.

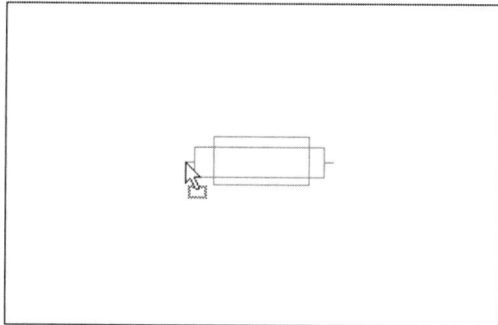

Figure 4–49. Symbol02 being dragged into the document.

5 Select and drag the symbol *Symbol05* into the document. (See Figure 4–49.)

6   Release the mouse button. You will find that the symbol stretches to fit the text in the symbol. (See Figure 4–50 and compare it with Figure 4–48.)

7   Save the file in a folder of your computer.

Figure 4–50. Symbol02 inserted.

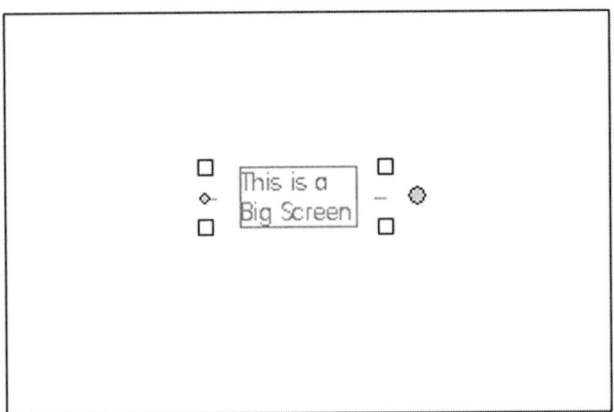

Gluing and Aligning a Symbol to Target Element

A symbol can be set to glue to and align with a target element when it is dropped into a document. Perform the following steps.

1   Open the symbol file *Symbol06.sym* from the *Chapter 4* folder of the companion CD-ROM.

2   Click on the Symbol Properties icon on the Symbol Authoring Tools toolbar. (See Figure 4–51.)

Figure 4–51. Symbol file opened.

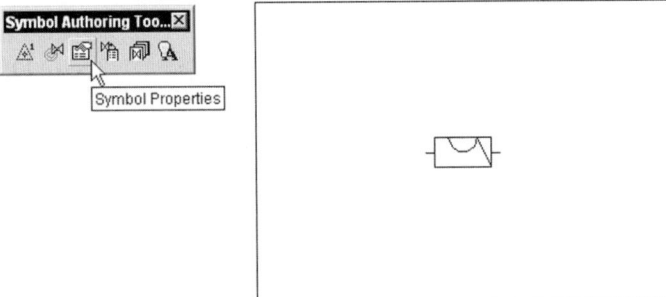

3   In the Behaviors tab of the Symbol Properties dialog box, check the *Glue to target object* and *Align parallel with target object* boxes and then click on the OK button. (See Figure 4–52.)

4 Select File > Save As to save the symbol in a folder of your computer.

5 Close the file.

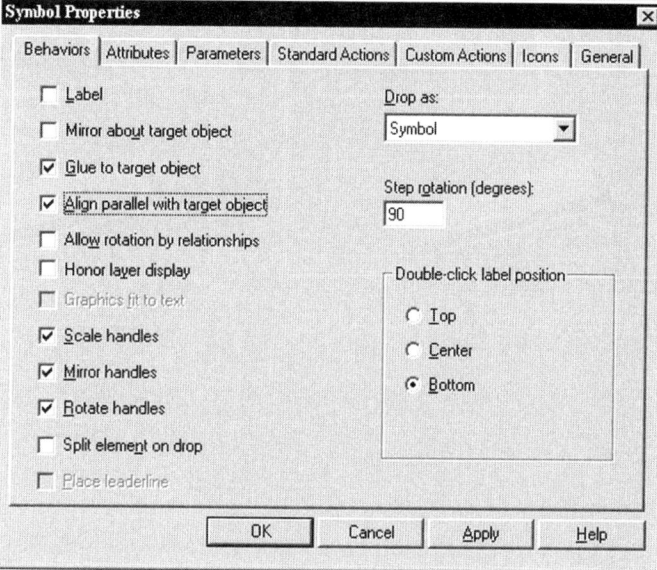

Figure 4–52.
Behaviors changed.

To appreciate how the symbol glues to and aligns with a target element, perform the following steps to insert the symbol.

1 Open the document file *Chapter4SymbolInsertion. igr*, if you already closed it.

2 Activate Sheet 2.

3 In the Symbol Explorer, browse the folder holding the custom symbols you constructed.

4 Select symbol *Symbol06* from the Symbol Explorer and drag it to endpoint A indicated in Figure 4–53.

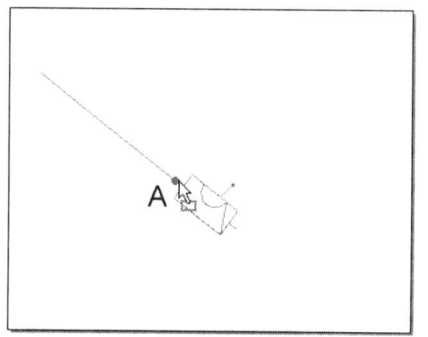

Figure 4–53. Symbol being inserted.

5 Release the mouse button. You will find a locking icon indicating that the symbol is glued to the attached object. (See Figure 4–54.)

6 To realize how the symbol is aligned and glued to the attached object, select the line segment to display the handles.

7 Select endpoint B indicated in Figure 4–55.

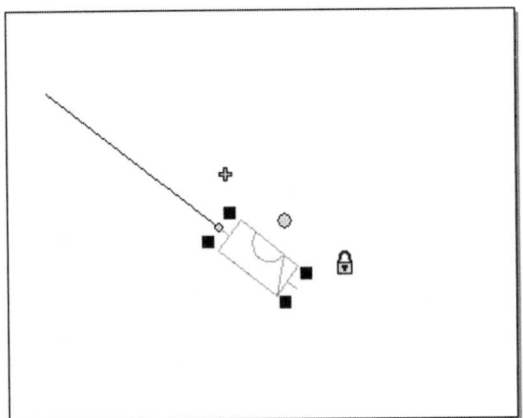

Figure 4–54. Symbol glued to the attached object.

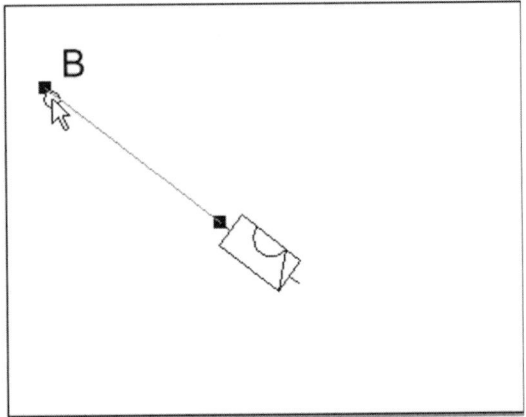

Figure 4–55. Endpoint of the attached line segment selected.

8 Hold down the mouse button and drag the endpoint of the line from B to C indicated in Figure 4–56.

Because the symbol is glued and aligned with the attached line, the symbol's location and orientation changed as well.

9 Select the symbol to display its handles.

10 Click on the lock icon D to change its status from locking position to unlocking position. (See Figure 4–57.)

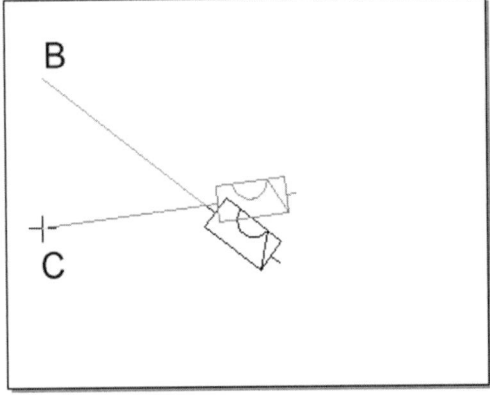

Figure 4–56. Endpoint of the line segment being moved.

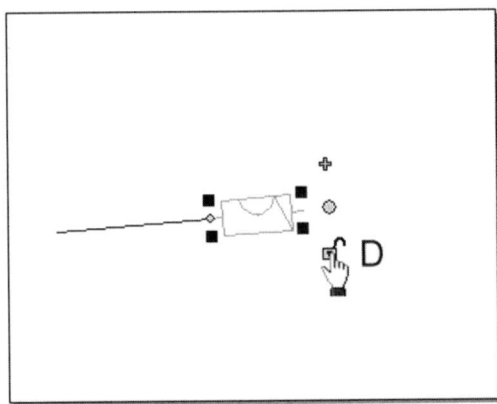

Figure 4–57. Locking icon's status changed.

11 With reference to Figure 4–58, select line segment E and drag it to location F. Because the symbol was unlocked, it does not move and change its orientation.

12 Save the document file.

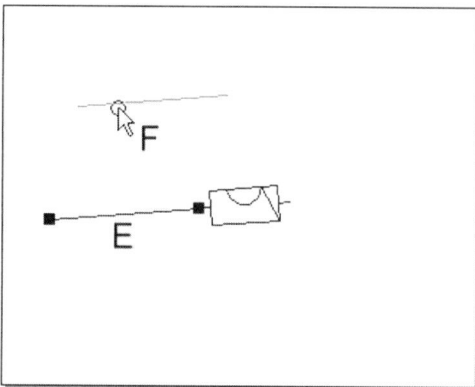

Figure 4–58. Line segment being moved.

Splitting the Target Element

Sometimes it is useful to have the target element split by a dropped symbol. Perform the following steps to modify other behavior aspects of a symbol.

1 Open the symbol file *Symbol03*.

2 Click on the Symbol Properties icon on the Symbol Authoring Tools toolbar.

3 In the Behaviors tab of the Symbol Properties dialog box, click the *Split element on drop* box and set step rotation to 5 degrees. (See Figure 4–59.)

4 Click on the OK button.

5 Save the symbol file.

To appreciate how the symbol split a target element upon dropping, perform the following steps.

1 Open the document file *Chapter4SymbolInsertion.igr*, if you already closed it.

2 Activate Sheet 3.

3 In the Symbol Explorer, browse the folder holding the custom symbols you constructed.

4 Select symbol *Symbol03* from the Symbol Explorer and drag it to location A indicated in Figure 4–60.

5 Release the mouse button. Note that the line segment is split.

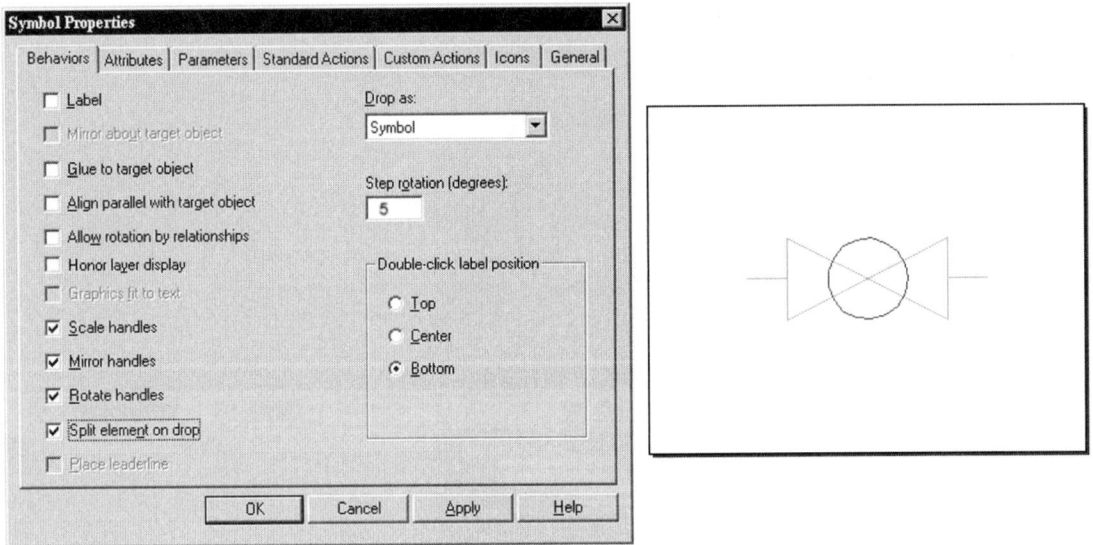

Figure 4–59. Symbol's properties being changed.

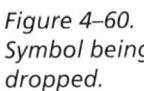

Figure 4–60. Symbol being dropped.

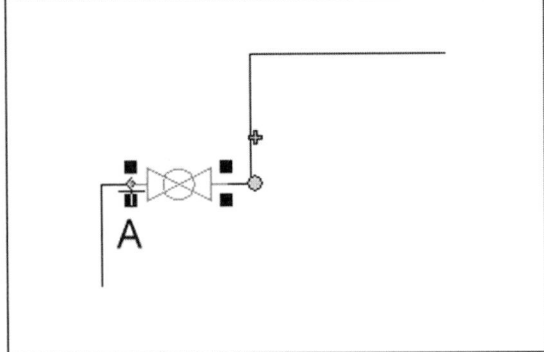

6 Select symbol *Symbol03* and drag it to location B indicated in Figure 4–61.

7 Do not release the mouse button. Press the ⇐ key or the ⇒ key to rotate the symbol.

8 Release the mouse button. The line segment is split and affected by the rotated symbol. (See Figure 4–62.)

9 Select Tools > Layer Groups. You will find that there is only one layer, the default layer, despite inserting a symbol with more than one layer.

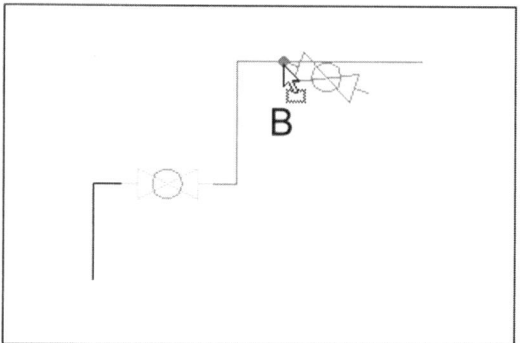

Figure 4–61. Symbol being rotated while inserting.

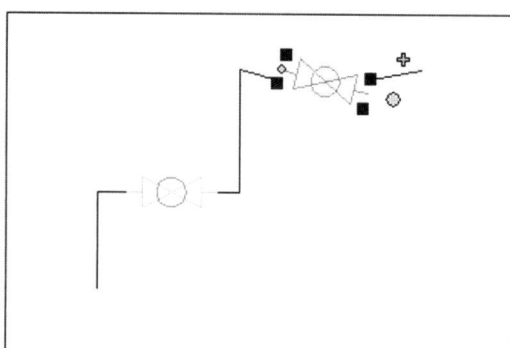

Figure 4–62. Symbol inserted.

Honoring Layer Settings

Layer settings in a symbol can be honored in the target document. Perform the following steps to further modify a symbol.

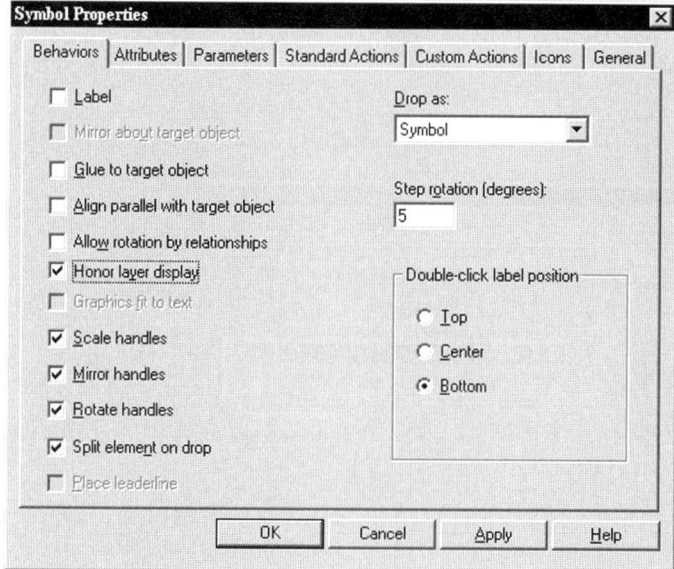

Figure 4–63. Behavior property changed.

1 Open the symbol file *Symbol03*, if you already closed it.

2 Click on the Symbol Properties icon on the Symbol Authoring Tools toolbar.

3 In the Behaviors tab of the Symbol Properties dialog box, click the *Honor layer display* option.

4 Click on the OK button. (See Figure 4–63.)

5 Save the symbol file.

To appreciate how the layer settings of a symbol are honored in the target document, perform the following steps.

1   Start a new document by using *Technical Drawing (Imperial).igr* as the template.

2   In the Symbol Explorer, browse the folder holding the custom symbols you constructed.

3   Select symbol *Symbol03* from the Symbol Explorer and drag it into the document.

4   Select Tools > Display Manager. (See Figure 4–64.) Layers from the symbol are created in the document.

5   Close the document file without saving.

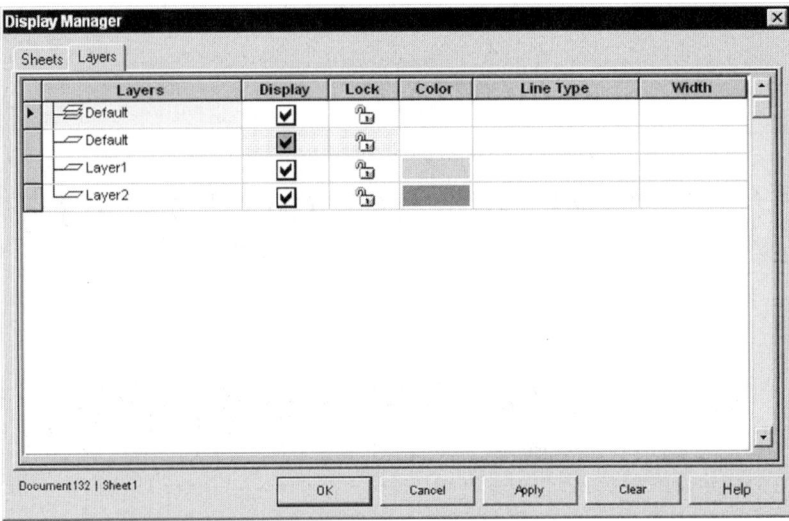

Figure 4–64.
Layers created.

Dropping as Discrete Objects

Instead of dropping a symbol as a unique symbol, you can drop a symbol into a target document as a set of discrete objects. Perform the following steps to further modify the symbol.

1   Open the symbol file *Symbol03*, if you already closed it.

2   Click on the Symbol Properties icon on the Symbol Authoring Tools toolbar.

3   In the Behaviors tab of the Symbol Properties dialog box, select Discrete Object from the *Drop as* pull-down list box. (See Figure 4–65.)

4 Click on the OK button.

5 Save the symbol file.

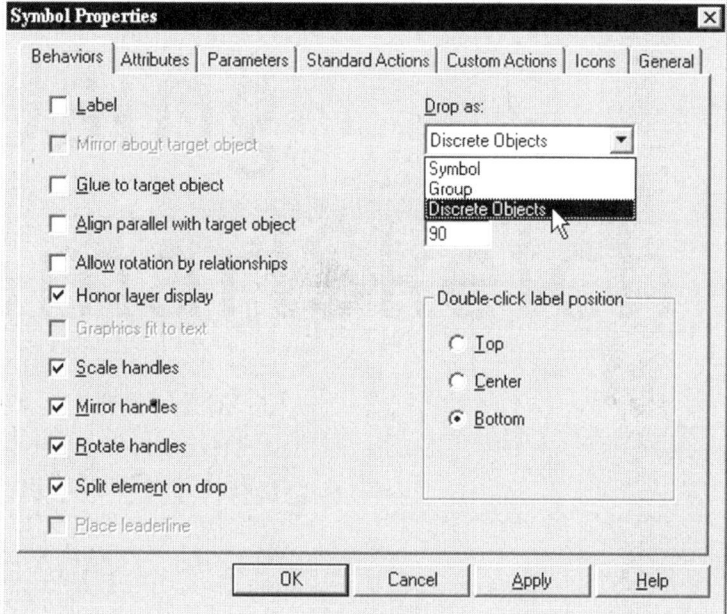

Figure 4–65.
Drop as discrete
objects selected.

To appreciate how symbols are dropped as discrete objects, perform the following steps to insert a symbol as discrete objects.

1 Start a new document file.

2 Insert *Symbo03* into the document.

3 You will find that the symbol is now inserted as discrete individual objects.

Mirroring While Dropped

To allow mirroring a symbol when it is being dropped in a target document, perform the following steps to construct a new symbol.

1 Open the symbol file *Symbol06*, if you already closed it.

2 Select File > Save As.

3 In the Save As dialog box, specify the symbol file name *Symbol07*.

Now the symbol file *Symbol06.sym* is closed and you will be working on the symbol file *Symbol07*.

4 Click on the Symbol Properties icon on the Symbol Authoring Tools toolbar.

5 In the Behaviors tab of the Symbol Properties dialog box, click the *Align parallel with target object* box and then the *Mirror about target object* box. (See Figure 4–66.)

6 Click on the OK button.

7 Save and close the symbol.

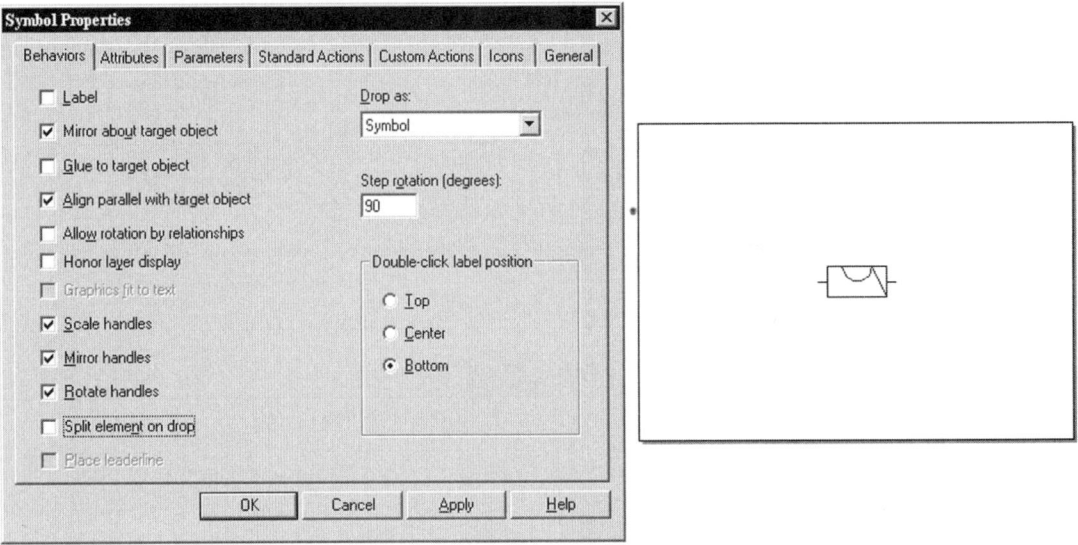

Figure 4–66. Behavior properties changed.

To appreciate how to mirror a symbol while dropping it into a document, perform the following steps.

1 Open the document file *Chapter4SymbolInsertion.igr*, if you already closed it.

2 Activate Sheet 4.

3 In the Symbol Explorer, browse the folder holding the custom symbols you constructed.

4 Select *Symbol07* and drag it to location A indicated in Figure 4–67.

5 With the mouse button held down, drag the symbol slightly up and down to mirror it. (See Figure 4–68.)

6 Click on A indicated in Figure 4–68 after you are satisfied with the symbol's orientation.

7 Save the document file in your computer and close it.

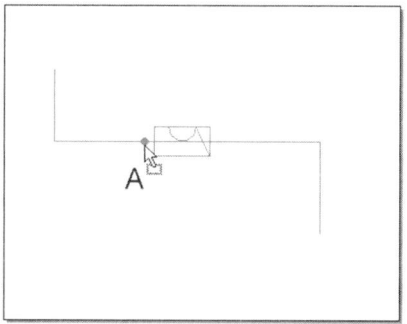

Figure 4–67. Symbol dragged to the connector.

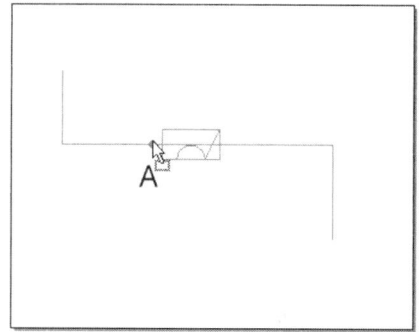

Figure 4–68. Symbol slightly dragged up and down to mirror it.

Defining Attribute Properties

You can add attributes to a symbol to provide textual meaning to it. Attributes can be a text string, a date, a number, or a Yes/No statement. Perform the following steps.

1 Open the symbol file *Symbol Attribute* from the *Chapter 4* folder of the companion CD-ROM.

2 Click on the Symbol Properties icon on the Symbol Authoring Tools toolbar.

3 Select the Attributes tab.

4 In the Attributes tab, type the text string *Furniture Type* in the Name box, select Text from the Type pull-down list box, and type *Desk* in the Value box. Click on the Add button. (See Figure 4–69.)

5 With reference to Figure 4–70, add three more attributes as follows:

Name	Value	Type
Purchase Date	June 2004	Date
Price in US$	200	Number
Refundable	Yes	Yes or No

6 Click on the OK button.

7 Save the symbol file in your computer and close the file.

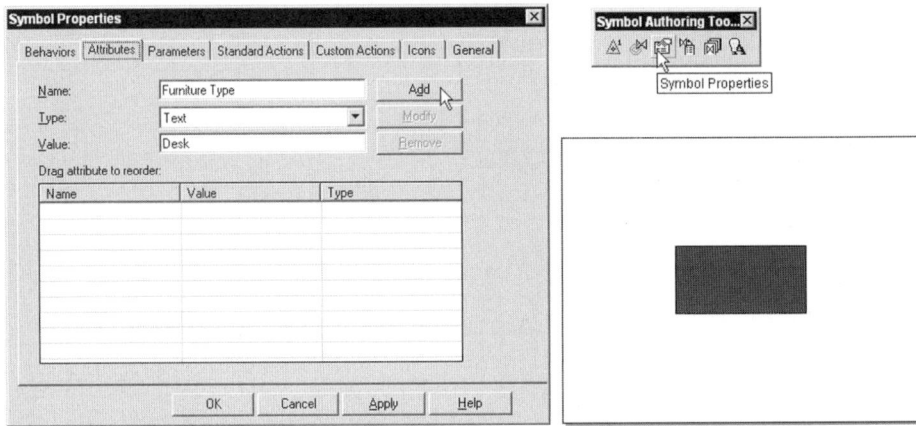

Figure 4–69. An attribute being defined.

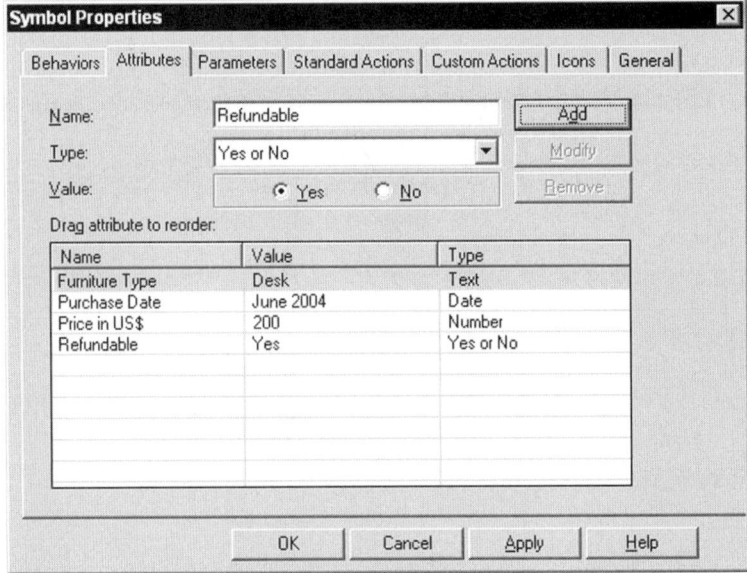

Figure 4–70.
Attributes
defined.

Perform the following steps to insert the symbol in a document.

1 Start a new document file using the template *Technical Drawing (Imperial).igr*.

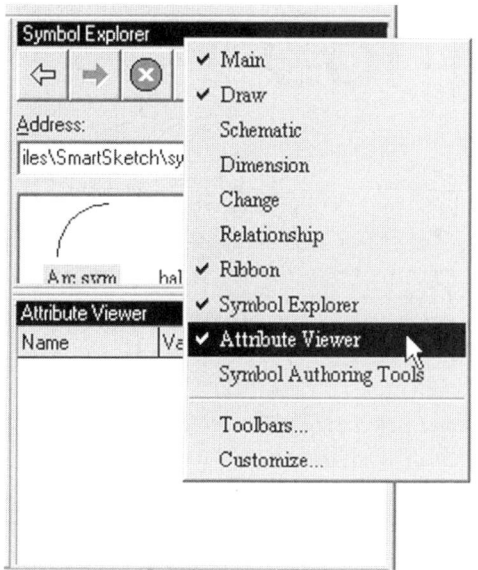

Figure 4–71. Displaying the Attribute Viewer.

2 Below the Symbol Explorer, there should be an Attribute Viewer. If it is not displayed there or anywhere in the screen, move the cursor to the title bar of the Symbol Explorer, right click, and click Attribute Viewer. (See Figure 4–71.)

3 In the Symbol Explorer, browse the folder holding the custom symbols you constructed.

4 Select and drag the symbol into the document. (See Figure 4–72.)

While the inserted symbol is selected, its attributes will be displayed in the Attribute Viewer. To change these attributes, type a new value in the Value column of the Viewer. You may also use the Symbol Properties dialog box, as follows.

5 Select the dropped symbol, if it is already unselected.

Figure 4–72. Attributes of a symbol.

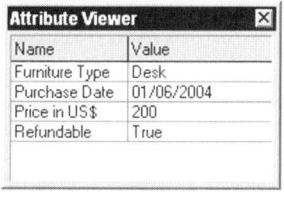

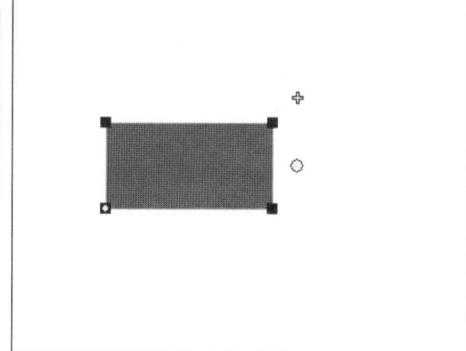

6 Select Edits > Properties.

7 Select the User tab of the Symbol Properties dialog box. (See Figure 4–73.)

8 Select an attribute from the Attributes list.

9 Type a new value in the Value box.

10 Click on the OK button to exit.

11 Close the document file without saving.

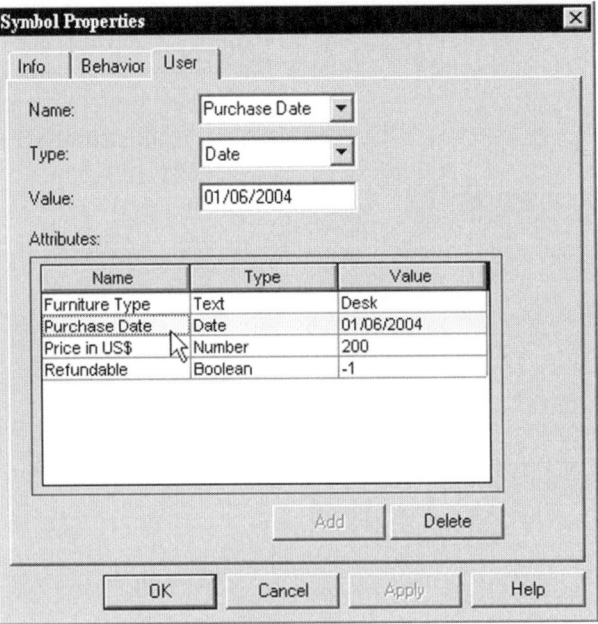

Figure 4–73.
User tab of
the Symbol
Properties
dialog box.

Defining Parameter Properties

The Parameters tab of the Symbol Properties dialog box concerns the driving dimensions in the symbol. As mentioned in Chapter 3, you have to select Tools > Maintain Relationships and click on the Driving/ Driven button in the dimension ribbon while constructing a dimension. You can include up to a maximum of four parameters in a symbol. Perform the followings steps.

1   Open the document file *Chapter4SymbolParameters.igr* from the *Chapter 4* folder of the companion CD-ROM.

2   Select Tools > Maintain Relationships, if there is no tick mark prefixing the menu item. This is mandatory for construction of driving dimensions.

3   Display the Dimension toolbar, if it is not displayed. Displaying the Dimension toolbar can be done through selecting View > Toolbars, clicking Dimension from the toolbars list in the Toolbars dialog box, and clicking on the OK button.

4   Construct four driving dimensions in accordance with Figure 4–74.

5   Click on the Select Tool icon on the Draw toolbar.

6   Select location A (indicated in Figure 4–75) and drag the mouse button to location B (indicated in Figure 4–75) to select the line segments and the dimensions.

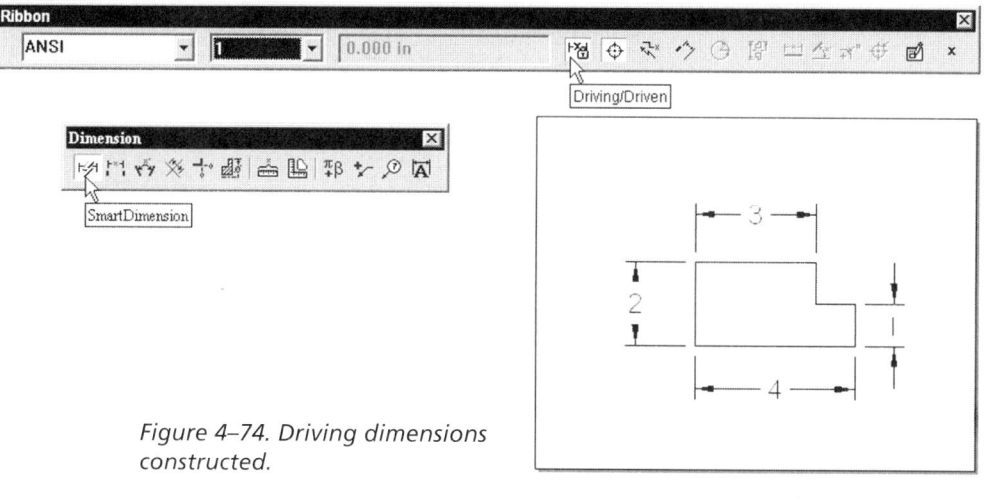

Figure 4–74. Driving dimensions constructed.

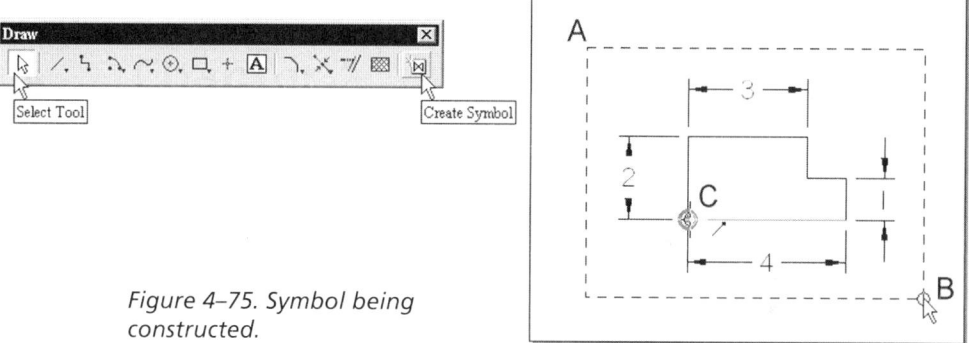

Figure 4–75. Symbol being constructed.

7 Click on the Create Symbol icon on the Draw toolbar.

8 Select endpoint C as the origin of the symbol.

9 In the Save As Symbol dialog box, specify a symbol file name *(Symbol Parameter)* and click on the Save button.

10 Close the document file without saving.

Perform the following steps to work on the symbol file.

1 Open the symbol file *Symbol Parameter* you just constructed.

2 Click on the Symbol Properties icon on the Symbol Authoring Tools toolbar.

3 In the Symbol Properties dialog box, select the Parameters tab. You will find four dimension parameters listed. (See Figure 4–76.) Note that the order of the dimensions displayed in the list may not be the same as yours.

4   Select the 4-inch dimension in the dialog box. Note that the corresponding dimension A in Figure 4–76 should be highlighted in red. If not, you probably have constructed a driven dimension rather than a driving dimension in the symbol. In that case, you have to go back to previous step 4 (on page 190) and redo the symbol.

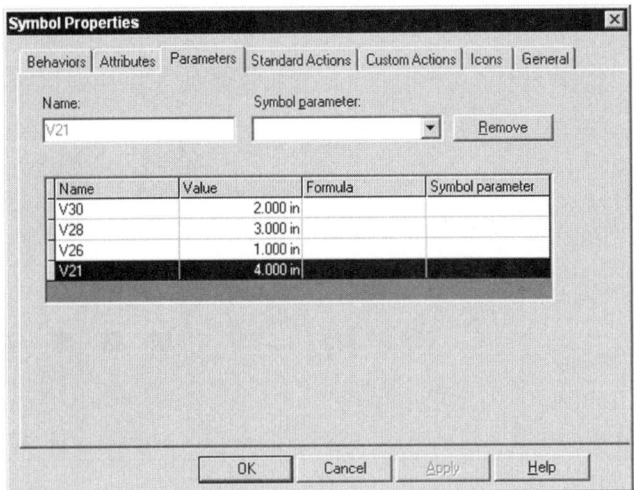
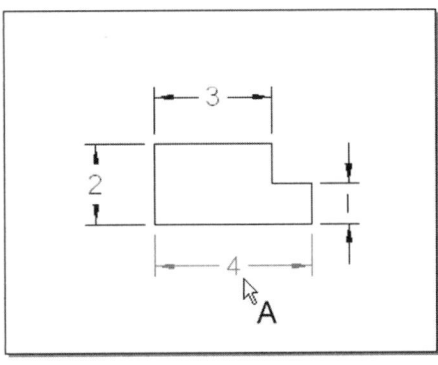

Figure 4–76. Dimensions listed in the Parameters tab of the Symbol Properties dialog box.

5   While the 4-inch dimension is selected, select Bottom from the Symbol Parameter pull-down list box. This sets the control to the bottom of the symbol. (See Figure 4–77.)

Figure 4–77. Bottom symbol parameter set.

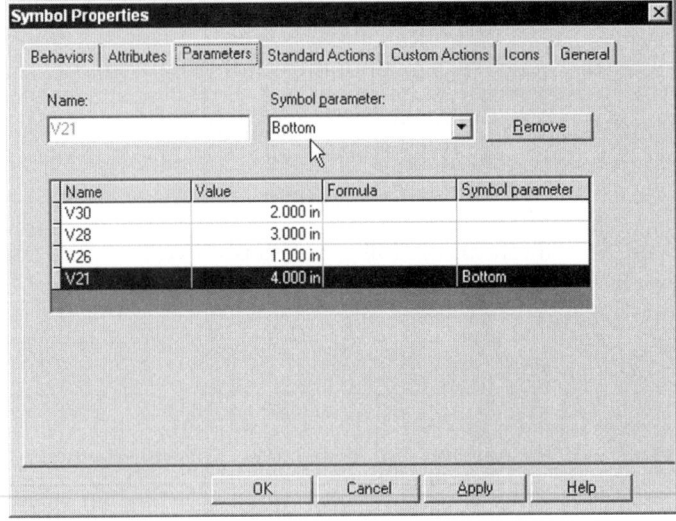

6 With reference to Figure 4–78, set the 1-inch dimension as right, the 3-inch dimension as top, and the 2-inch dimension as left.

7 Save and close the symbol file.

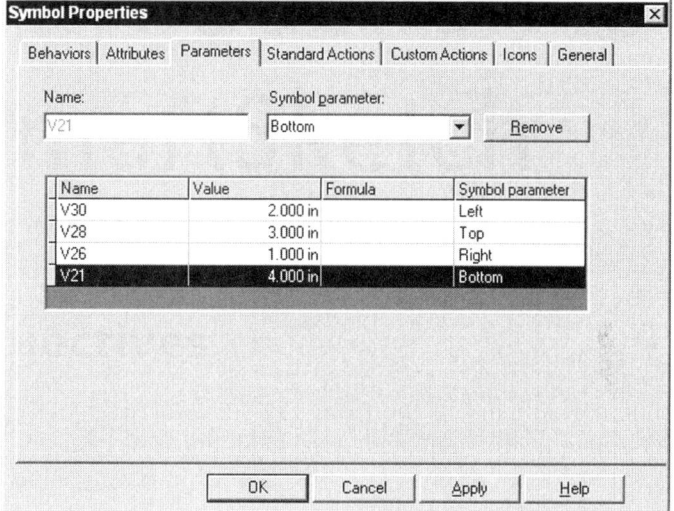

Figure 4–78.
Three symbol
parameters
established.

Perform the following steps to insert the symbol in a document.

1 Start a new document file using the template *Technical Drawing (Imperial).igr.*

2 In the Symbol Explorer, browse the folder holding the custom symbols you constructed.

3 Select the symbol *Symbol Parameters* from the Symbol Explorer. As can be seen in Figure 4–79, the Attribute Viewer displays the symbol's parameters and their dimension values in terms of name and value in two columns.

4 Select and drag the symbol into the document.

5 Modify the symbol's size by changing the values of the parameters in the Attribute Viewer. (See Figure 4–80.)

Because there is a scale handle in the symbol, you may also change the overall size of the symbol by selecting the dragging the scale handle. To avoid confusion due to scaling, you may disable the scale handle of the symbol by clearing the Scale Handles box of the Behavior tab of the Symbol Properties dialog box.

6 Close the document file without saving.

*Figure 4–79.
Symbol
parameters
and values
displayed in
the Attribute
Viewer.*

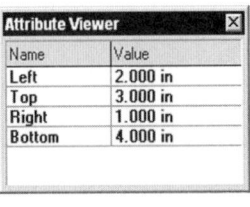

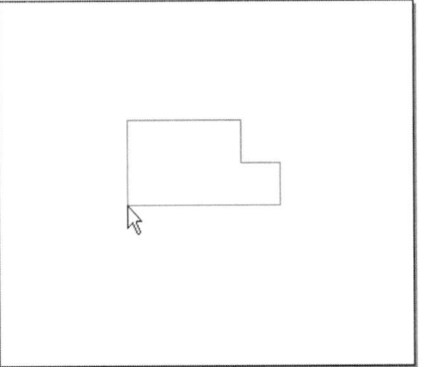

*Figure 4–80.
Symbol's size
being modified.*

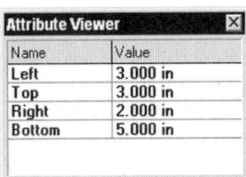

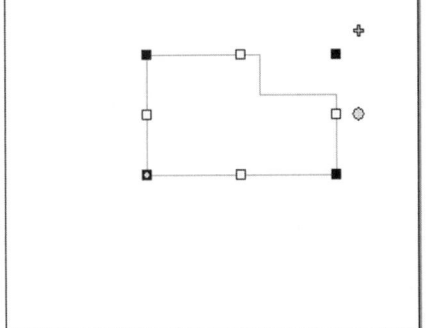

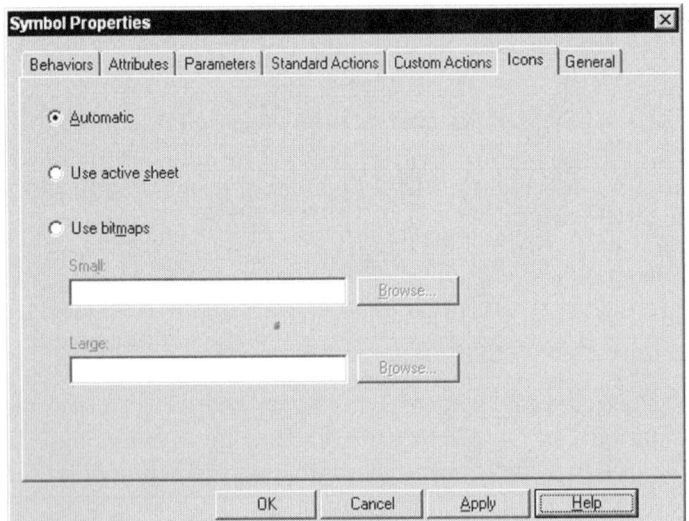

Figure 4–81. Icons tab of the Symbol Properties dialog box.

Defining Icons

You can define the symbol icon that will be displayed in the Windows Explorer and the Symbol Explorer by using the Icons tab of the Symbols Properties dialog box. As shown in Figure 4–81, you can define an icon via three options: *Automatic, Use active sheet, and Use bitmaps.* The *Automatic* option uses the elements and objects in the current window to construct the icon. The *Use active sheet* option saves the latest change to the symbol icon. The *Use bitmaps* option enables you to select a bitmap file as the icon.

Defining Standard Actions, Custom Actions, and General Properties

The Standard Actions tab and Custom Actions tab require preconstructed program files with an extension of *.dll*, *.ocx*, or *.exe* in Visual Basic. The General tab requires preconstructed help files with an extension of *.hlp*.

The Standard Actions tab, shown in Figure 4–82, defines actions to be taken while a symbol is being dropped into a document, when a dropped symbol is double clicked, and when you select Edit > Properties.

The Custom Actions tab, shown in Figure 4–83, enables you to define custom actions to be taken when you right click and select a custom command from the shortcut menu.

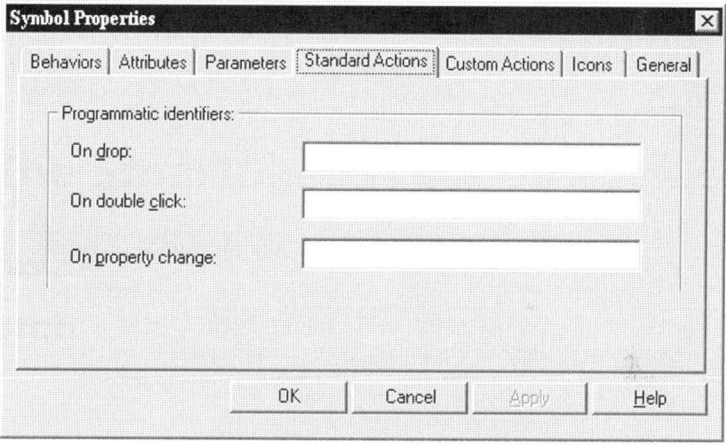

Figure 4–82.
Standard
Actions tab.

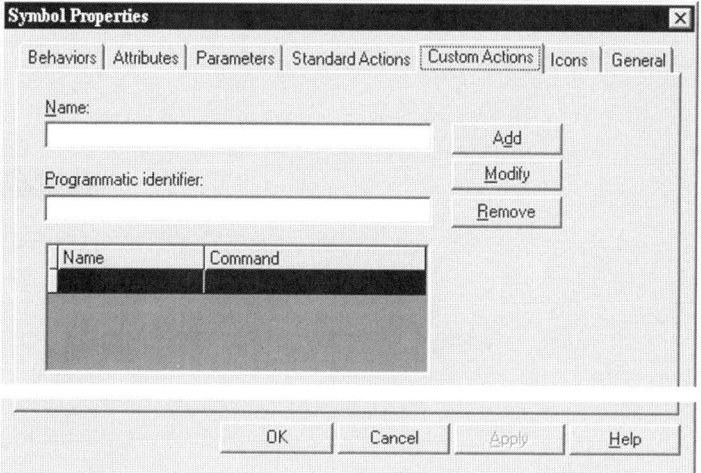

Figure 4–83.
Custom
Actions tab.

The General tab, shown in Figure 4–84, specifies a special help document that delineates special information about the symbol.

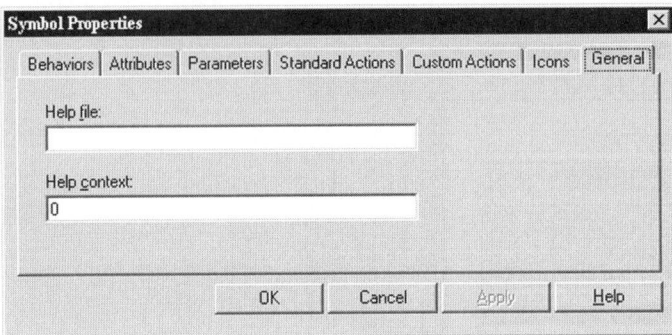

Figure 4–84.
General tab.

Using Lookup Tables

If you already have a Visual FoxPro file and wanted to use the table's data to associate with the parameters of a symbol, you use the Lookup Table command from the Symbol Authoring Tools toolbar. Figure 4–85 shows the Lookup Table dialog box.

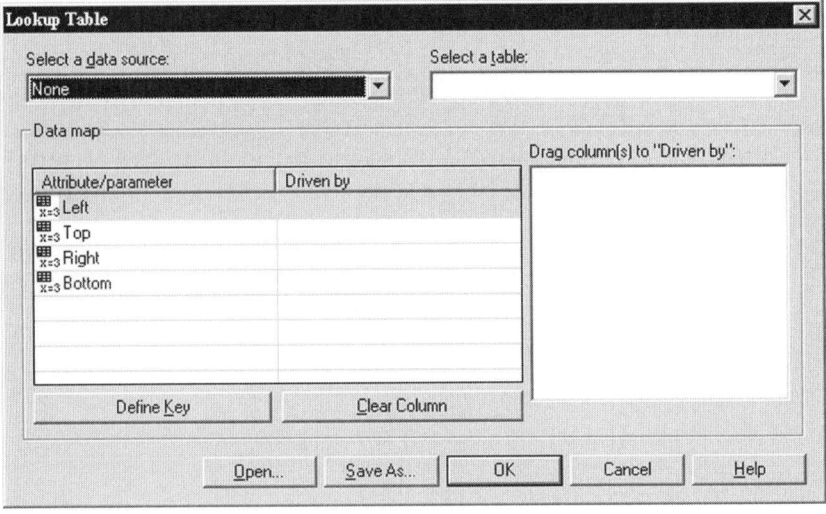

Figure 4–85.
Lookup Table
dialog box.

Defining Different Representations

Instead of constructing individual symbol files to represent different symbol shapes, you can use a single symbol file with a number of representations in it. Perform the following steps to construct a symbol.

1 Open the document file *Chapter4SymbolRepresentation.igr* from the *Chapter 4* folder of the companion CD-ROM.

2 Click on the Select Tool icon on the Draw toolbar.

3 With reference to Figure 4–86, click on location A and drag the mouse to location B.

4 Click on the Create Symbol icon on the Draw toolbar.

5 Select endpoint C to indicate the symbol origin.

6 In the Save As Symbol dialog box, specify a symbol file name *(Symbol Representation)* and click on the Save button.

7 Close the document file without saving.

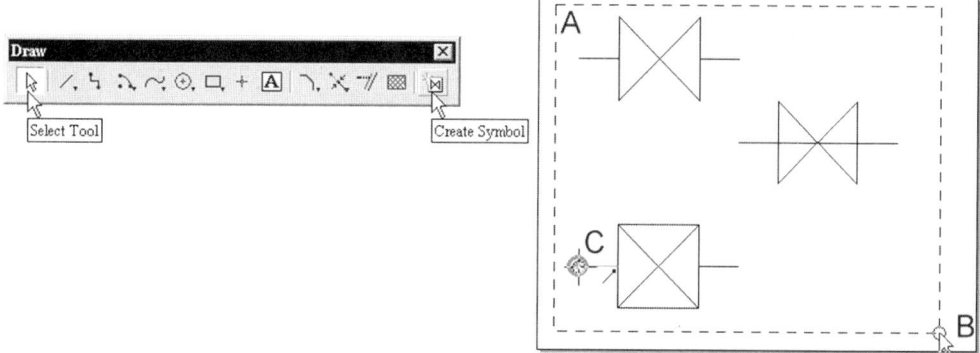

Figure 4–86. Objects selected for making a symbol.

Perform the following steps to work on the symbol file.

1 Open the symbol file *Symbol Parameter* you just constructed.

2 Click on the Select Tool icon on the Draw toolbar.

3 With reference to Figure 4–87, click on location A and drag to location B.

4 Click on the Group icon on the Change toolbar. Selected elements are put into a group.

5 Repeat steps 3 and 4 twice to construct two more groups. (See Figures 4–88 and 4–89.)

6 Click on the Symbol Representation icon on the Symbol Authoring Tools toolbar.

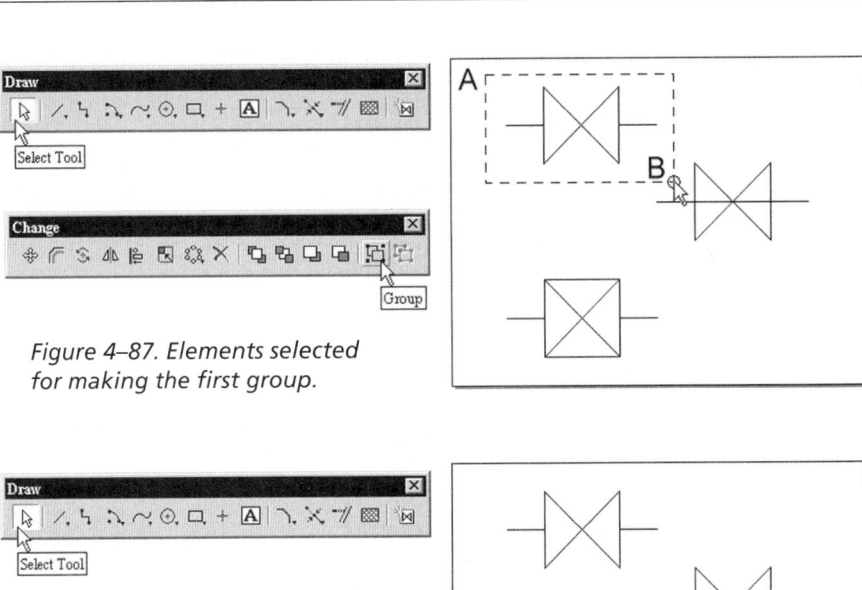

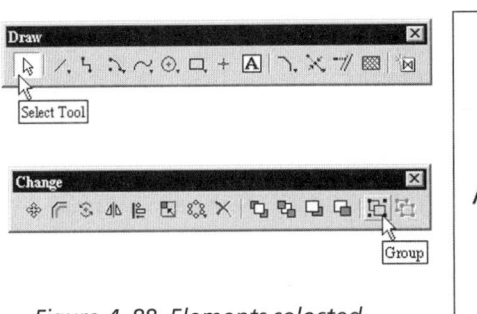

Figure 4–87. Elements selected for making the first group.

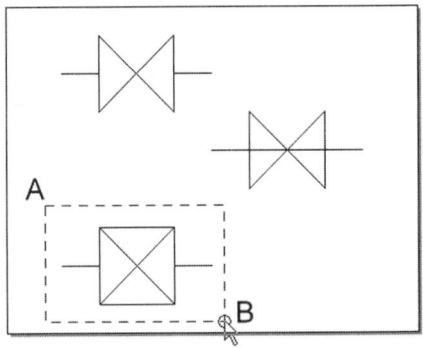

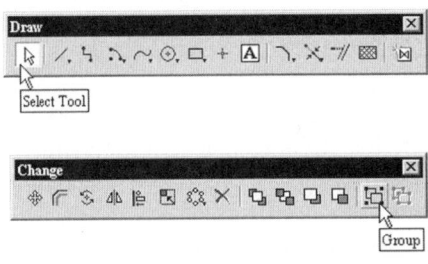

Figure 4–88. Elements selected for making the second group.

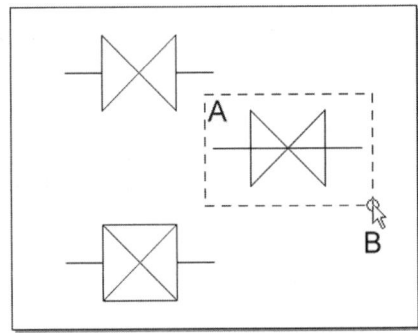

Figure 4–89. Elements selected for making the third group.

7 Select A indicated in Figure 4–90. In the Group Name box of the Define Symbol Representation dialog box, you will find a group name generated by the system. Note that the name may be different from yours. However, the name here is unimportant.

8 In the Representation Name box, type the text string *Symbol A*. This is the name of the first representation of the symbol.

9 Click on the Add button. A representation is constructed.

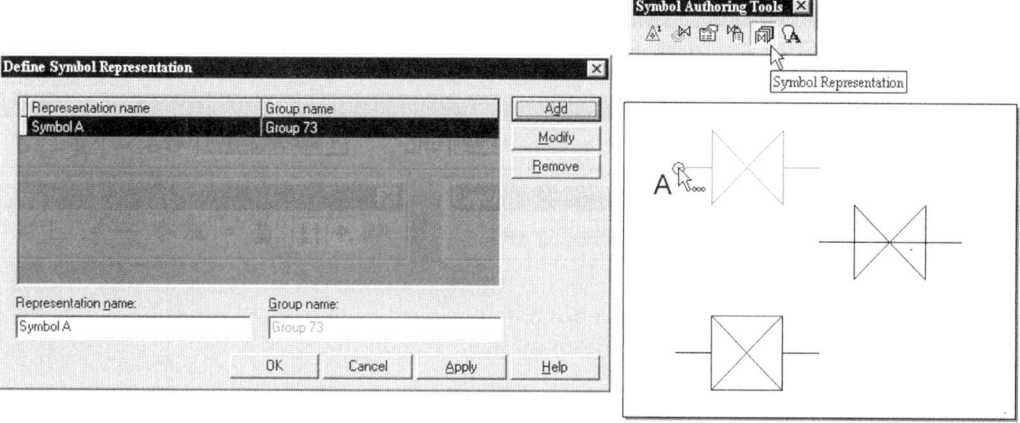

Figure 4–90. First representation being made.

10 Select B indicated in Figure 4–91 and type the text string *Symbol B* in the Representation Name box of the Define Symbol Representation dialog box.

11 Click on the Add button. The second representation is added.

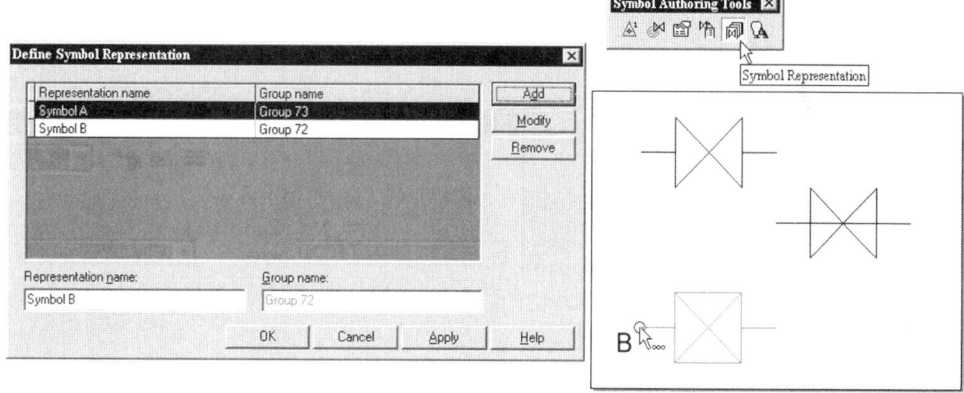

Figure 4–91. Second representation being made.

12 Select C indicated in Figure 4–92.

13 Type the text string *Symbol C* in the Representation Name box.

14 Click on the Add button.

15 Click on the OK button.

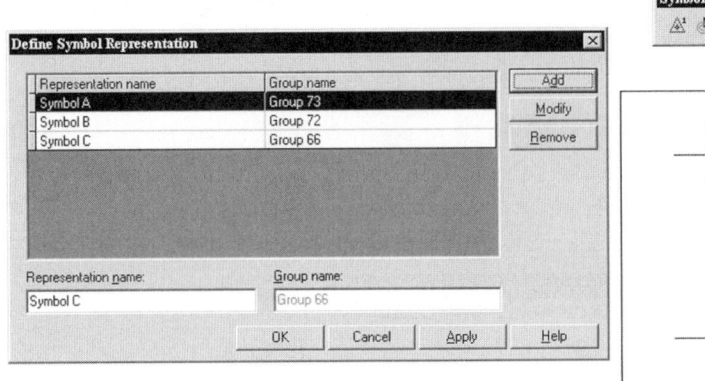

Figure 4–92. Third representation being made.

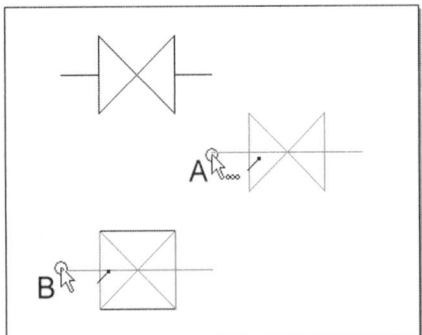

Figure 4–93. Third group being placed over the second group.

16 Select A indicated in Figure 4–93 and drag it to endpoint B. This moves a symbol over anther symbol.

17 Select A indicated in Figure 4–94 and drag it to endpoint B. Three symbols now overlap. (See Figure 4–95.)

18 Save and close the symbol file.

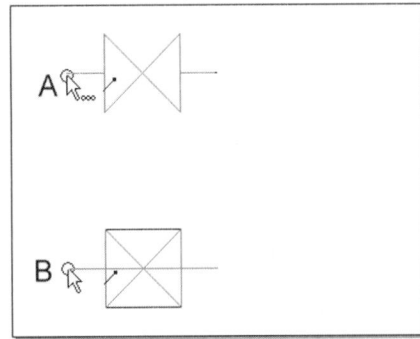

Figure 4–94. First group being placed over the second group.

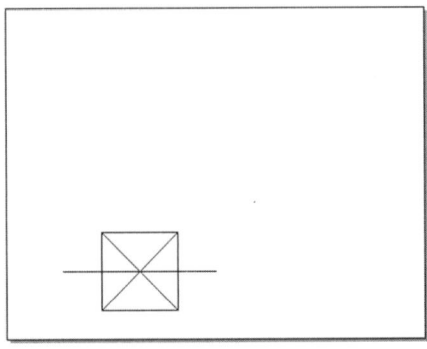

Figure 4–95. Three groups of elements placed together.

Perform the following steps to use the symbol with multiple representations.

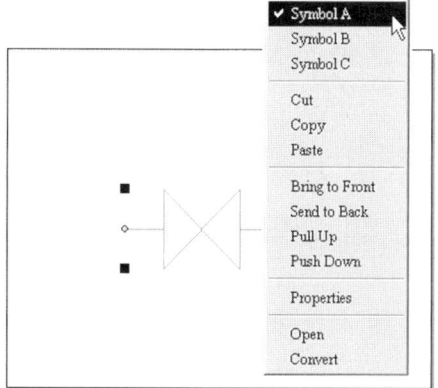

Figure 4–96. Three representations of a symbol to choose from.

1 Start a new document file using the template *Technical Drawing (Imperial).igr.*

2 In the Symbol Explorer, browse the folder holding the custom symbols you constructed.

3 Select the symbol *Symbol Representation* and drop it into the document.

4 Select the inserted symbol and right click. You will find three representations, together with other options. Clicking one of the representations enables you to use different versions of the symbol. (See Figure 4–96.)

5 Close the document file without saving.

Constructing a Smart Label

A Smart Label relates to an attribute of a symbol, showing the attribute's value. Perform the following steps to construct a Smart Label.

1 Open the symbol file *Symboµl Smart Label* from the *Chapter 4* folder of the companion CD-ROM.

2 Click on the Symbol Properties icon on the Symbol Authoring Tools toolbar.

3 In the Behaviors tab of the Symbol Properties dialog box, click the boxes Label and *Glue to target object*. (See Figure 4–97.)

4 Click on the OK button to close the dialog box.

5 Click on the Edit Smart Text icon on the Symbol Authoring Tools toolbar.

6 In the Smart Text Editor dialog box, select *item one* from the Item pull-down list box and select String from the Format pull-down list box.

7 Type the string *Furniture Type* in the Property box. Note that this text string must be identical to the attributes of the other symbol that this symbol attaches to.

8 In the Value box, type the string *Desk*. (See Figure 4–98.)

9 Click on the Insert Field button. This will put the objects entered in the SmartText field into the Existing Text box of the dialog box.

10 Select all objects in the Existing Text Box.

11 Click on the Font button.

12 Set the font size to 20 points.

13 Click on the OK button. A Smart Text is constructed.

14 Save the symbol file in a folder of your computer.

15 Close the symbol file.

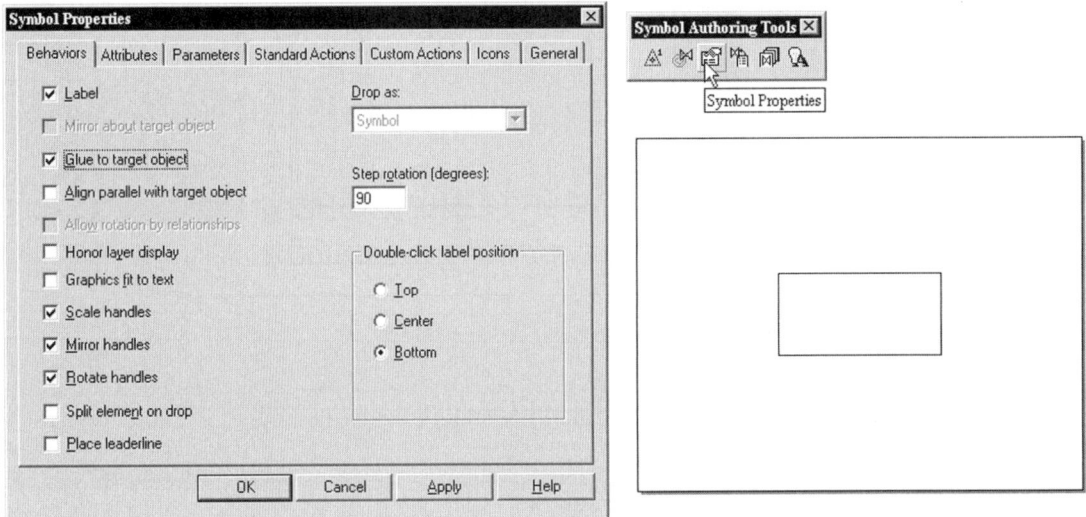

Figure 4–97. Symbol file opened and behaviors changed.

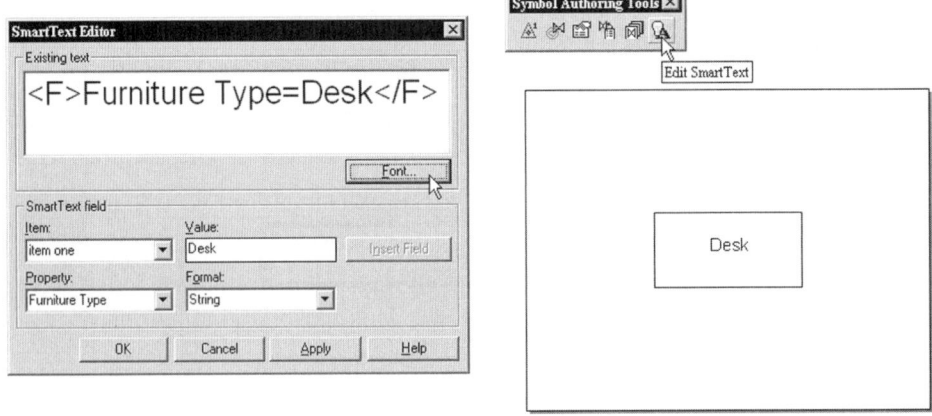

Figure 4–98. Smart Text being constructed.

To appreciate how to attach a Smart Label to a symbol, perform the following steps.

1 Start a new document file using the template *Technical Drawing (Imperial).igr.*

2 In the Symbol Explorer, browse the *Chapter 4* folder of the companion CD-ROM.

3 Select the symbol *Symbol for Label* and drop it into the document.

4 Select the dropped symbol, if it is deselected. You will find four attributes in the Attribute Viewer. (See Figure 4–99.)

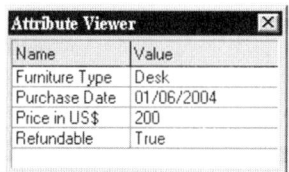

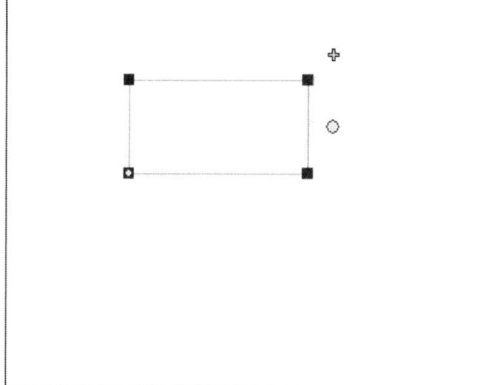

Figure 4–99. Symbol dropped into a document.

5 In the Symbol Explorer, browse the folder in which you saved the symbol file *Symbol Smart Label.*

6 With reference to Figure 4–100, select and drag the symbol *Symbol Smart Label* into the document. Drag the cursor A (with Smart Label being attached) to endpoint B of the dropped symbol.

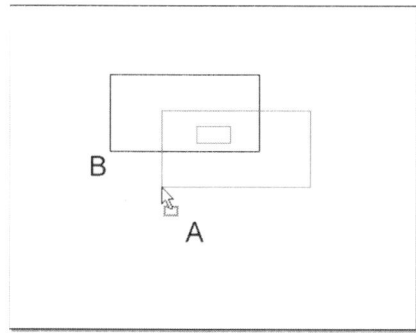

Figure 4–100. Smart label being dragged to a dropped symbol.

After the Smart Label is attached to the dropped symbol, a lock icon will be displayed, indicating that the label is locked to the symbol. If the Attribute Viewer is displayed, the label's value will be displayed. (See Figure 4–101.)

7 Type the string *Stool* in the Value box of the Attribute Viewer. Note in Figure 4–101 that the label is changed.

8 Click on the Push Down icon on the Change toolbar. This will bring the label below its attached symbol.

9   Select the symbol. The Attribute Viewer now shows the symbol's attribute. Note in Figure 4–102 that the value for Furniture Type is changed.

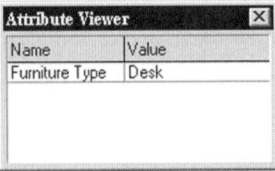
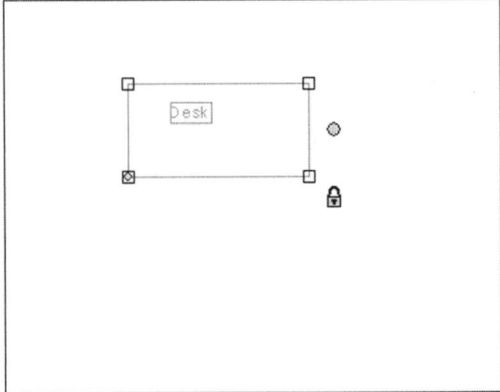

Figure 4–101. Label locked to a symbol.

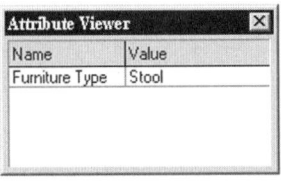
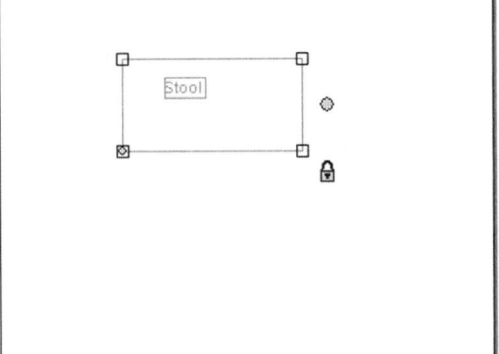

Figure 4–102. Value of the label changed.

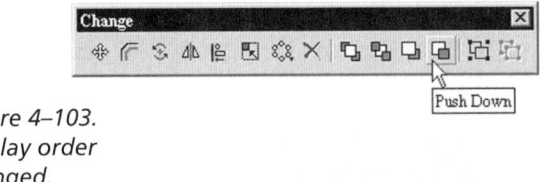
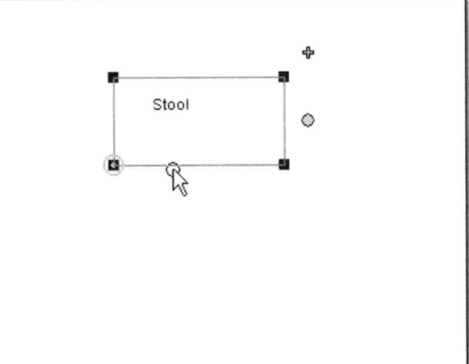

Figure 4–103. Display order changed.

10 Change the value of the Furniture Type to Chair. Note that the label changes as well. (See Figure 4–103 and 4-104.)

11 Close the document without saving.

Figure 4–104. Label changed.

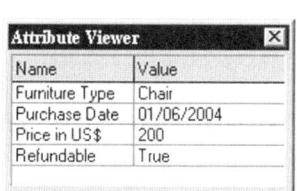

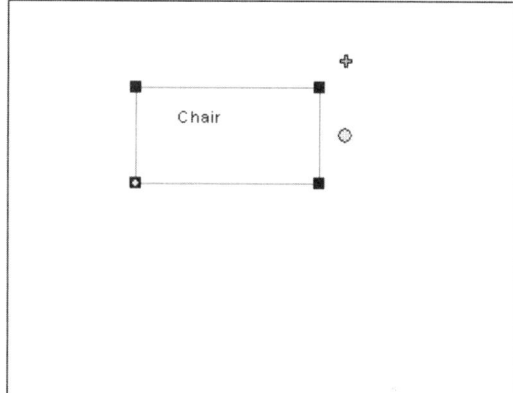

■ ■ ■ ■   **Summary**

The use of symbols in document construction enhances productivity by shortening production time and reusing drawing objects stored in the library. Symbols can be constructed from scratch, by modifying an existing symbol or by combining two or more symbols.

To insert a symbol in a document, you use the Symbol Explorer, through which you drag and drop symbols from a symbol library. The Symbol Explorer's functions are very similar to those of the Windows Explorer in many respects, including browsing the World Wide Web.

A dropped symbol can be embedded in the target document or linked to the source symbol in the library. An embedded symbol is saved in the document and has no relationship with the source symbol. On the other hand, a linked symbol is always associated with the source symbol in such a way that if the source symbol changes the linked symbol also changes.

To customize a symbol, you use the Symbol Authoring tools. You can add Smart Points in a symbol to manage the symbol's connect points, drop point, and drag points. Although the origin point has to be defined while making the symbol, you can change the origin point afterward. You can decide how a symbol will behave when it is dropped into a target document. To give added meaning to a symbol, you can add attributes to a symbol. With a parametric symbol, you can resize a symbol by manipulating the parameters.

If you have good programming knowledge, you can go one step further to associate a symbol to programs, executing standard actions and custom actions, as well as providing custom help. To help identify a symbol, you manage the icon used to depict the symbol. In a single symbol file, you can have multiple symbols to choose from when inserted into a target document. A symbol can be constructed as a Smart Label that associates with a dropped symbol, thus displaying the associated attribute values.

■ ■ ■ ■ Review Questions

1   What are the main functions of the Symbol Explorer and the Attribute Viewer?

2   How many ways can you construct a symbol?

3   How are Smart Points defined in a symbol?

4   Explain how behaviors of a symbol can be modified.

5   How can multiple representations of a symbol be defined?

6   Explain how a Smart Label is constructed.

CHAPTER 5

Drafting III

■ ■ ■ ■ ## Objectives

The goals of this chapter are to explore how to construct detail views and isometric views, as well as the use of variables in a document. The chapter also explores inserting hyperlinks, Windows objects, and raster images in a document. After studying this chapter, you should be able to:

- ❑ Construct detail views and isometric views
- ❑ Manipulate variables
- ❑ Insert hyperlinks, Window Objects, and raster images

Overview

In order to provide a clearer close-up view to details of a document, you insert detail views. To represent a 3D object on a 2D plane, you use isometric drawings. To maintain relationships among the parameters of a document, you manipulate variables. In addition to inserting SmartSketch objects in your document, you can insert hyperlinks, Windows objects, and raster images.

■ ■ ■ ■ ## Constructing Detail Views and Isometric Views

In order to better see the details of a drawing, you insert a detail view. To better show the details of an object viewed in a 3D, you construct isometric views. Isometric views are 2D drawings simulating the effect of a 3D drawing. To construct an isometric drawing, you need special tools.

Detail View

To appreciate how a detail view can be produced, perform the following steps.

1   Open the document file *Chapter5Detail.igr* from the *Chapter 5* folder of the companion CD-ROM.

2   Select Insert > Detail View.

3   In the Insert Detail View ribbon, click on the *Circular detail* button, if it is not already selected.

4   Select endpoint A indicated in Figure 5–1 and click location B to describe a circle.

5   Click location C indicated in Figure 5–2. A detail view is constructed. (See Figure 5–3.)

6   Save the file in a folder of your computer and close the file.

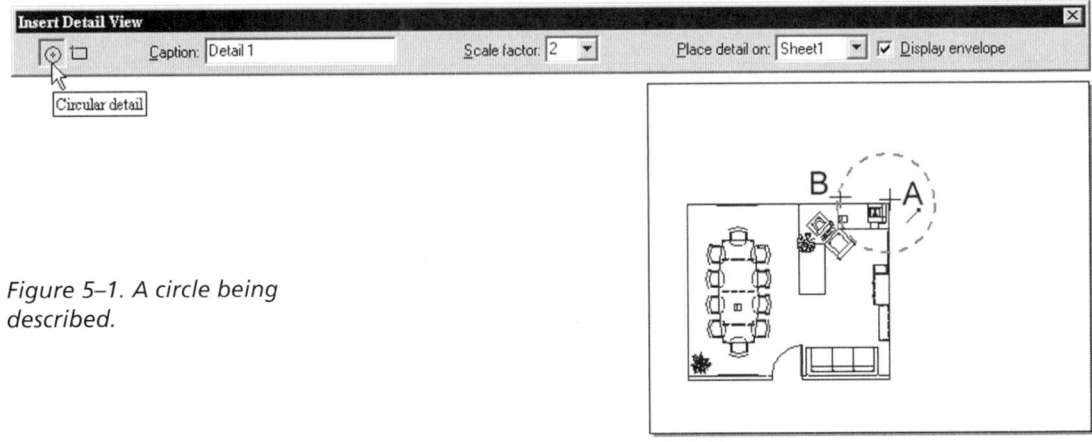

Figure 5–1. A circle being described.

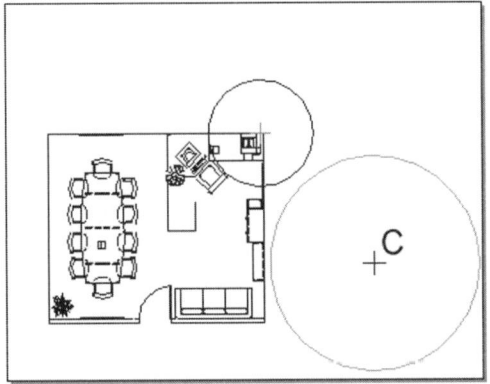

Figure 5–2. Detail view location being determined.

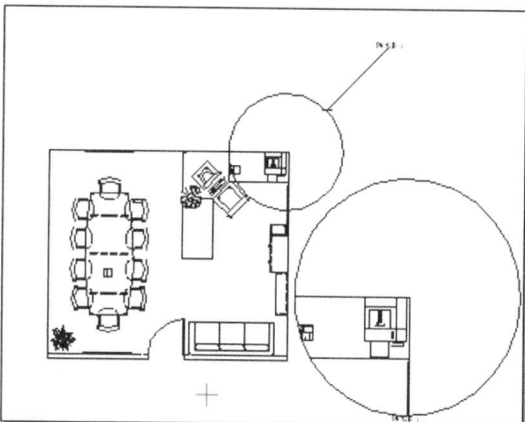

*Figure 5–3.
Detail view
constructed.*

Isometric Views

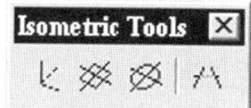

*Figure 5–4. Isometric
Tools toolbar.*

To construct an isometric view, you need to use the Technical Illustration template together with its associated Isometric Tools toolbar, shown in Figure 5–4. Details of the options available in the toolbar are delineated in Table 5–1.

Table 5–1 Isometric Toolbar Options and Their Functions

Options	Function
Isometric Line	Constructs an isometric line. In essence, it is a line command with the line's direction restricted to one of the three isometric axes.
Isometric Rectangle	Constructs an isometric rectangle on one of the three isometric planes. In essence, it produces a parallelogram.
Isometric circle	Constructs an isometric circle on a 2D plane that looks like a 3D circle. In essence, it produces an ellipse in one of the three isometric planes.
Segmented Style	Applies a line type, line weight, or color to part of an element.

Perform the following steps to construct and isometric drawing.

1   Start a new document using the template file *Technical Illustration (Imperial).igr.*

2   Turn on PinPoint.

3   Click on the Isometric Line icon on the Isometric Tools toolbar.

4   Click on location A indicated in Figure 5–5 to specify the first point of the isometric line.

5   Click on location B indicated in Figure 5–5 to specify the second point of the first isometric line.

6   Click on location C indicated in Figure 5–5 to specify the endpoint of the second isometric line.

7   Right click. Two isometric lines are constructed.

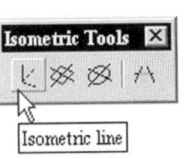

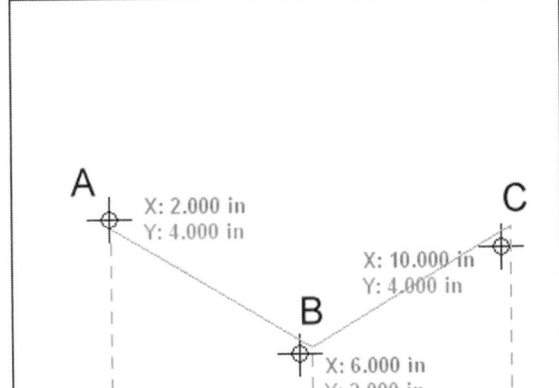

*Figure 5–5.
Isometric lines
constructed.*

8   Click on the *Isometric rectangle* icon on the Isometric Tools toolbar.

9   Click Top on the Isometric Rectangle ribbon.

10 Click location A indicated in Figure 5–6 to specify the first point of the rectangle.

11 Click location B indicated in Figure 5–7. Together with point A constructed in the previous step, an edge of the rectangle is defined.

12 Select location C indicated in Figure 5–8 to specify the other edge of the rectangle.

13 Click on the Isometric Circle icon on the Isometric Tools toolbar.

14 Select Top on the Isometric Circle ribbon.

15 Select location A indicated in Figure 5–9 to specify the center of the isometric circle.

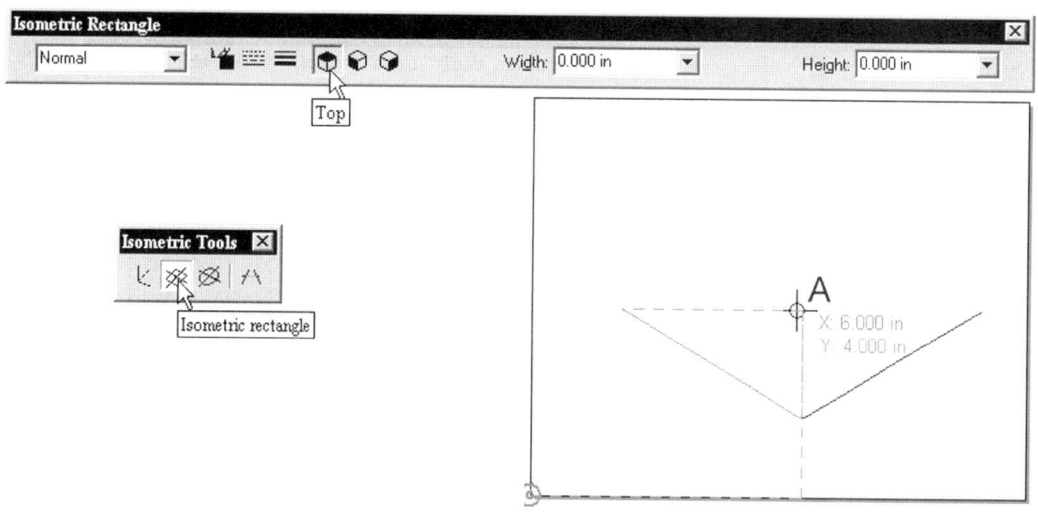

Figure 5–6. First point of the rectangle defined.

Figure 5–7. An edge of
the rectangle defined.

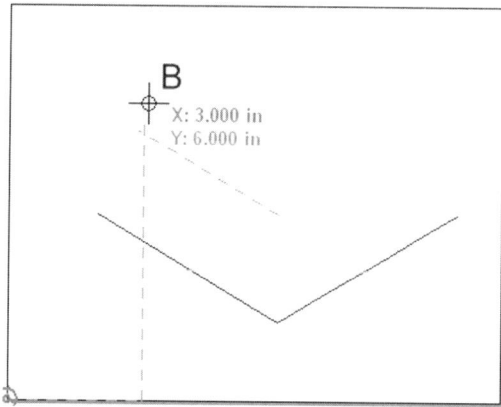

Figure 5–8. Rectangle's
other edge defined.

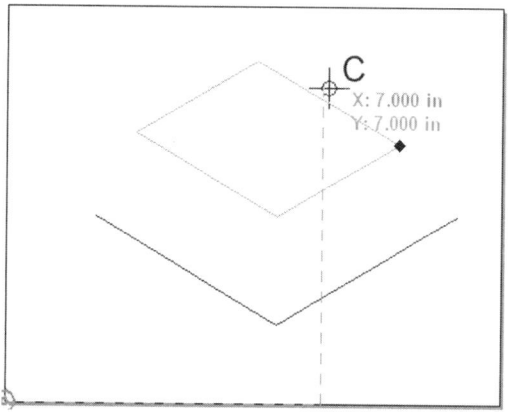

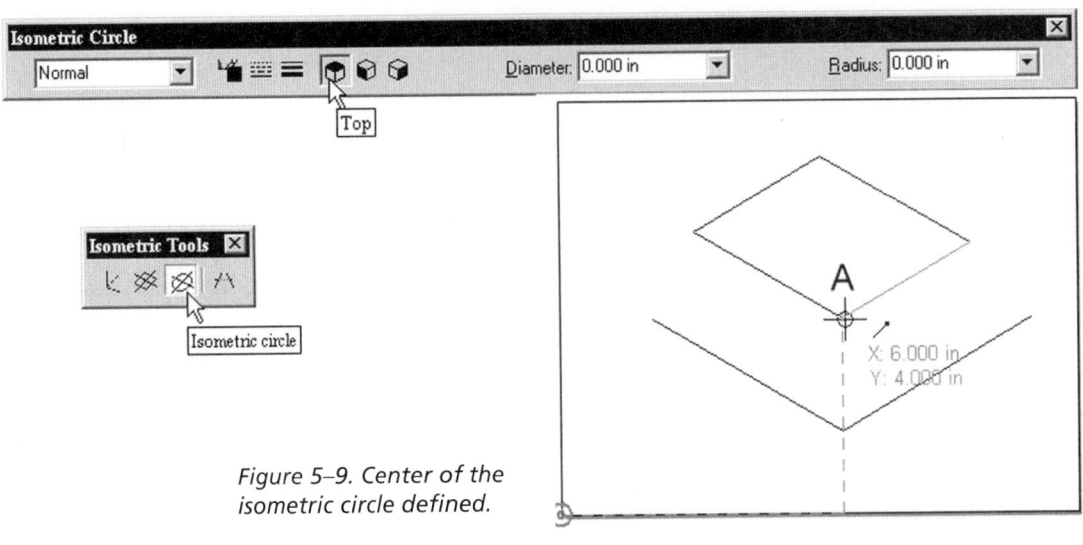

Figure 5–9. Center of the isometric circle defined.

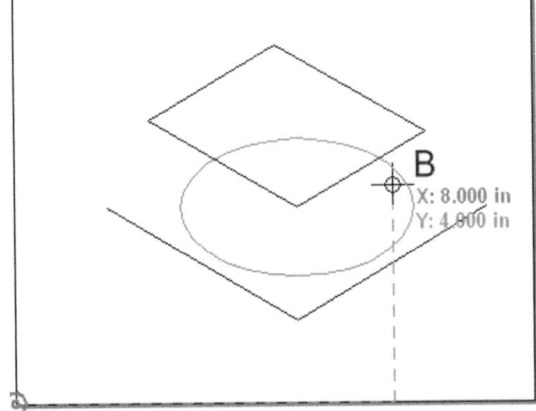

Figure 5–10. Isometric circle constructed.

16 Select location B indicated in Figure 5–10 to specify a point on the isometric circle.

17 Click on the *Segmented style* icon on the Isometric Tools toolbar.

18 Select Dashed on the Isometric Segmented Style ribbon.

19 Select location A indicated in Figure 5–11. Segment BAC is changed to dashed line.

20 Save the file as *Chapter5Isometri.igr* and then close the file.

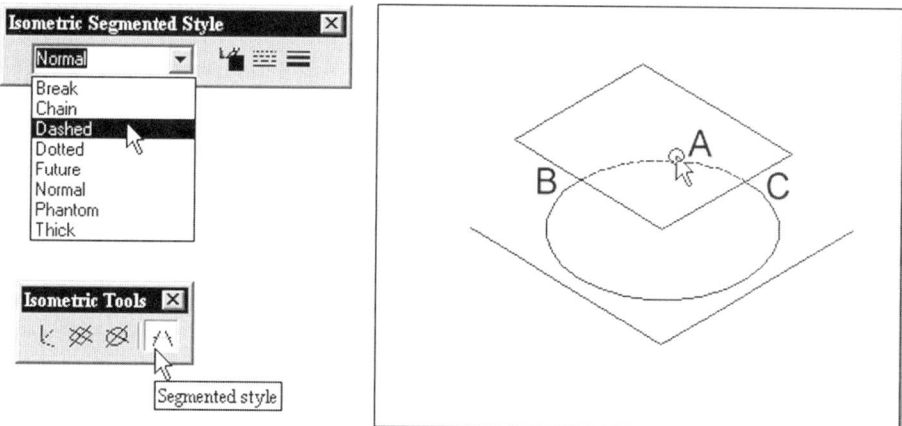

Figure 5–11. Line segmented with a different line type.

■ ■ ■ ■ Manipulating Variables

You can display, define, and manipulate design variables and functional relationships between the variables. There are two types of variables: user and system. User variables, as the name implies, are defined by the end user constructing the SmartSketch document. On the other hand, system variables are generated automatically when dimensions are added to the document. To manipulate variables, you use the variable table, accessible by selecting Tools > Variables. Perform the following steps to manipulate the variables of the document.

1   Open the file *Chapter5Variable* from the *Chapter 5* folder of the companion CD-ROM.

2   Select Tools > Variables to display the Variable Table dialog box, shown in Figure 5–12.

Depending on the filter setting, the variable table may or may not display any variables. On top of the Variable Table dialog box there are, from left to right, a box showing the types of variables and a number of buttons. The Tick button confirms the changes made to the table and the Cross button cancels the previous change. The button A is the filter button. The F_x button is used for bringing out the Function Wizard for constructing complex functions. The ? button is the help button. Continue with the following steps.

3   Click on the filter button (A shown in Figure 5–12) to display the Filter dialog box.

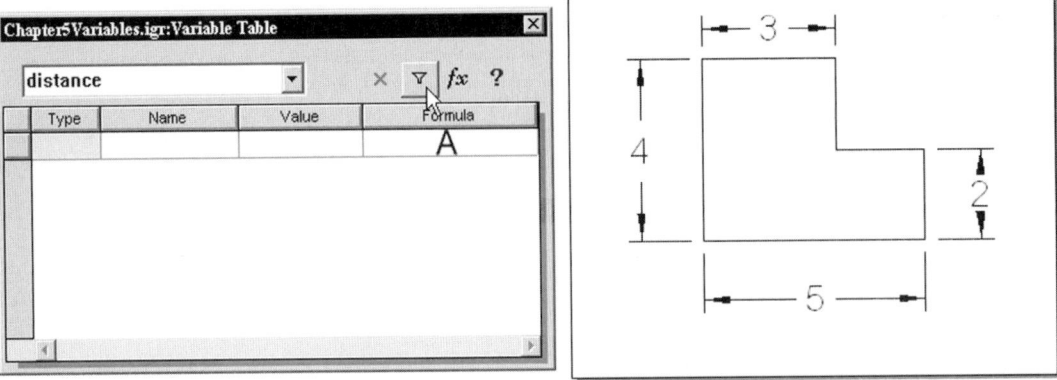

Figure 5–12. Variable Table dialog box displayed.

4 In the Filter dialog box, shown in Figure 5–13, click on Both and then on the OK button.

Figure 5–13. Setting the filter.

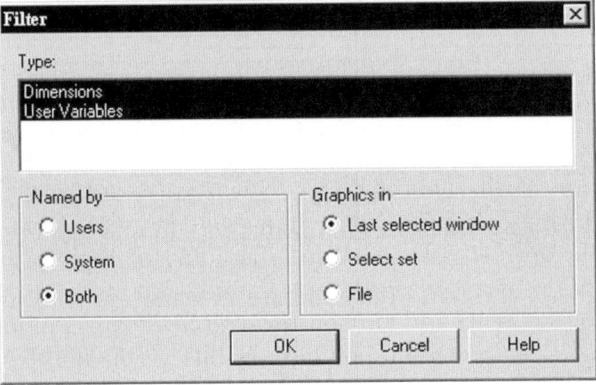

Now all variables, including the system variables depicting the four dimensions in the document, are displayed. Note in Figure 5–14 that the Name column is automatically generated.

5 With reference to Figure 5–15, change the value of a variable to 1 inch. As a result, the drawing is modified.

6 With reference to Figure 5–16, type the text string *V152*2* in field A. After that, the value column of the row is grayed out and the dimension B is changed. In the text string, *V152* is the name of variable C. Because the name is generated by the system, you should refer to the actual variable name as seen in your computer display. The meaning of the formula is that this variable has the value of another variable multiplied by two.

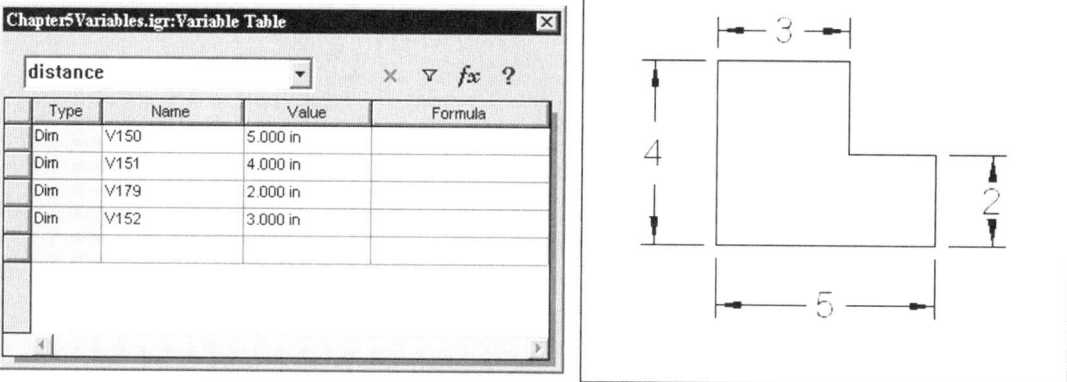

Figure 5–14. System dimension variables displayed.

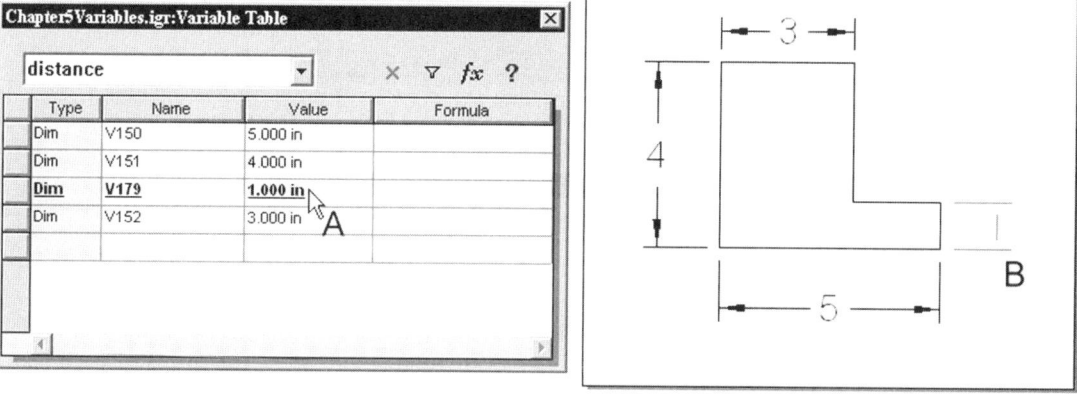

Figure 5–15. Variable modified.

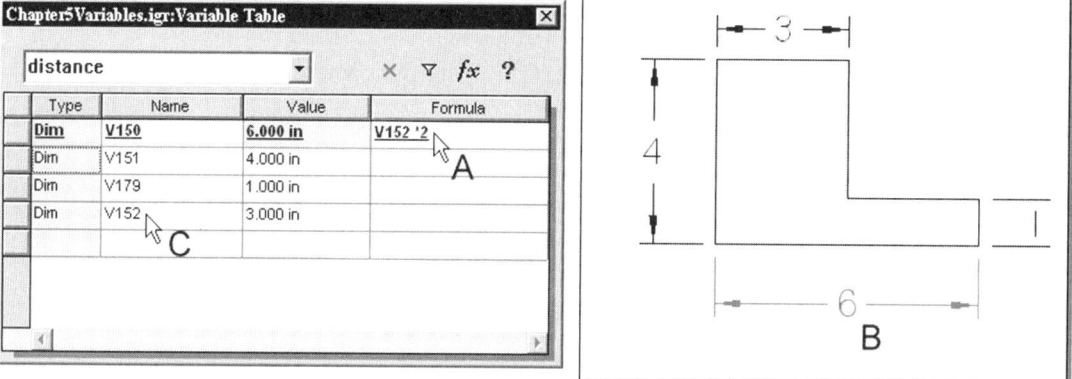

Figure 5–16. A variable set as the function of another variable.

To appreciate how to construct a user variable, perform the following steps.

1　With reference to Figure 5–17, type the text string *height* in the blank row under the Name column and input a 3-inch value under the Value column. A variable called *height* with a value of 3 inches is constructed in the variable table.

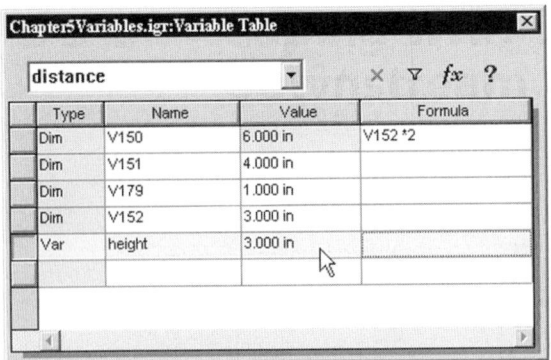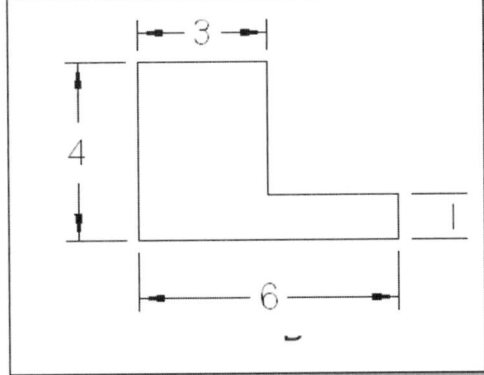

Figure 5–17. User variable constructed.

2　With reference to Figure 5–18, type the variable name *height* in field A. All at once, the variable V151 takes on the value of the variable *height* and dimension B associated to the variable changes as well.

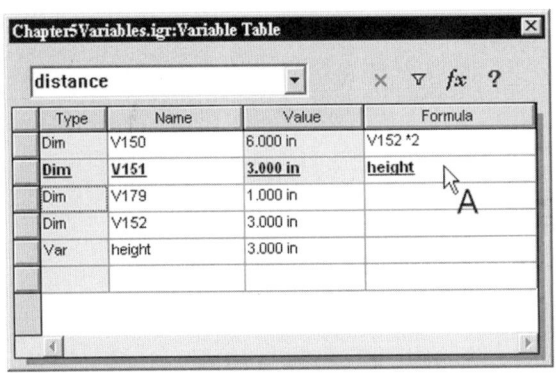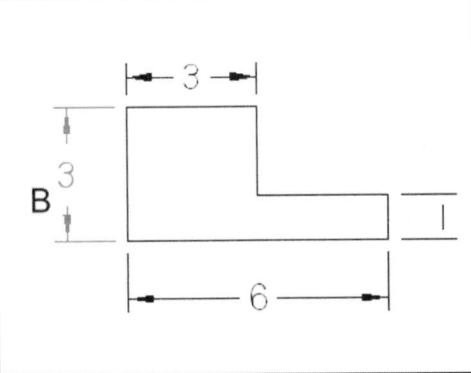

Figure 5–18. System variable as a function of a user variable.

To appreciate how to use an Excel spreadsheet to control the variables in a document, perform the following steps.

1 Open the Excel file *Variable* from the *Chapter 5* folder of the companion CD-ROM and save it in a folder of your computer.

2 As seen in Figure 5–19, the spread sheet has two fields: 4 and *in*. Together, they mean "4 inches."

3 Select these two fields.

4 Select Edit > Copy from the Excel menu.

5 With reference to Figure 5–20, click on field A, right click, and select Paste Link. This links the selected data of the spreadsheet to a field of the variable table.

6 The value of the variable now takes on the value in the spreadsheet. As a result, the dimension also changes. (See Figure 5–21.)

Figure 5–19. Excel spreadsheet.

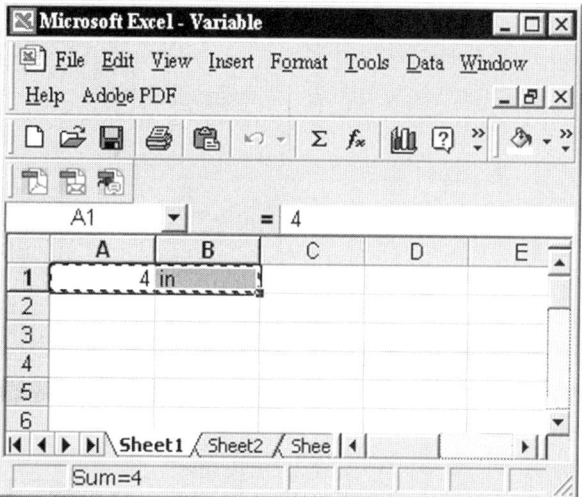

Figure 5–20. Pasting link.

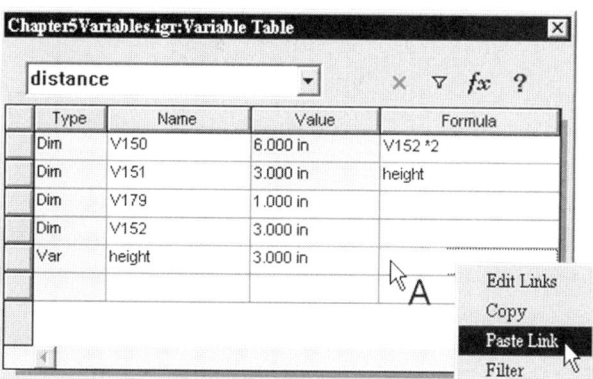

7   Change the value of field A1 in the spreadsheet to *2*. As seen in Figure 5–22, the variable changes and the dimension changes.

8   Save the files in your computer and close the files.

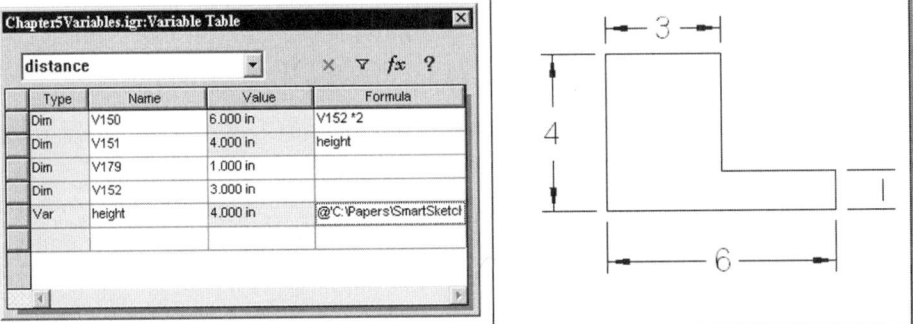

Figure 5–21. Variable linked to a spreadsheet.

Figure 5–22.
Spreadsheet
modified.

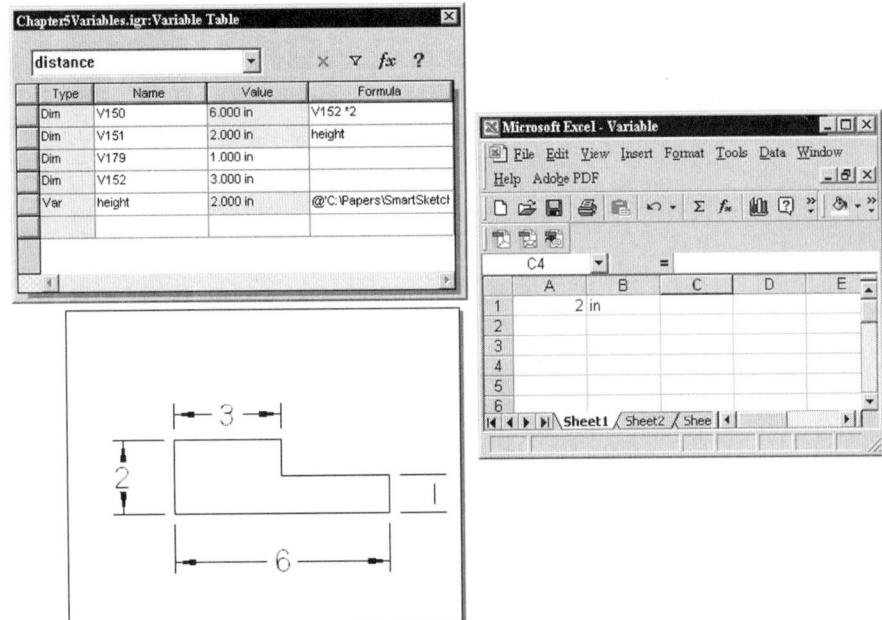

Adding a Hyperlink to a Document

You can add hyperlinks to objects in a SmartSketch document. Perform the following steps. To share design information, you can use the Internet as a sharing platform.

1 Open the document file *Chapter5Hyperlink.igr* from the *Chapter 5* folder of the companion CD-ROM.

2 Select Insert > Hyperlink or select Hyperlink on the Main toolbar and select A. (See Figure 5–23.)

3 In the Add Hyperlink dialog box, shown in Figure 5–24, type the text string *http://www.smartsketch.com* in the source box, type the text string *Smart Sketch* in the Name box, click on the Add button, and click on the OK button.

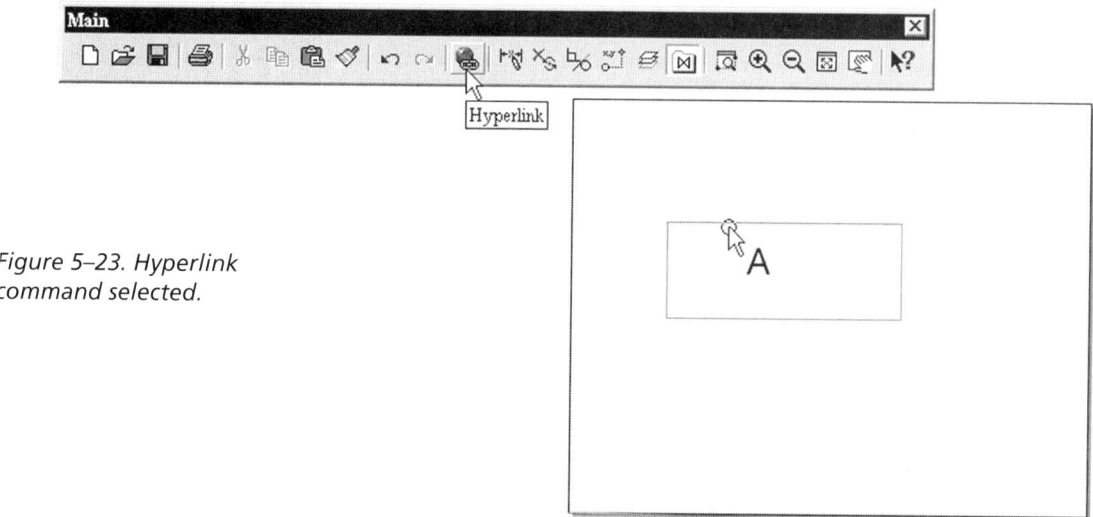

Figure 5–23. Hyperlink command selected.

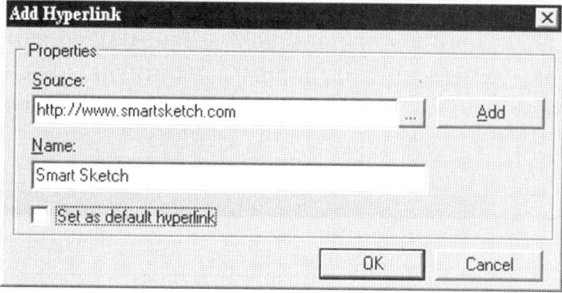

Figure 5–24. Hyperlink being added.

To see how a hyperlink works, continue with the following steps.

4 Click on the Hyperlink icon on the Main toolbar.

5 Move the cursor over A in Figure 5–25. A tooltip displays, indicating that there is a hyperlink.

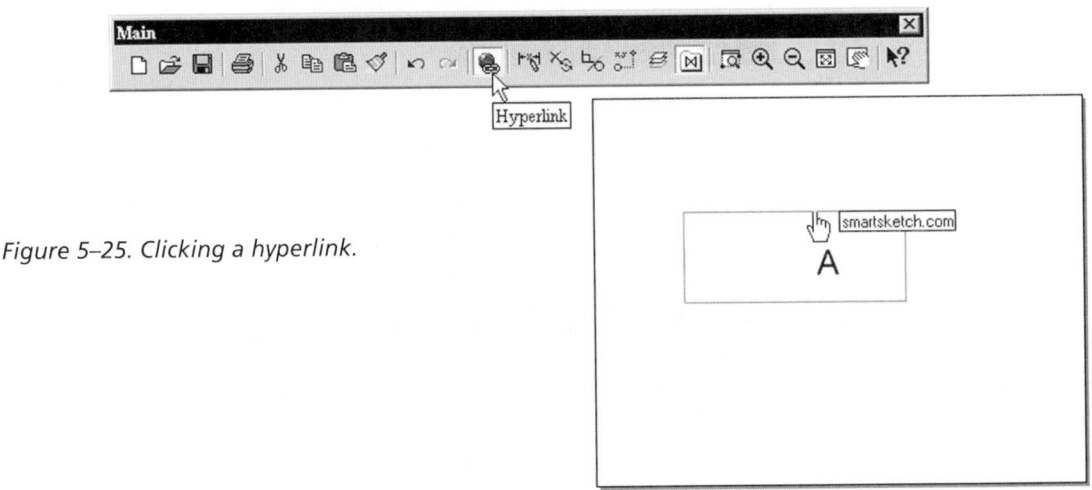

Figure 5–25. Clicking a hyperlink.

6   Click on A. Internet Explorer is activated and the web site is visited.

7   Save the file in a folder of your computer and close the file.

■ ■ ■ ■   Object Linking and Embedding (OLE)

Because SmartSketch is fully compatible with OLE-compliant applications, you can transfer text, numbers, sound bites, or other graphics between SmartSketch documents and other Microsoft applications, as well as embedding a file or linking to a file.

Embedding

There are two ways to insert a file: embed and link. Embedding embeds a file into a document. After embedding, the embedded files form independent versions of the original source file. Therefore, changes in the original files will have no effect on the embedded files. Because the embedded file is saved in the target document, the document's file size increases. However, the end user of the SmartSketch document does not have to gain access to the source file.

To edit an embedded file, the application that is used to construct the embedded file has to be installed in the computer. Opening the embedded file can be done by double clicking it.

To insert an object, select Insert > Object. In the Insert Object dialog box, clear the Link box, if it is already selected.

Create New and Create from File

Object files that are embedded in a SmartSketch document or linked to a SmartSketch document can be constructed from scratch or can be selected from an existing file. (See Figures 5–26 and 5–27.)

Figure 5–26. Create from an existing file.

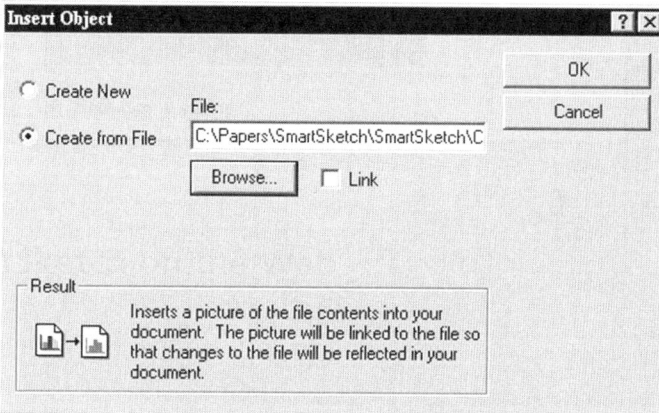

Figure 5–27. Create new file.

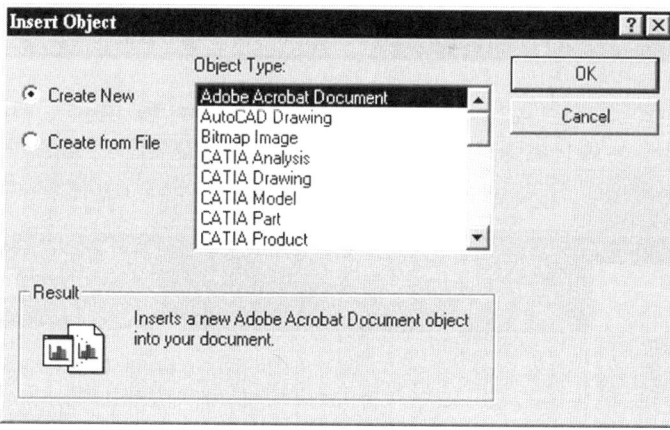

Linking

Instead of embedding a file, you can link a SmartSketch document file to another file. Because linking does not copy the file into the target Smart-Sketch document, file size increase will be insignificant, saving the linking information instead of the entire object file. To link a file, select Insert > Object and click on the Link box in the Insert Object dialog box.

Using the Image Integrator Tools

Apart from constructing drawing elements in a document, you can incorporate raster images by using the image integrator tools. Using the tools, you can view and manipulate images. You can work on all or part of an image in regard to cutting, moving, and erasing. You can also adjust contrast and brightness, and invert color values. Image file types that can be processed are GIF, JPG, BMO, TIF, CAL, PC, RLE, COT, CIT, TG4, CRL, CMP, and RGB. Images in the format of lines of pixels are called raster images. Objects in a raster image can be manipulated only by individual pixels. To work on raster images, you use the Image Integrator toolbar, shown in Figure 5–28.

Figure 5–28. Image Integrator toolbar.

The image integrator toolbar contains 14 options. Details are outlined in Table 5–2.

Table 5–2 Image Integrator Toolbar Options and Their Functions

Option	Function
Inset Image	Inserts an image into the document
Save Selected Images	Saves the changes made to selected images' source files
Image Undo	Undoes the last change made to an image
Image Redo	Redoes the last undone operation to an image
Rectangular Select Area	Selects a rectangular area of an image by describing two diagonal points to depict a rectangle
Polygonal Select Area	Selects a polygonal area of an image by selecting a number of points to depict the vertices of a polygon
Contrast and Brightness	Adjusts the contrast and/or brightness of an image
Invert	Inverts the values of an image, thus producing a negative image

Option	Function
Fill	Fills a selected area with a fill color
Position	Positions an image in a document
Image Properties	Displays the properties of an image
Image Erase	Erases the selected image or selected edited area of an image
Speckle Remove	Removes all unwanted areas of speckle in a binary image
Multi-point Warp	Two-dimensionally transforms a source area to fit to a destination area

To familiarize yourself with the image integrator tools, perform the following steps.

1 Start a new document using *Technical Drawing (Imperial).igr* as the template.

2 Select Tools > Add-Ins.

3 In the Add-In Manager dialog box, select the Image Integrator option under *Available add-ins* and then click on the OK button. (See Figure 5–29.)

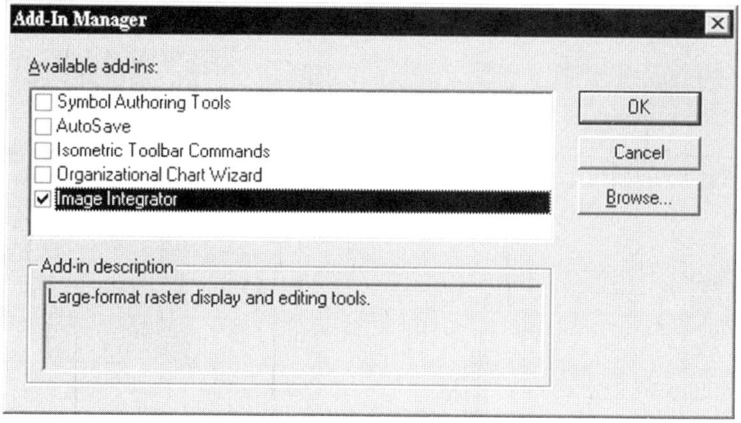

Figure 5–29. Selecting the Image Integrator add-in program.

4 Select View > Toolbars.

5 In the Toolbars dialog box, click on ImageIntegrator and click on the OK button. The ImageIntegrator toolbar is displayed.

6 Click on the Insert Image icon on the ImageIntegrator toolbar or select Insert > Image.

7 In the Insert Image dialog box, select the image *SSimage01.tif* from the *Chapter 5* folder of the companion CD-ROM.

8 Select location A indicated in Figure 5–30 to position the image.

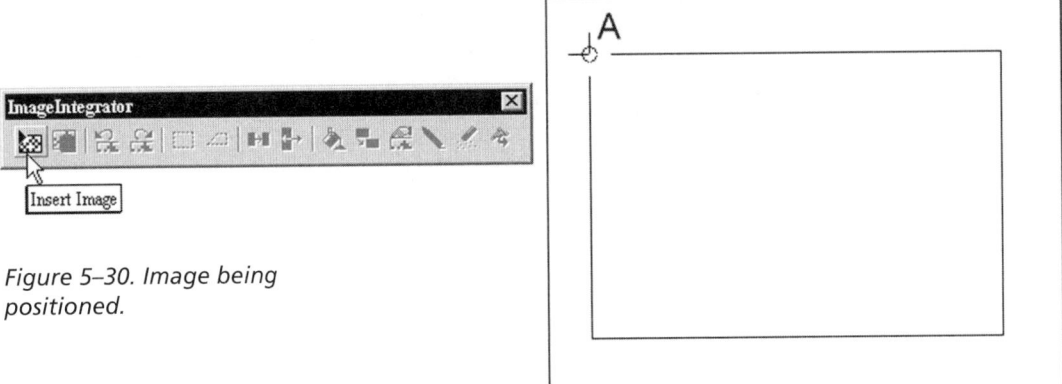

Figure 5–30. Image being positioned.

9 Click on the Select Tool icon on the Draw toolbar and select the inserted image. (See Figure 5–31.)

Figure 5–31. Inserted image selected.

10 Click on the Contrast and Brightness icon on the ImageIntegrator toolbar.

11 In the Contrast and Brightness dialog box, click the crosshairs at A and drag it to location B indicated in Figure 5–32 to increase the contrast and brightness values.

12 Click on the Image Undo icon on the ImageIntegrator toolbar to undo the last command. (See Figure 5–33.)

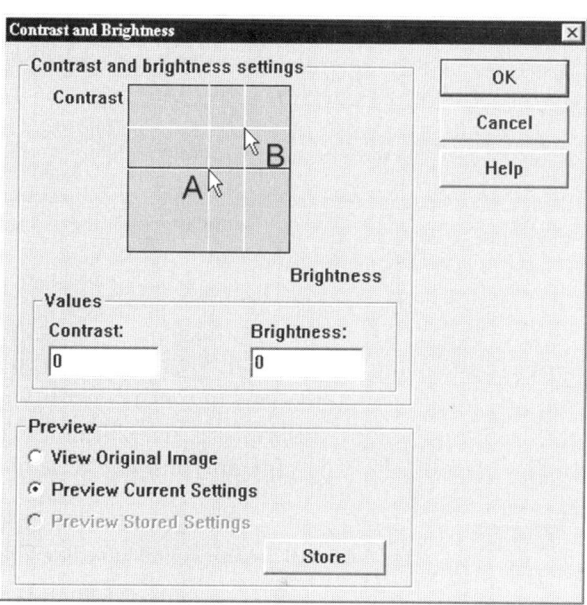

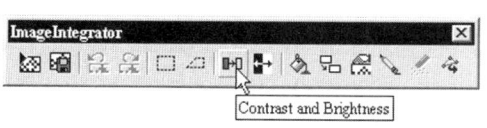

Figure 5–32. Contrast and brightness of the image being adjusted.

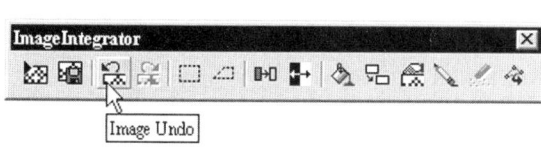

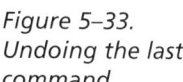

Figure 5–33. Undoing the last command.

13 If you want to redo the last undone operation, you can click on the Redo icon on the ImageIntegrator toolbar. (See Figure 5–34.) However, do not redo the last undo command.

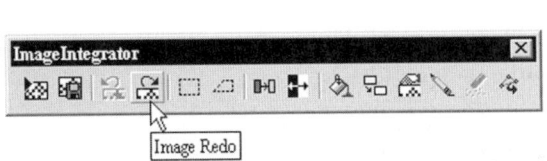

Figure 5–34. Redoing the last undone command.

14 Click on the Invert icon on the ImageIntegrator toolbar to reverse the color value of the image. (See Figure 5–35.)

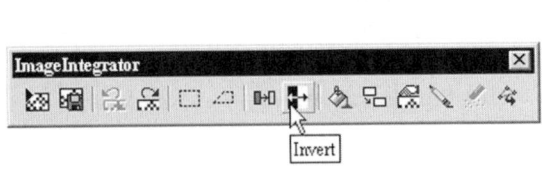

Figure 5–35. Reversing the color values.

15 Click on the Undo icon and then the Position icon on the ImageIntegrator toolbar.

16 Select the image and move it to a new location. (See Figure 5–36.)

17 While the Position command is active, select the S1 icon to scale and rotate the image. (See Figure 5–37.)

18 Click on the Polygonal Select Area icon on the ImageIntegrator toolbar.

19 Click locations A, B, C, and D indicated in Figure 5–38.

20 Right click.

Figure 5–36. Image moved
to a new location.

Figure 5–37. Image
repositioned.

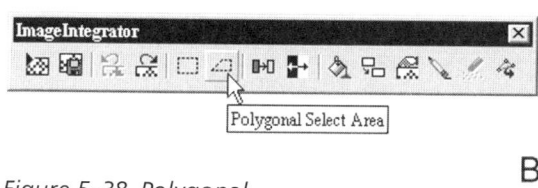

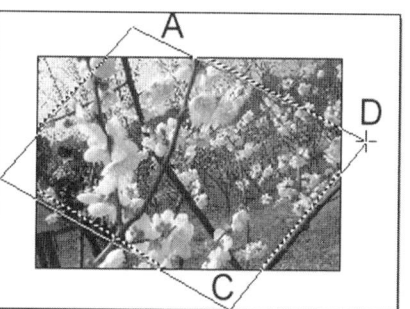

Figure 5–38. Polygonal
area selected.

21 Click on the Fill icon on the ImageIntegrator toolbar. (See Figure 5–39.)

22 In the Fill dialog box, shown in Figure 5–40, click on the Edit button.

23 In the Color dialog box, select a color and click on the OK button. (See Figure 5–41.)

Figure 5–39. Fill command selected.

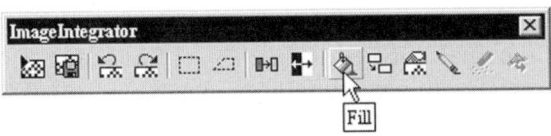

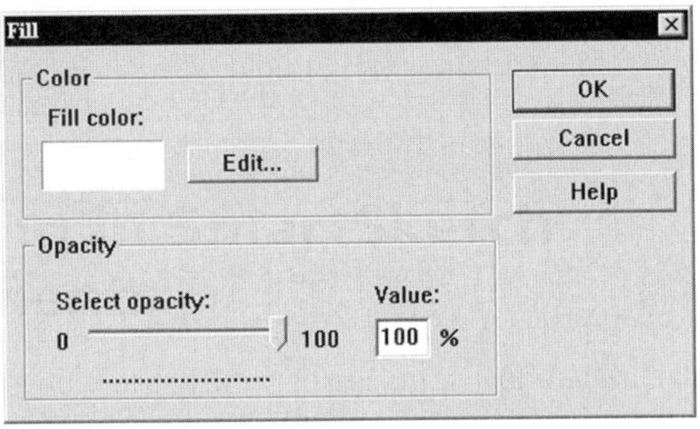

Figure 5–40. Fill dialog box.

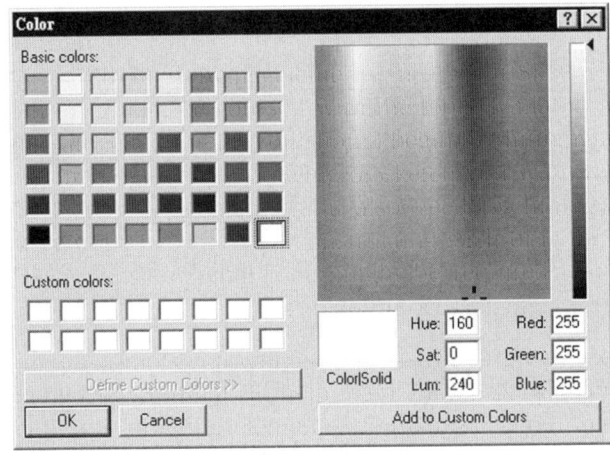

Figure 5–41. Color dialog box.

24 Click on the OK button in the Color dialog box and in the Fill dialog box click on the Proceed without Undo button. The selected area is filled. (See Figure 5–42.)

25 To display the Image Properties dialog box, click on the Image Properties icon on the Image Integrator toolbar. (See Figure 5–43.)

26 If you want to save the changes to the source image, click on the Save Selected Image(s) icon on the ImageIntegrator toolbar. (See Figure 5–44.) However, do not click on this icon unless you want to change the source image.

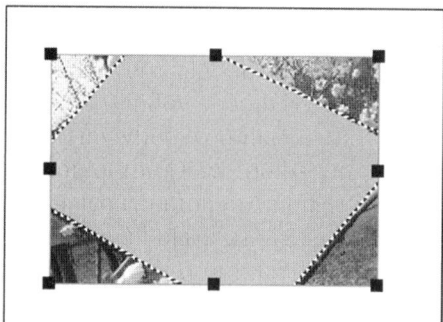

Figure 5–42.
Selected
polygonal
area filled.

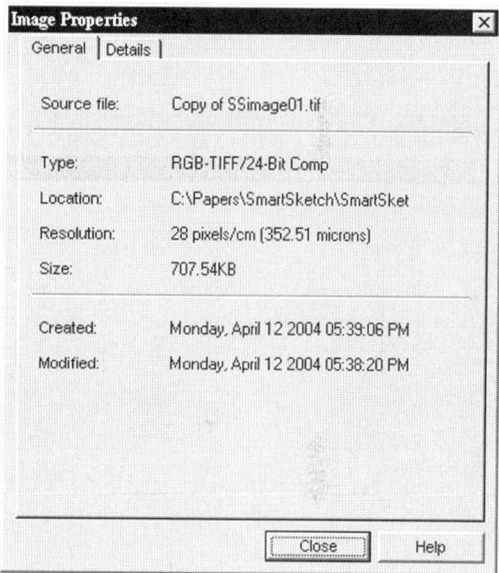

Figure 5–43. Image
Properties dialog
box.

Figure 5–44. Saving
changes to source
image.

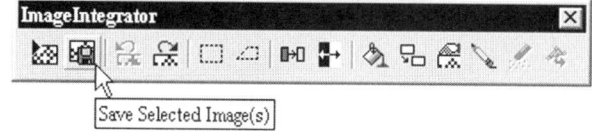

27 If you want to delete the image, click on the Image Erase icon on the ImageIntegrator toolbar. (See Figure 5–45.)

28 Close the file without saving.

Perform the following steps to transform a raster image.

Figure 5–45. Erasing the image.

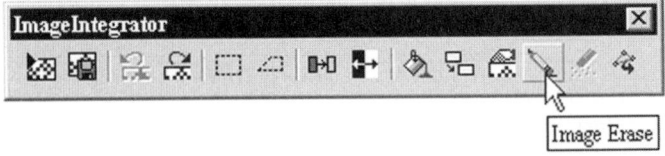

1 Open the document file *Chapter5Warp.igr* from the *Chapter 5* folder of the companion CD-ROM. In the document file, you will find two identical images and a rectangle.

2 Click on the Select Tool icon on the Draw toolbar and select image A indicated in Figure 5–46.

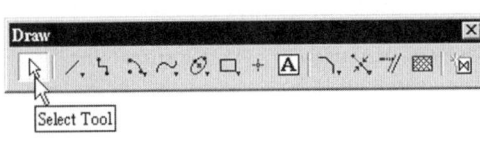

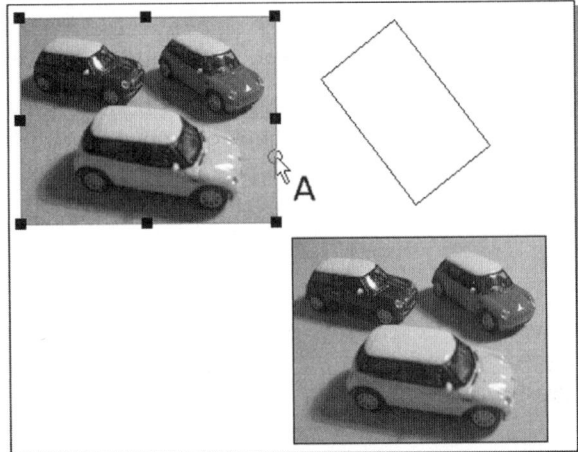

Figure 5–46. An image selected.

3 Click on the Multi-point Warp icon on the ImageIntegrator toolbar. (See Figure 5–47.)

4 Select 1st Order Polynomial from the ribbon.

5 Click on corner S1 and then T1, depicting the source corner and the target corner.

6 Click on corner S2 and T2.

7 Click on corner S3 and T3.

8 Click on the Finish button on the ribbon. The image is transformed. (See Figure 5–48.)

9 Save and close your file.

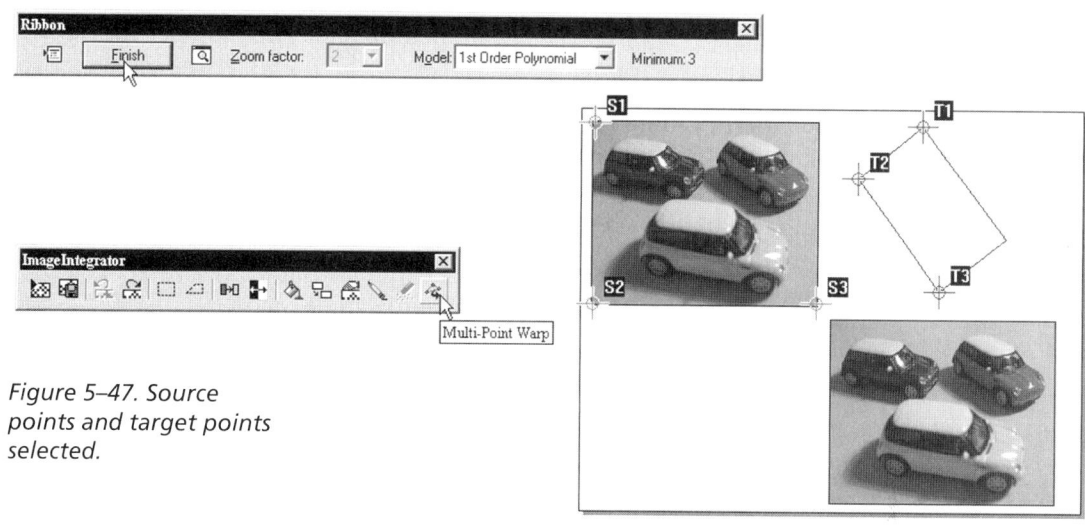

Figure 5–47. Source points and target points selected.

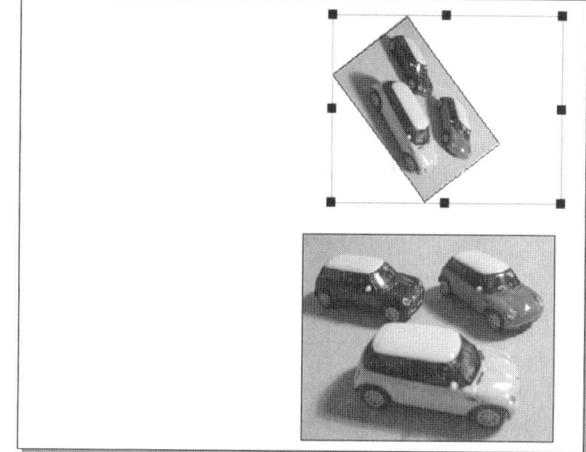

Figure 5–48. Image transformed.

■ ■ ■ ■ Summary

Because display scale set in a sheet applies to all drawing elements on the sheet, you need to add a detail view with a different display scale to magnify certain parts of the drawing. It is understood that SmartSketch is a 2D drafting tool. To produce a 2D isometric drawing, you need special tools to construct isometric lines, isometric circles, and isometric rectangles on three isometric planes. To add reality to a 2D isometric drawing, you need to change portions of the drawing elements to dashed lines.

There are two types of variables, user and system. To manipulate the variables in a document and to link it to an external spreadsheet, you use the variable table.

In addition to insertion of symbols, you can insert hyperlinks, Window objects, and images in a document. To give added information to a document, you can add hyperlinks as well. Insertion of Window objects makes it possible to incorporate other data in a SmartSketch file. For example, you can insert a Word document to provide a detail description about the SmartSketch document. After insertion of an image, you can modify its color values and save the changes to the source image file.

Review Questions

1   What is reason for adding a detail view to a drawing?

2   What are the four types of drawing objects you can construct in an isometric view?

3   Distinguish between user variables and system variables.

4   Briefly explain how a hyperlink can be added.

5   What is the difference between linking and embedding a Window object in a SmartSketch document?

6   How many ways can an inserted image be modified? What are they?

CHAPTER 6

Schematic Diagrams and Precision Drawings

■ ■ ■ ■ ## Objectives

The goal of this chapter is to explore the use of various schematic diagrams and precision drawing templates and their associated symbols. After studying this chapter, you should be able to:

❐ Produce schematic diagrams

❐ Produce precision drawings

Overview

Generally speaking, you can construct two major types of documents: schematic diagrams and precision drawings. Schematic diagrams are primarily constructed using symbols, connectors, and text boxes. Precision drawings are used to represent real-world objects on a 2D drawing sheet. You use various drawing elements available from the Draw toolbar. Naturally, you also use symbols wherever necessary to enhance drawing productivity. As a consolidation chapter to the use of Smart-Sketch as a tool in computer-aided drafting, this chapter introduces the use of various templates, including general diagramming templates, AEC solutions templates, electrical diagramming templates, mechanical engineering templates, and process diagramming templates.

■ ■ ■ ■ Schematic Diagrams

Schematic diagrams, as the name implies, are diagrams depicting a system or a flow in a simplified way. As such, objects are usually represented by using symbols, rather than precise drawings delineating details. For example, you would use a schematic if you wanted to construct a diagram explaining how a certain product is made from raw material up to the final state using various types of machines. Instead of producing a detail drawing delineating the individual machines, you would use symbols (probably in the form of rectangles with text boxes inside the rectangles) to convey what types of machines they are. Then you use connectors to join the symbols. Naturally, you might use a lot of text boxes to further explain the flow of the process.

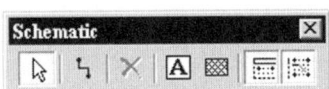

Figure 6–1. Schematic toolbar.

To produce schematic diagrams, you need the Schematic toolbar, shown in Figure 6–1. This toolbar will display automatically after you select a schematic diagram template. If it is not displayed, you can use the Toolbar dialog box accessible from View > Toolbars.

The Schematic toolbar contains seven options, the functions of which are outlined in Table 6–1.

Table 6–1 Schematic Toolbar Options and Their Functions

Option	Function
Select Tool	Selects objects. The cursor is changed to the arrow-shaped selection pointer with a circle at the end of the cursor indicating the locate zone.
Connector	Draws a series of line segments to connect two elements.
Delete	Deletes selected drawing objects.
Text Box	Constructs a text box, which is a rectangular element consisting of text and symbols.
Fill	Fills a closed boundary with a selected pattern or solid color.
Grid Display	Turns on/off grid display.
Grid Snap	Activates/deactivates snapping to grid points.

The types of templates to be deployed depend on what type of schematic diagrams you are going to construct. Typically, there are templates for general diagramming, electrical diagramming, and process diagramming.

To cope with Imperial standard and metric standard, each type of diagram provides two types of templates. The Imperial templates use imperial units in decimal inches, ANSI sheet sizes, ANSI or ASA dimensioning, and Arial, ANSI, or Architectural text. On the other hand, metric templates use metric units in millimeters, ISO sheet sizes, ISO dimensioning, and Arial or ISO text.

General Diagramming

General diagramming concerns atlas mapping, basic diagramming, directional mapping, flowcharts, network diagram, office layout, organization charts, and workflow diagrams. Figure 6–2 shows the general diagramming templates available.

Figure 6–2. General diagramming templates.

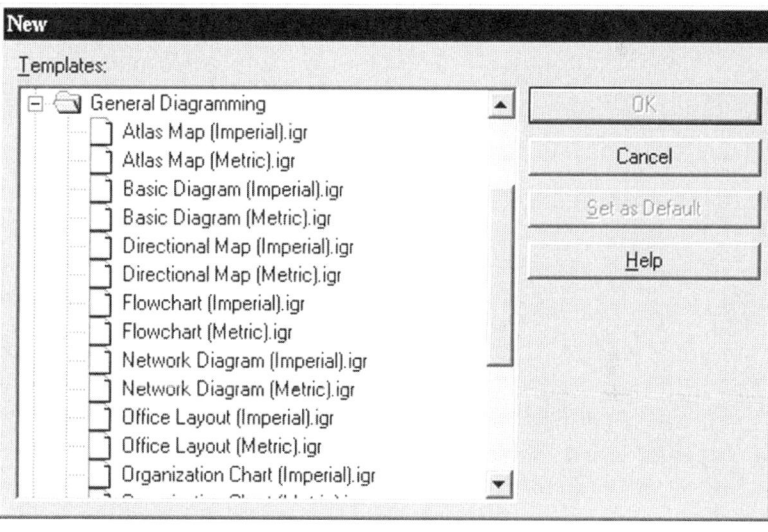

Atlas Mapping

As the name implies, Atlas mapping templates are mainly used for constructing atlas mappings. To start with, you select a template from the *General Diagramming* subfolder of New dialog box. Depending on the units of measurement you are going to use, you select the Atlas Map file *(Imperial).igr* or *(Metric).igr*.

To use the symbols available, you click on the Explore Elsewhere button of the Symbol Explorer and select *US States* from the *Atlas Map* subfolder, shown in Figure 6–3. Figure 6–4 shows the symbols available from the *US States* subfolder, and Figure 6–5 shows the SmartSketch sample document.

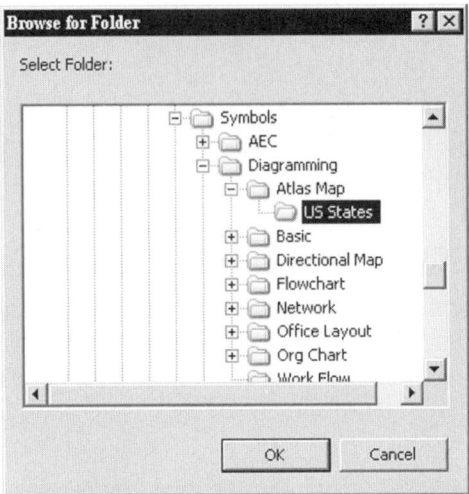

Figure 6–3. Symbols for U.S. States.

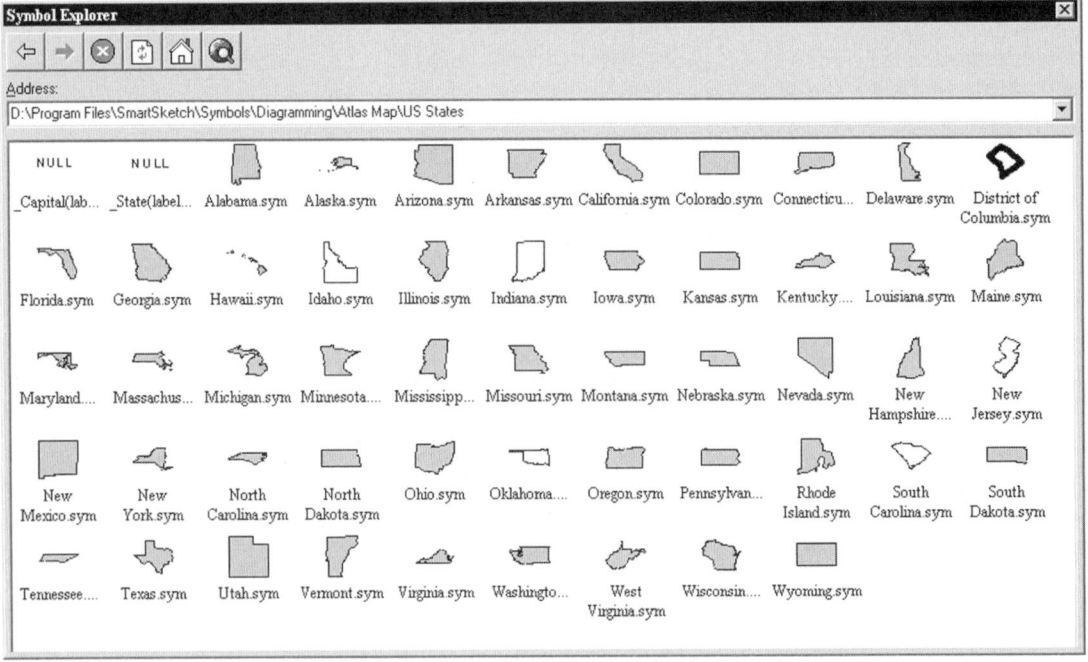

Figure 6–4. Symbol Explorer showing U.S. States.

Figure 6–5. SmartSketch sample atlas mapping showing the Unites States of America.

Basic Diagramming

Figure 6–6. Basic diagramming symbols.

To produce business diagrams, you use the *Basic Diagramming* template from the *General Diagramming* subfolder. Figure 6–6 shows various types of boxes for depicting ideas and facts to be placed in a business diagram. To join the basic boxes, you use the connectors in conjunction with the arrows and terminators shown in Figure 6–7. To further supplement a business diagram, you use the miscellaneous box symbols shown in Figure 6–8.

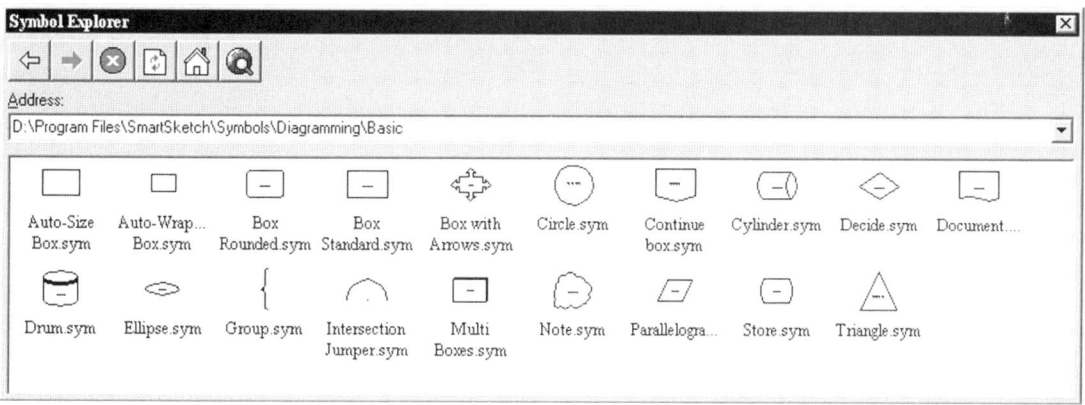

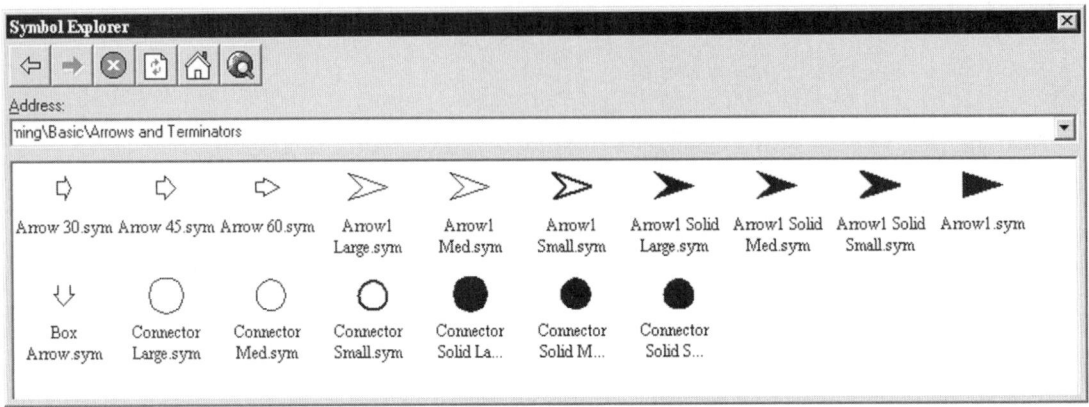

Figure 6–7. Arrows and terminators for basic diagramming.

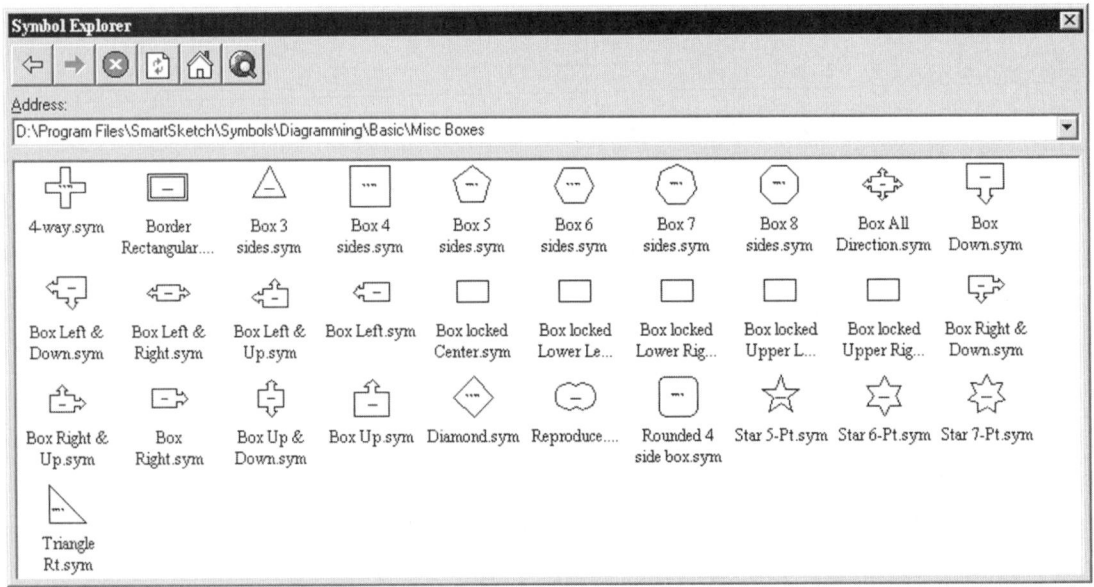

Figure 6–8. Miscellaneous boxes for basic diagramming.

Directional Mapping

To produce a map to illustrate directions, you use the *Directional Map* template. There are two sets of symbols available. Figure 6–9 shows various directional symbols, and Figure 6–10 shows various landmarks for inserting into a directional map. A typical directional map is shown in Figure 6–11.

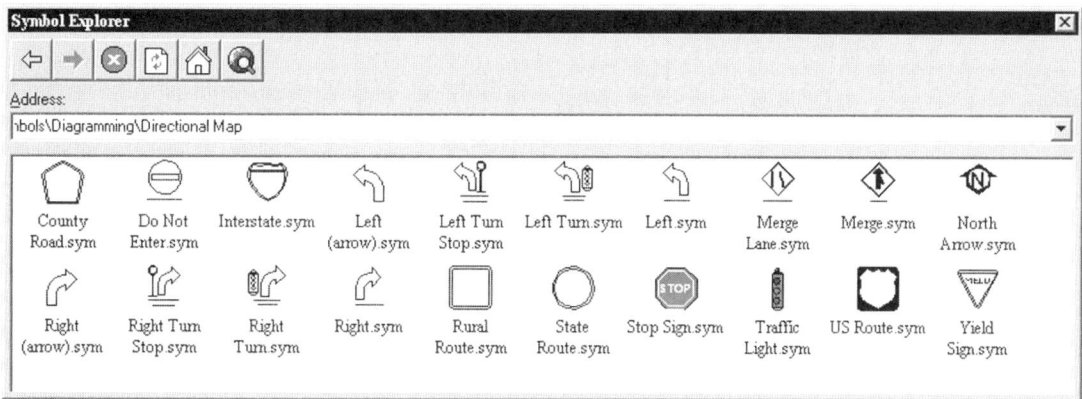

Figure 6–9. Directional mapping symbols.

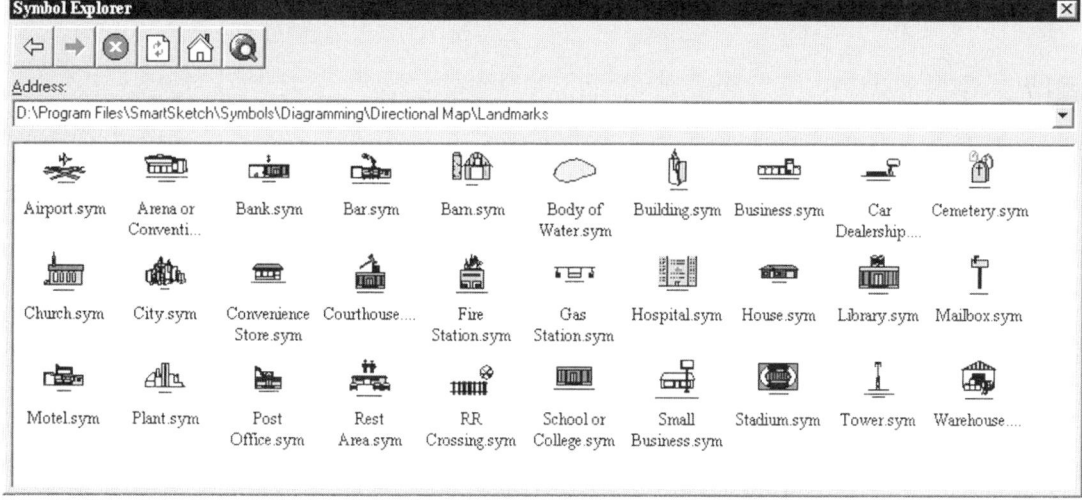

Figure 6–10. Landmark symbols for directional mappings.

Figure 6–11. SmartSketch sample directional mapping document.

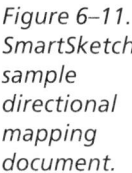

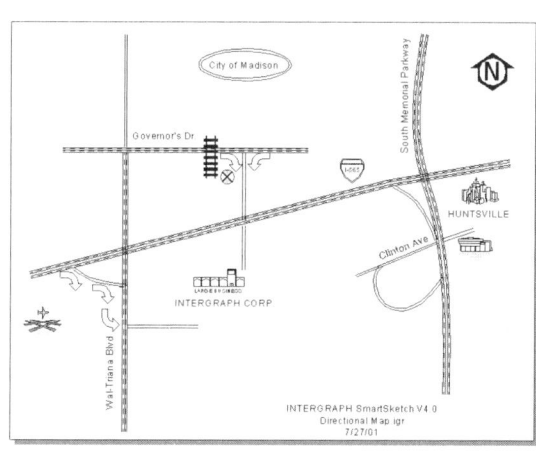

Flowcharts

Flowcharts depict process flows such as how a product is to be made or how a certain operation is to be carried out. Five sets of symbols are available. They are depicted in Figures 6–12 through 6–16.

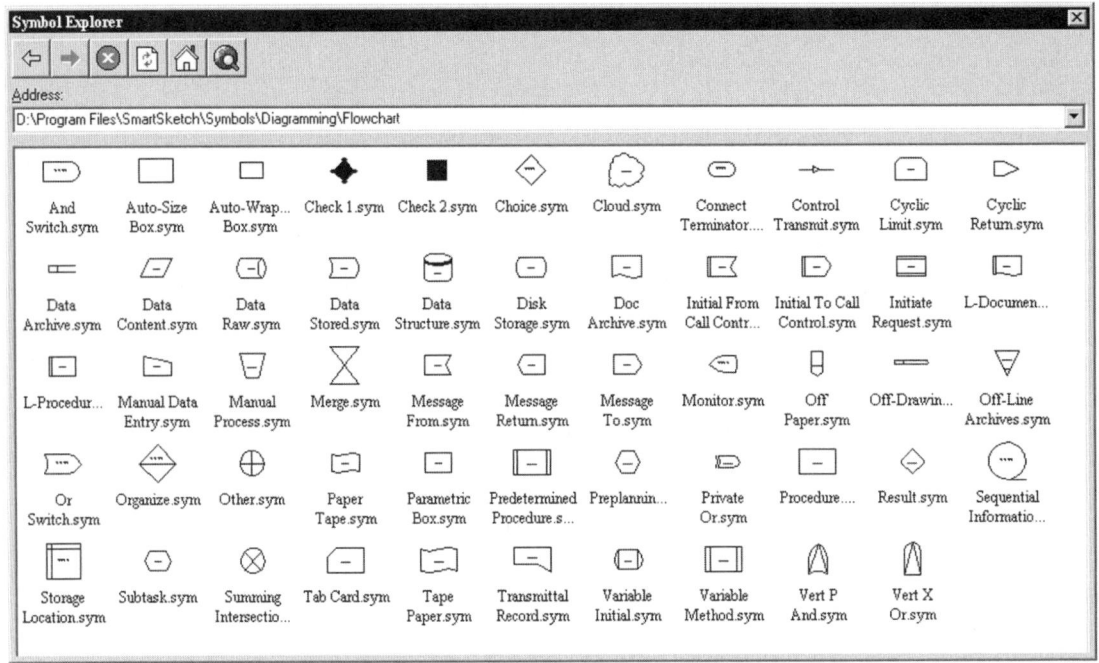

Figure 6–12. Flowchart symbols.

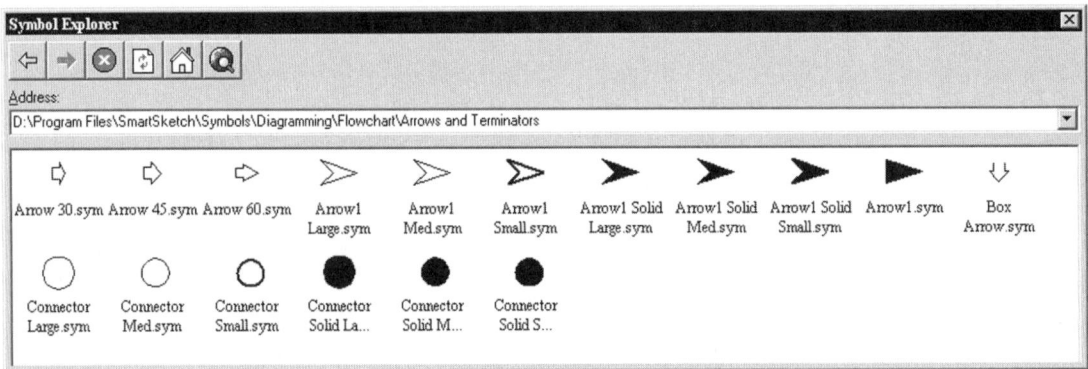

Figure 6–13. Arrows and terminators for flowcharts.

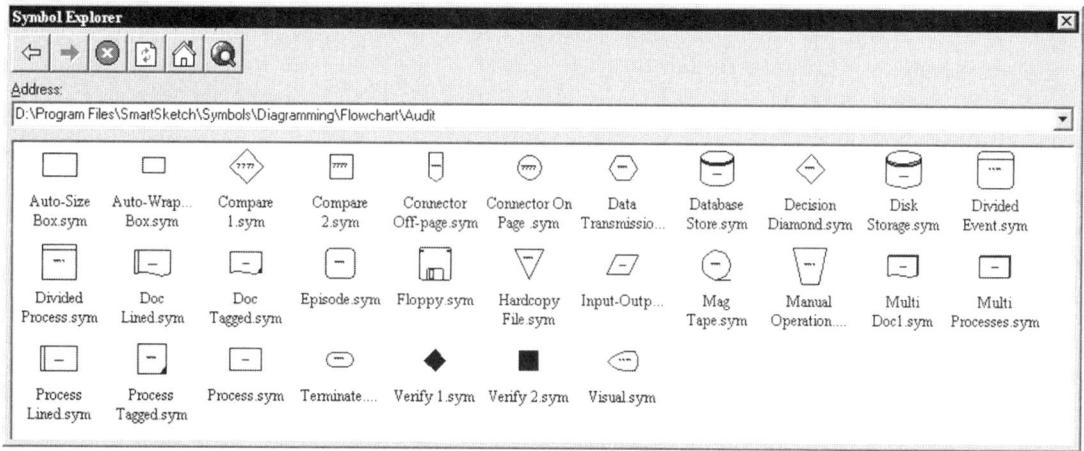

Figure 6–14. Audit symbols for flowcharts.

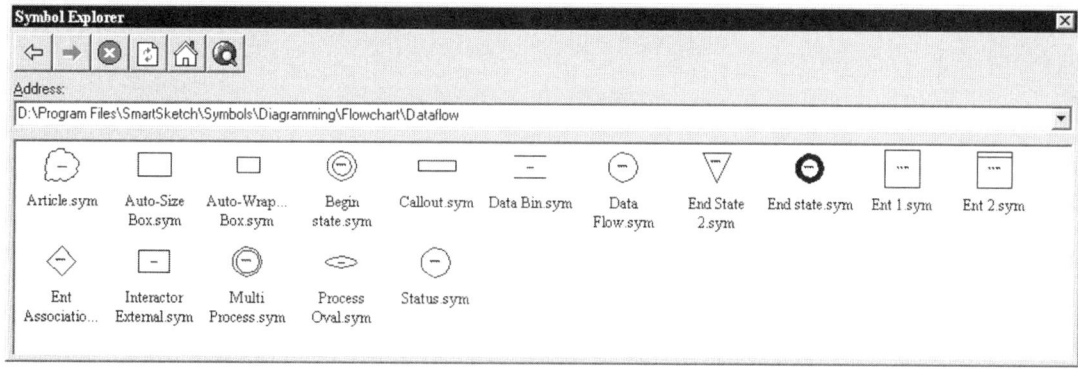

Figure 6–15. Dataflow symbols for flowcharts.

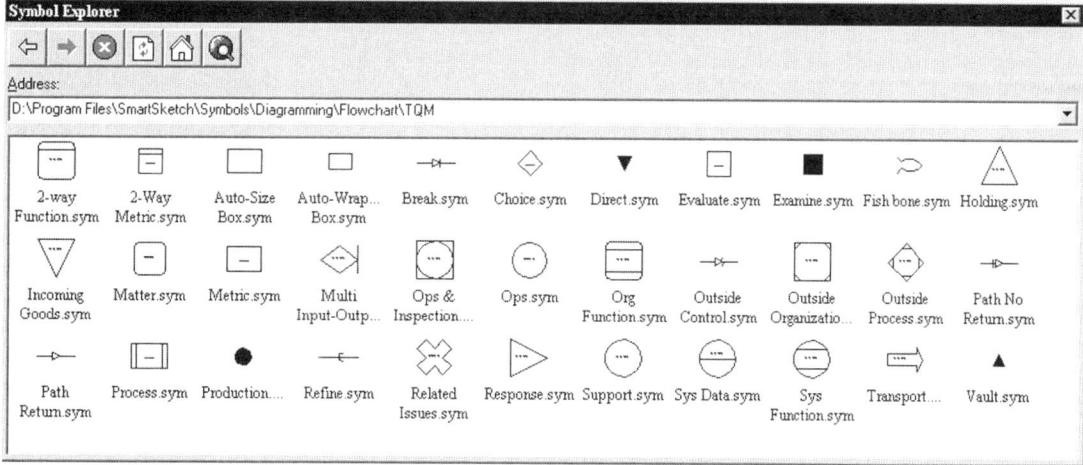

Figure 6–16. Total quality management symbols for flowcharts.

Network Diagrams

Network diagrams depict computer networking. There are a lot of symbols available in a number of folders under the subfolder *Network*. Figures 6–17 through 6–19 show some of the symbols available, and Figure 6–20 shows a sample network diagram.

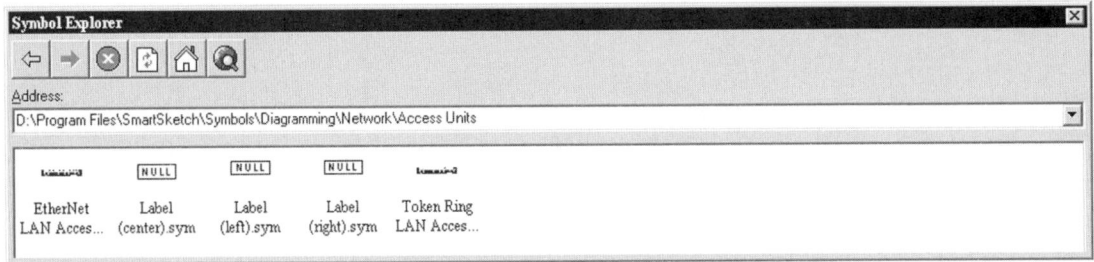

Figure 6–17. Access unit symbols for network diagrams.

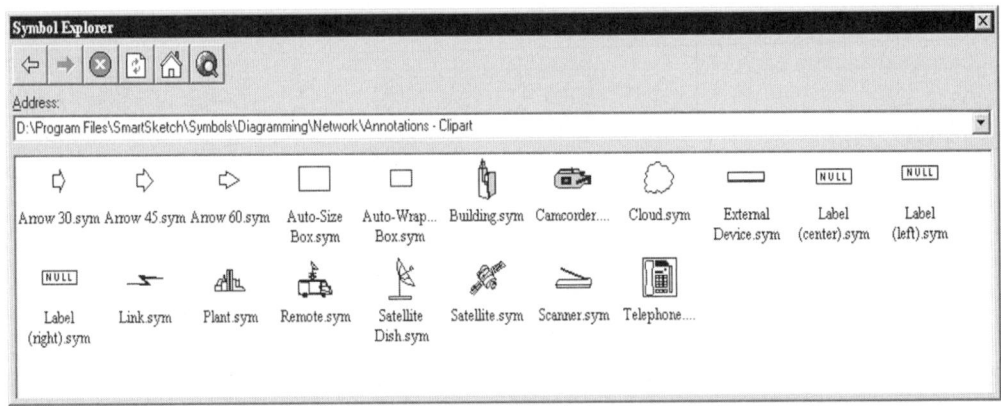

Figure 6–18. Annotation symbols for network diagrams.

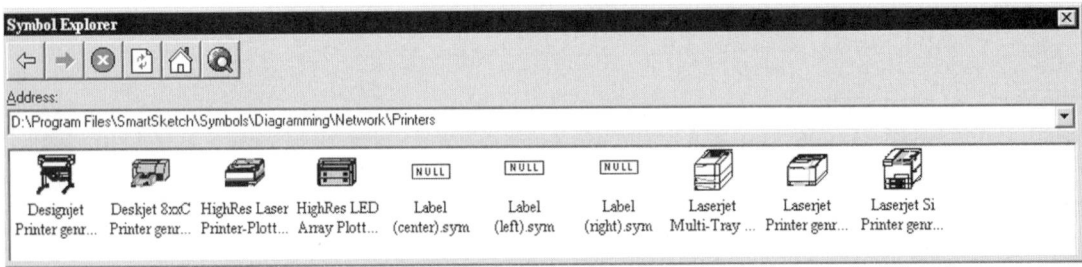

Figure 6–19. Printer symbols for network diagrams.

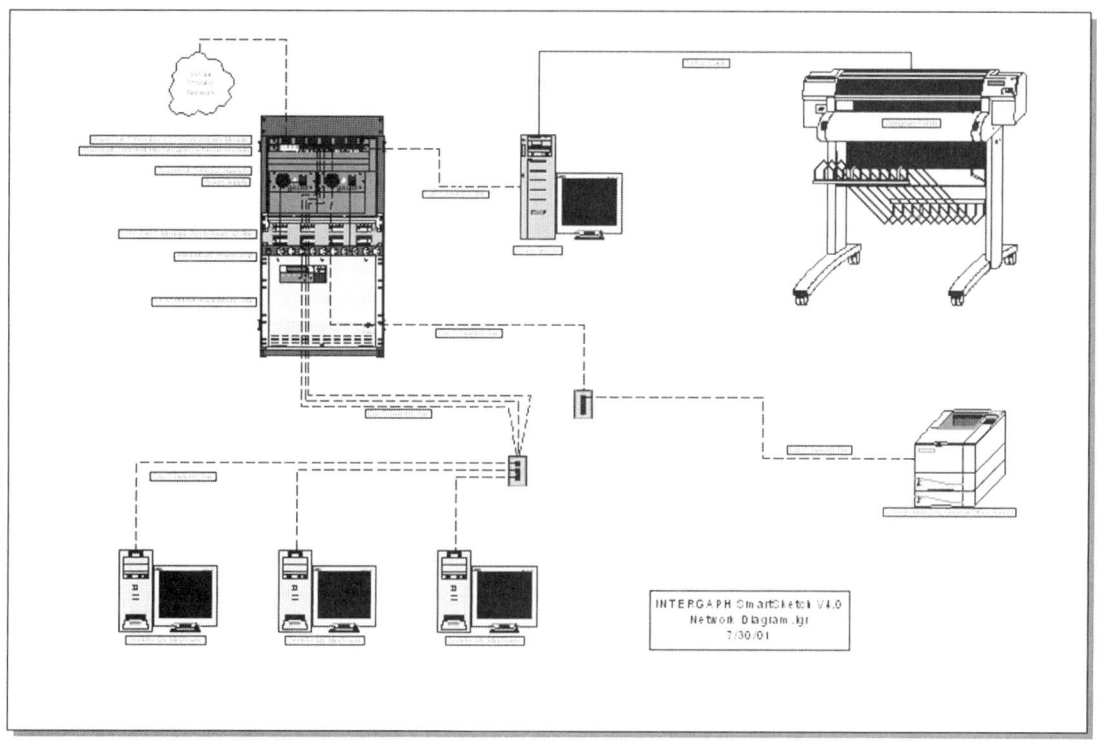

Figure 6–20. SmartSketch sample network diagram.

Office Layouts

Figure 6–21. Office layout symbols.

To produce an office layout, you use the office layout template together with the symbols shown in Figures 6–21 through 6–23. A SmartSketch sample office layout is shown in Figure 6–24.

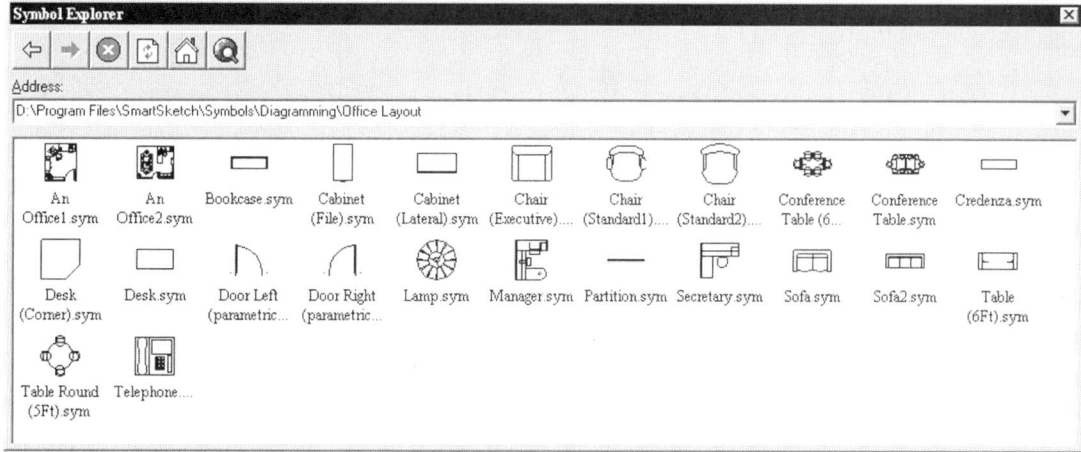

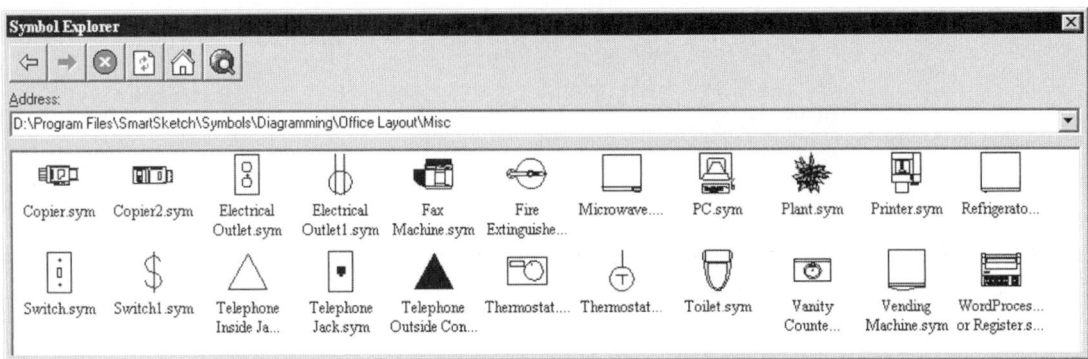

Figure 6–22. Miscellaneous symbols for office layout diagrams.

Figure 6–23. Report symbols for office layout diagrams.

Figure 6–24. SmartSketch sample office layout diagram.

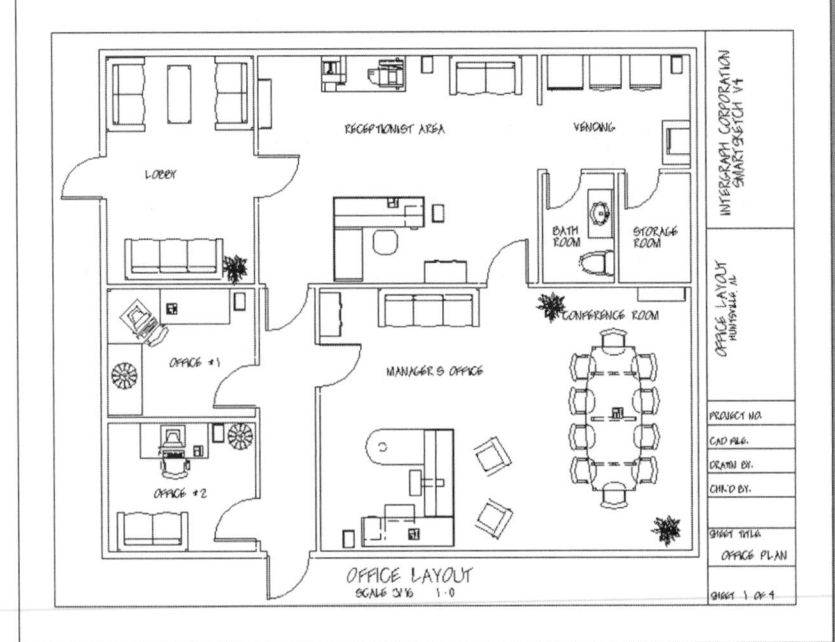

Organizational Charts

To construct an organizational chart for a company or an organization, you use the organization chart template and its associated symbols shown in Figures 6–25 through 6–27. Figure 6–28 shows a SmartSketch sample organizational chart.

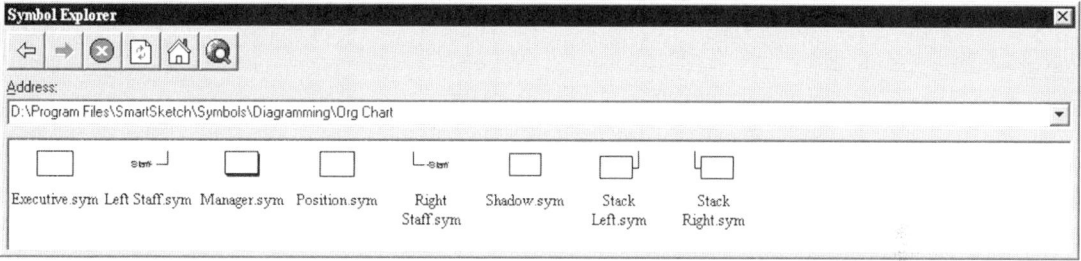

Figure 6–25. Organizational chart symbols.

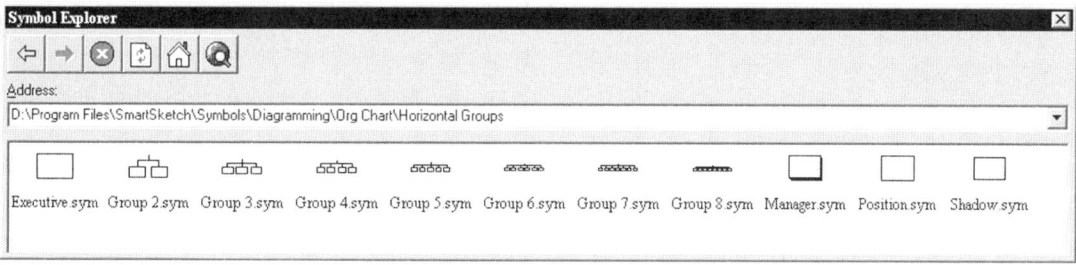

Figure 6–26. Horizontal grouping symbols for organizational charts.

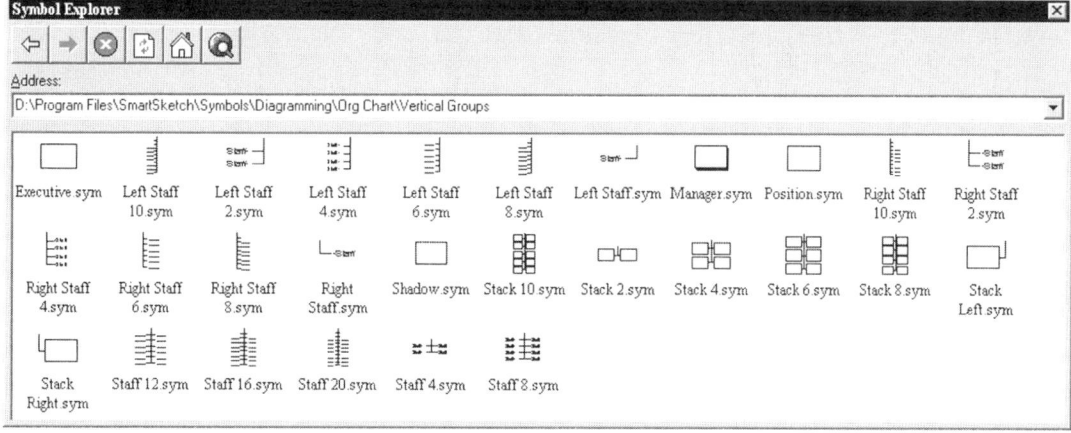

Figure 6–27. Vertical grouping symbols for organizational charts.

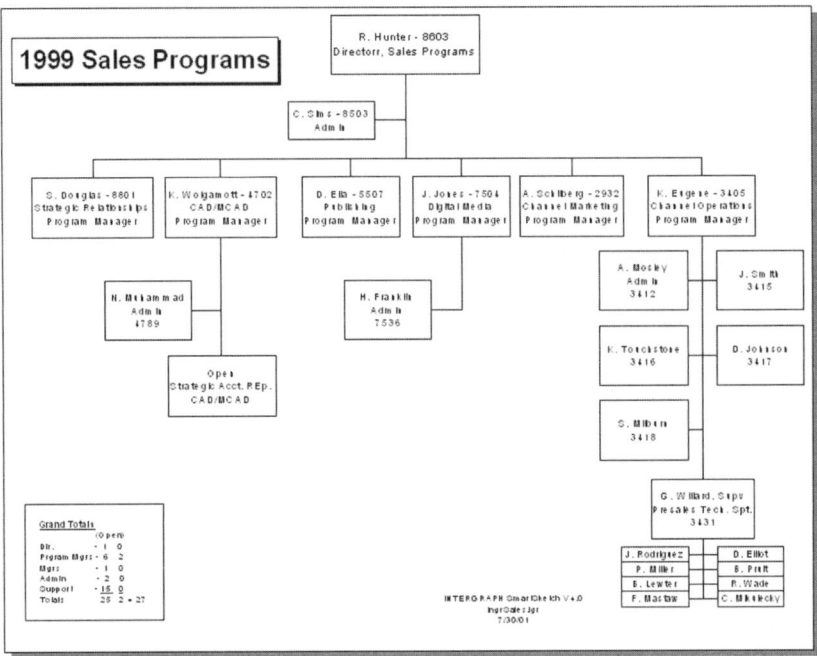

Figure 6–28. SmartSketch sample organizational chart.

Figure 6–29. Workflow symbols.

Workflow Diagrams

To construct workflow diagrams, you use the workflow diagram template and its associated symbols shown in Figure 6–29.

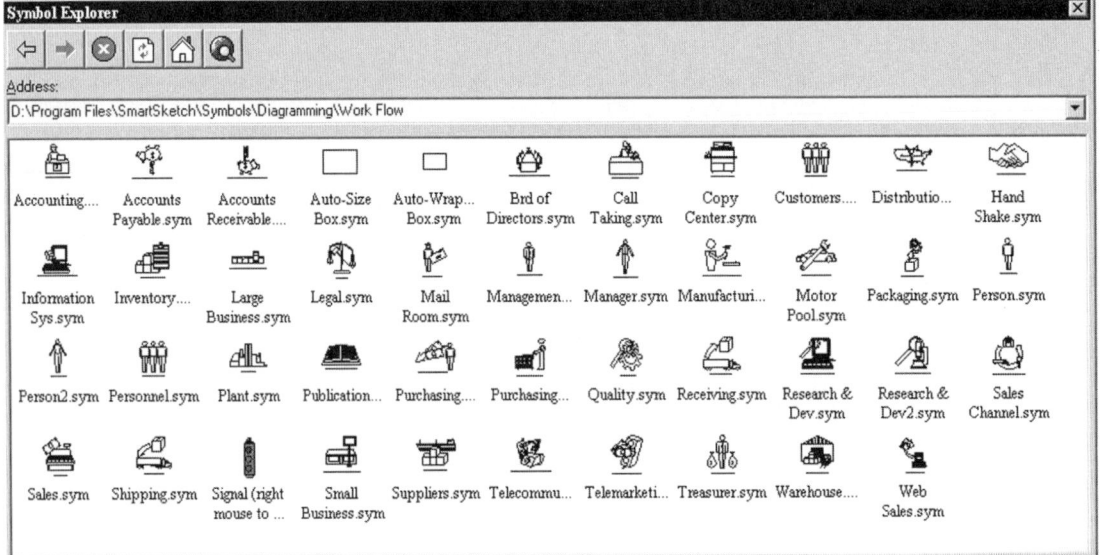

Electrical Diagramming

Electrical schematic diagrams can be quickly produced by using the control loop or electrical schematic templates shown in Figure 6–30, together with the repertoire of symbols available. Symbols include circuit protectors, contacts and relays, electron tubes, fundamental items, voltage symbols, logic gates, qualifying symbols, rotating mach, semiconductors, signaling, switches, terminals and connectors, and transformers and inductors. Some of the symbols are illustrated in Figures 6–30 through 6–33.

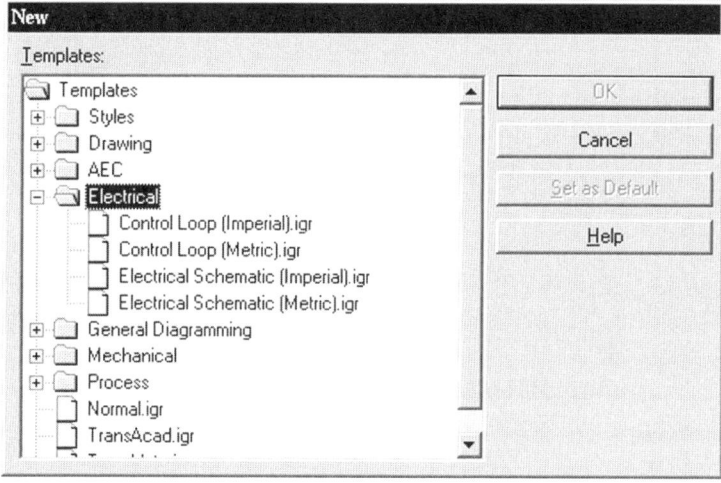

Figure 6–30. Electrical diagram templates.

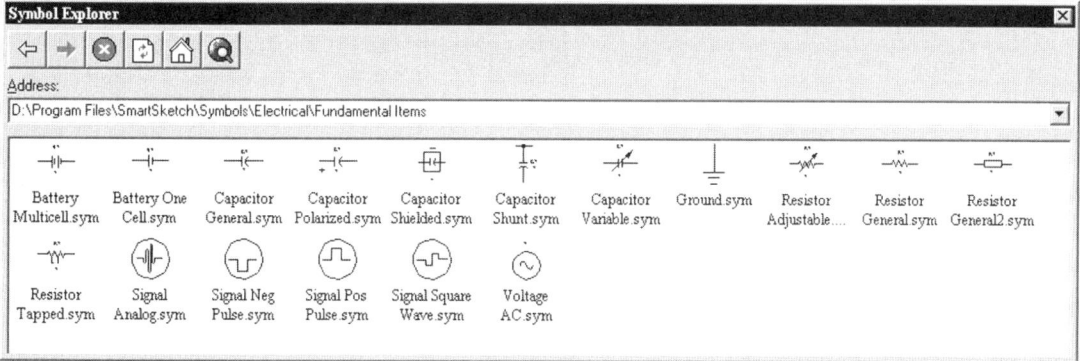

Figure 6–31. Symbols for fundamental items.

Figures 6–34 and 6–35 show a sample control loop diagram and an electrical diagram.

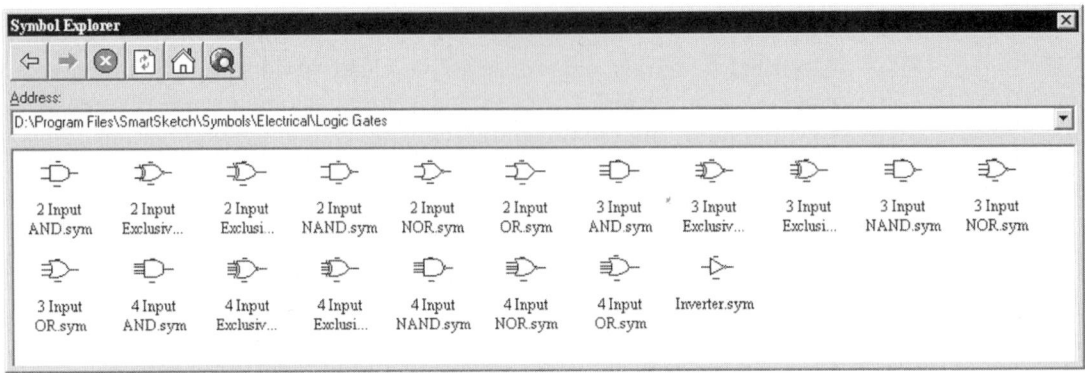

Figure 6–32. Logic gate symbols.

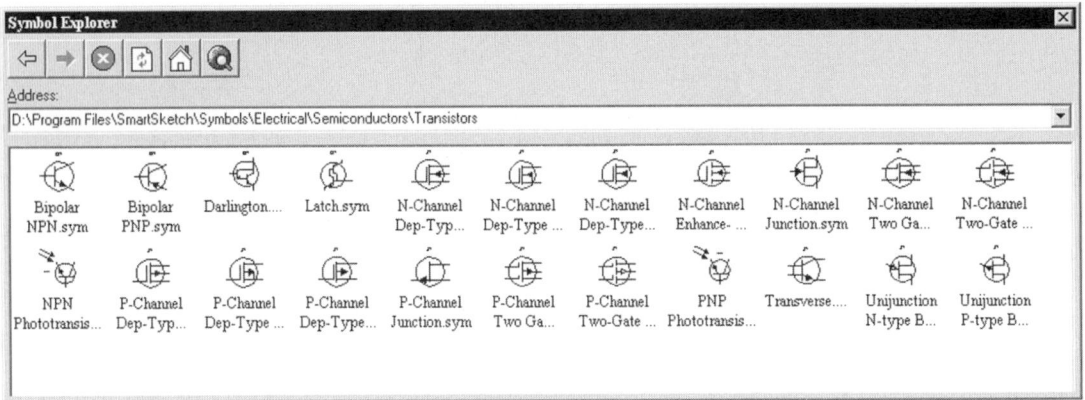

Figure 6–33. Transistor symbols.

*Figure 6–34.
SmartSketch
sample
control loop
diagram.*

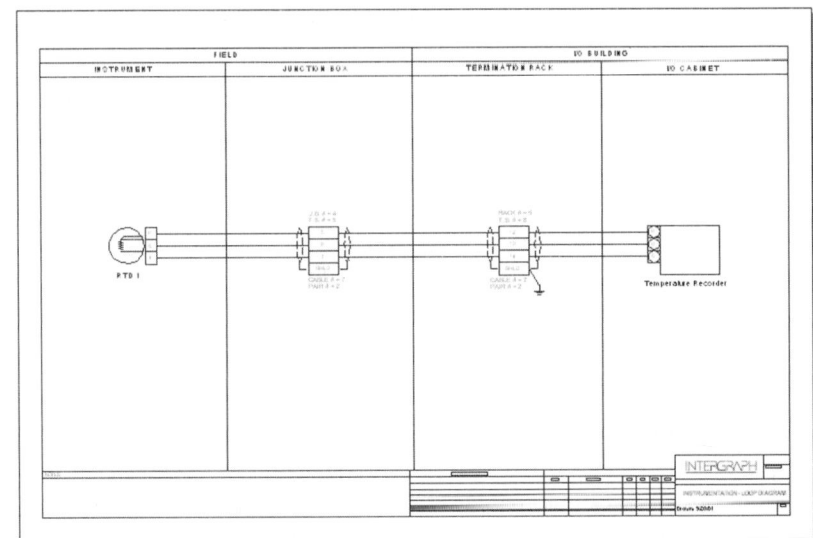

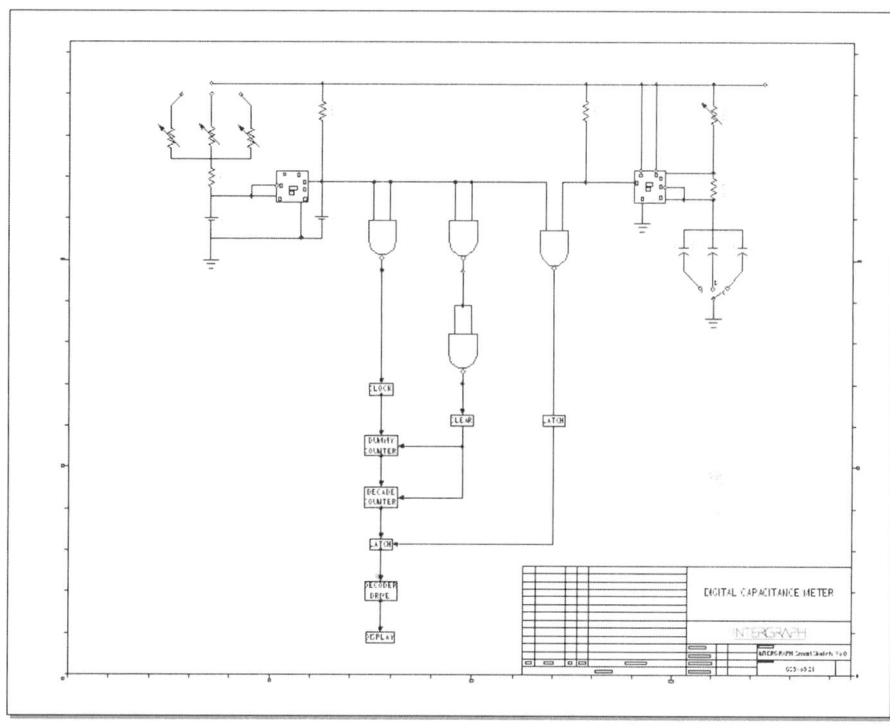

Figure 6–35. SmartSketch electrical diagram.

Process Diagramming

To produce process diagrams, you use the process diagram templates shown in Figure 6–36 and their associated symbols.

Figure 6–36. Process diagram templates.

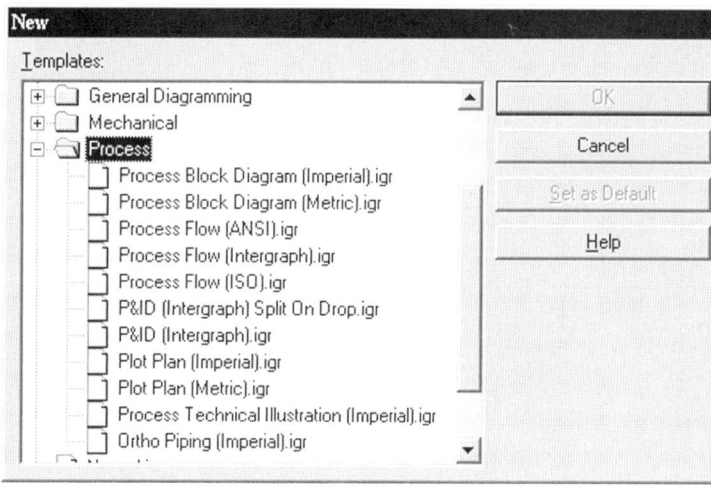

There are six types of process diagrams, including ortho piping process diagrams, P&ID process diagrams, plot plans, process block diagrams, process design workflow diagrams, and control loop diagrams. Some of the symbols available are illustrated in Figures 6–37 through 6–42, and SmartSketch sample process diagrams are depicted in Figures 6–43 through 6–48.

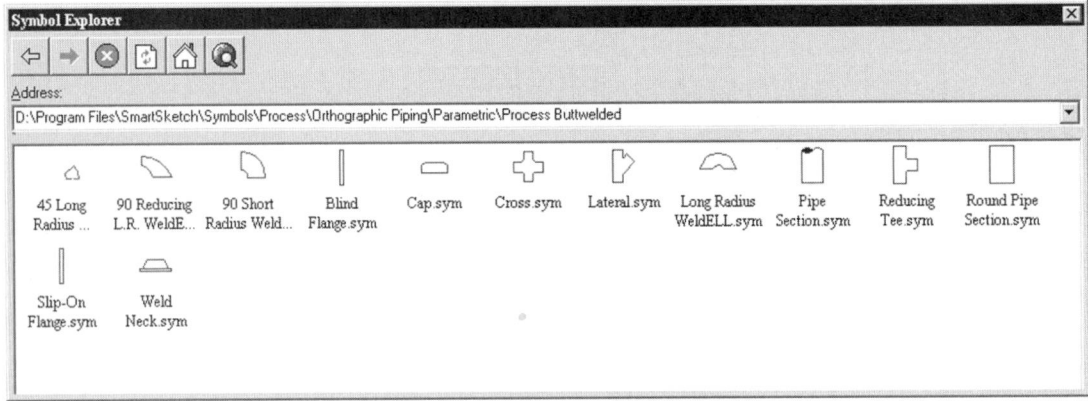

Figure 6–37. Process butt-weld symbols.

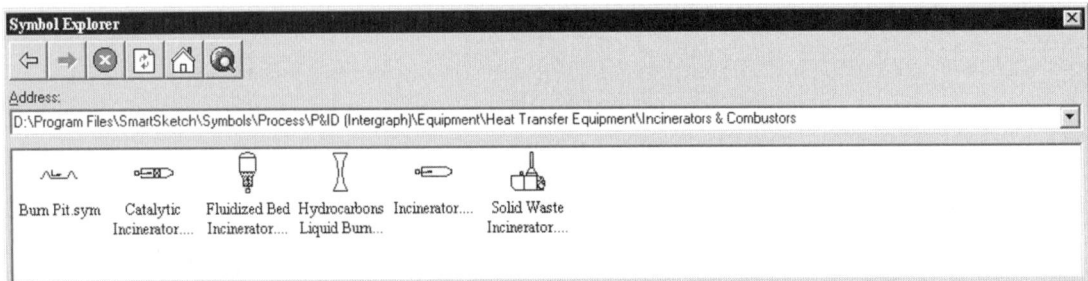

Figure 6–38. Incinerators and combustion equipment symbols.

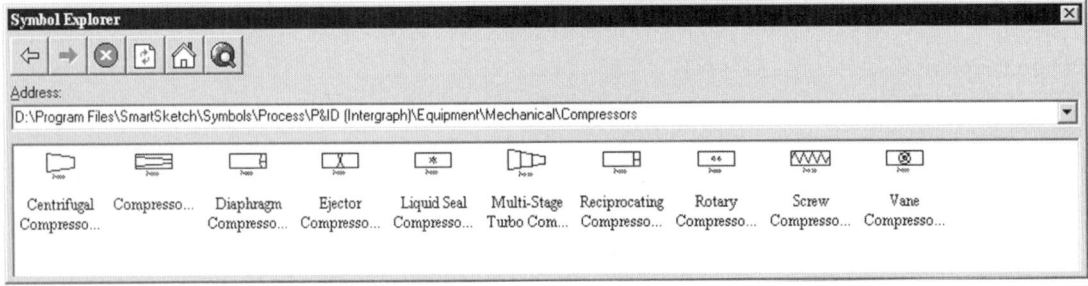

Figure 6–39. Mechanical compressor symbols.

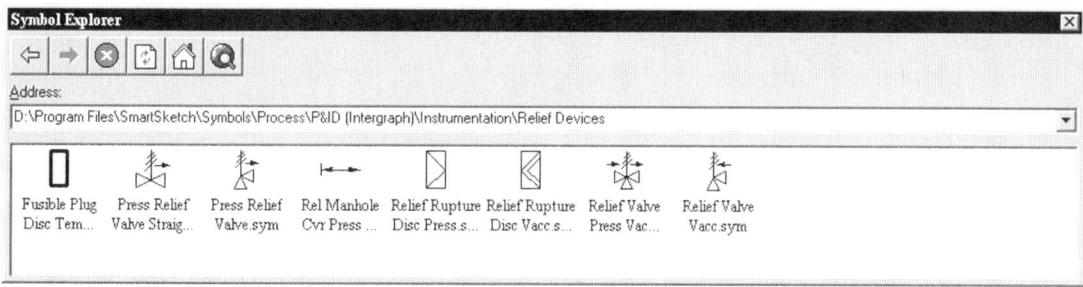

Figure 6–40. Relief device symbols.

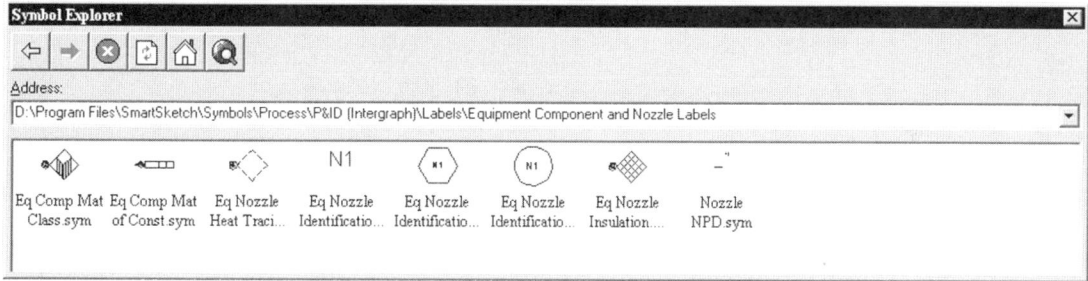

Figure 6–41. Equipment component and nozzle labels.

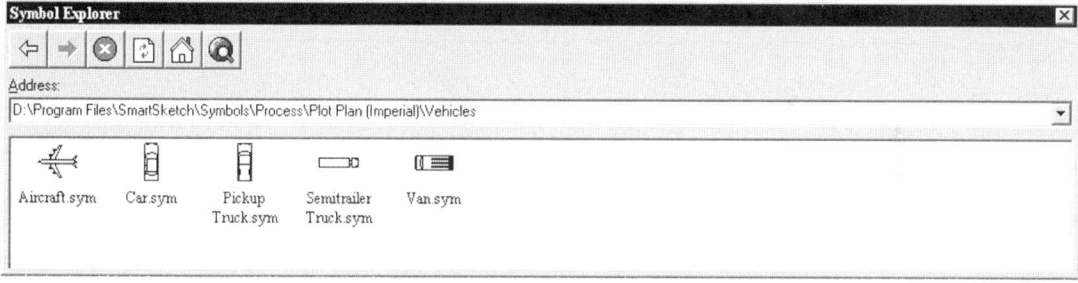

Figure 6–42. Vehicle symbols.

*Figure 6–43.
Ortho piping
process diagram.*

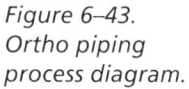
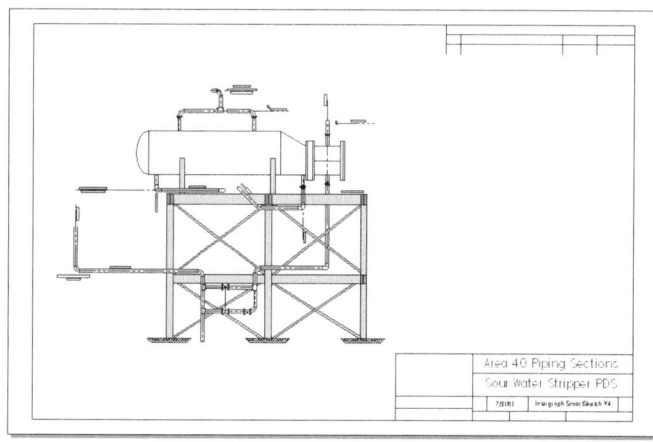

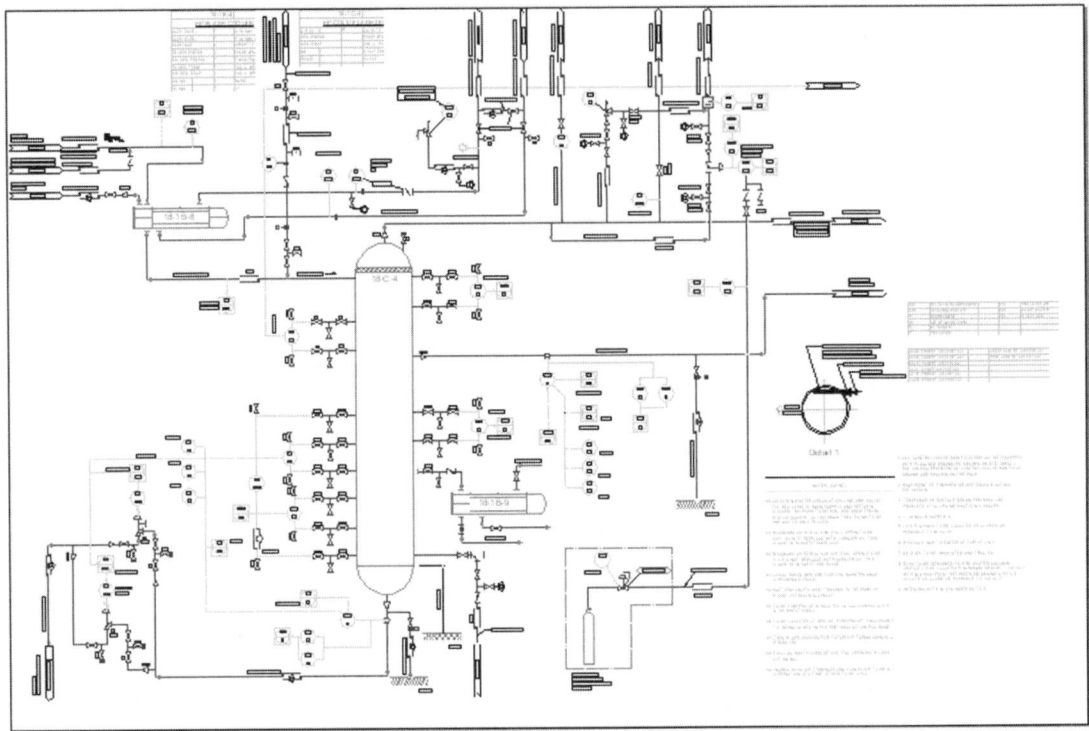

Figure 6–44. P&ID process diagram.

*Figure 6–45.
Plot plan.*

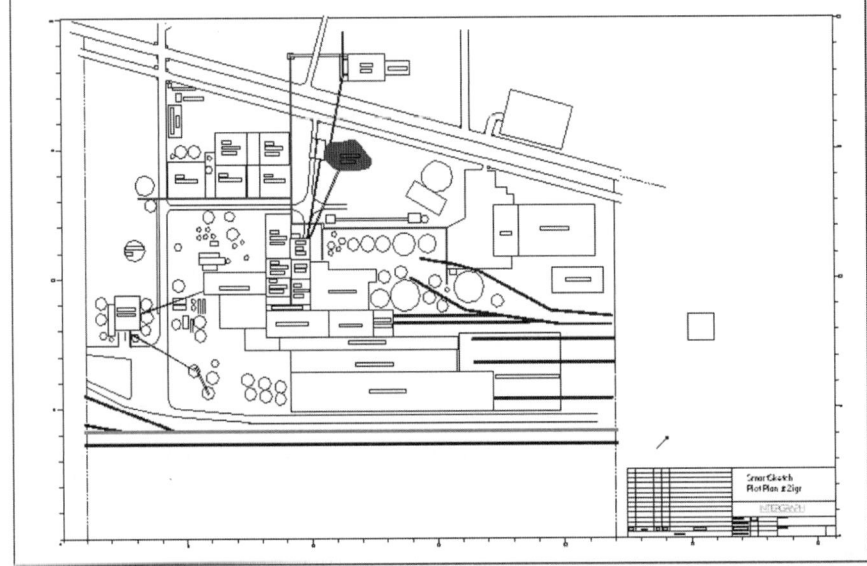

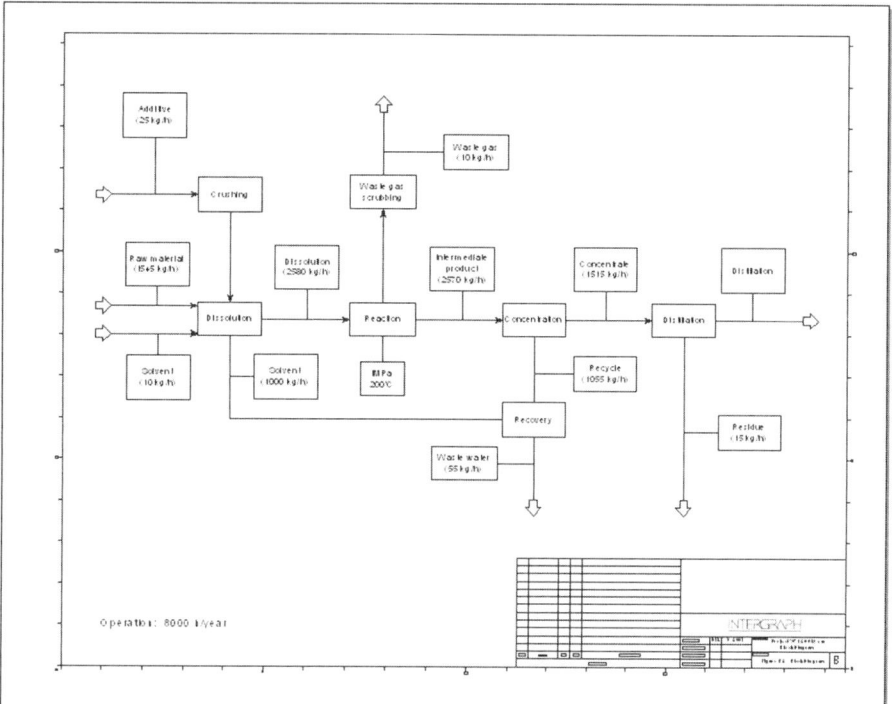

Figure 6–46. Process block diagram.

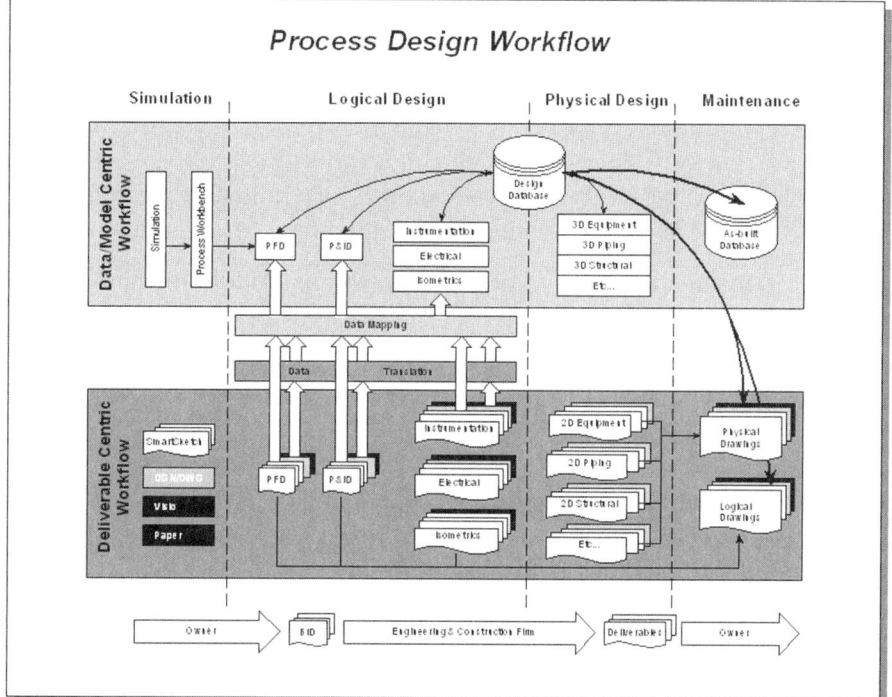

Figure 6–47. Process design workflow diagram.

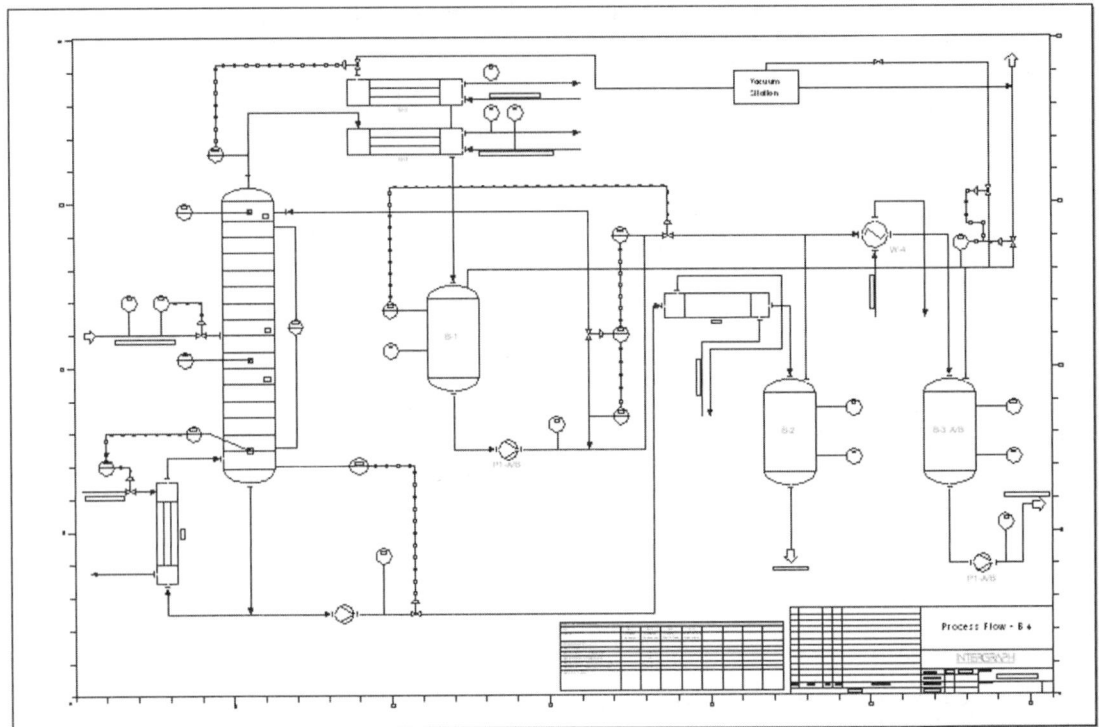

Figure 6–48. Process flow diagram.

■ ■ ■ ■   Precision Drawings

There are two major types of precision engineering drawings. They are AEC (architectural, electrical, and construction) solution drawings and mechanical engineering drawings. These 2D drawings are used to depict floor plans and orthographic engineering projection drawings. Unlike schematic diagrams, in which you deploy substantial amount of symbols, you normally use the drafting tools depicted in Chapters 2 and 3 and symbols as may be necessary. Similar to constructing schematic diagrams, you also need an appropriate template with either Imperial units or metric units.

AEC Solutions

AEC solution drawings depict architectural, building, or civil engineering entities. To produce AEC drawings, you select a template from a set of templates shown in Figure 6–49.

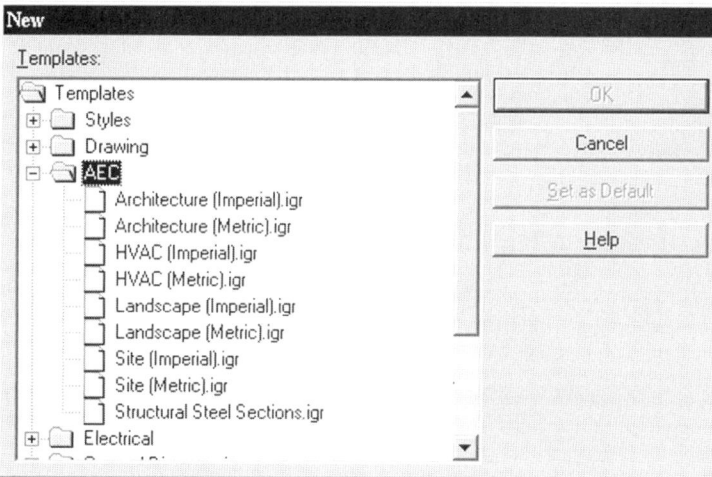

Figure 6–49.
AEC drawing
templates.

To enhance the production of AEC drawings, a lot of symbols are provided. Some of these symbols are illustrated Figures 6–50 through 6–60.

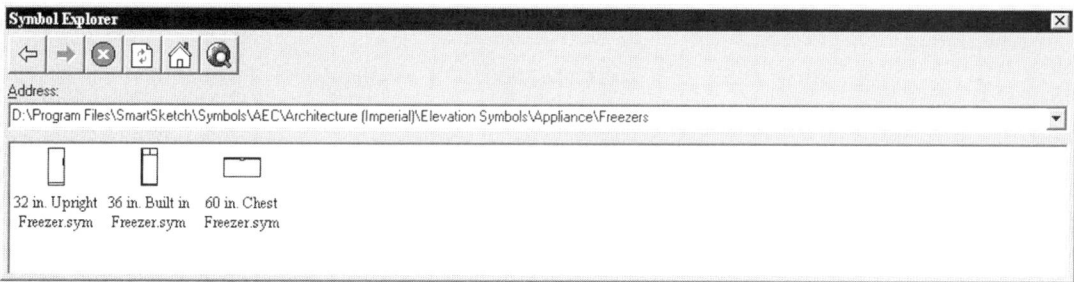

Figure 6–50. Freezer symbols.

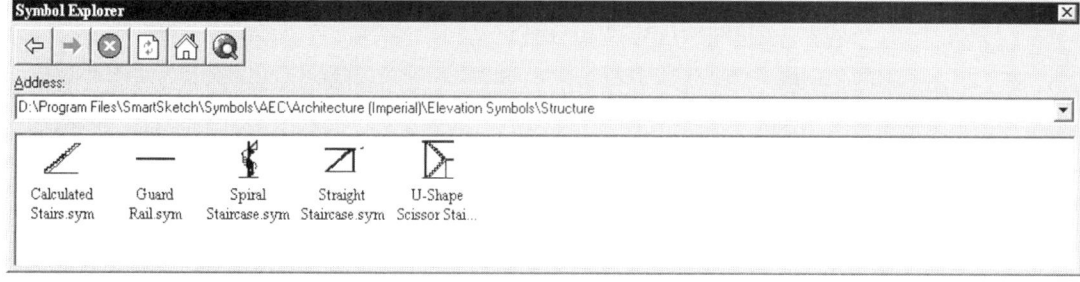

Figure 6–51. Structure symbols.

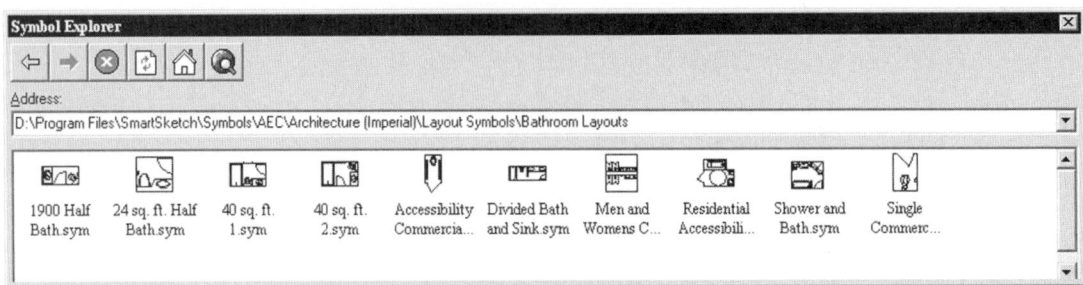

Figure 6–52. Bathroom layout symbols.

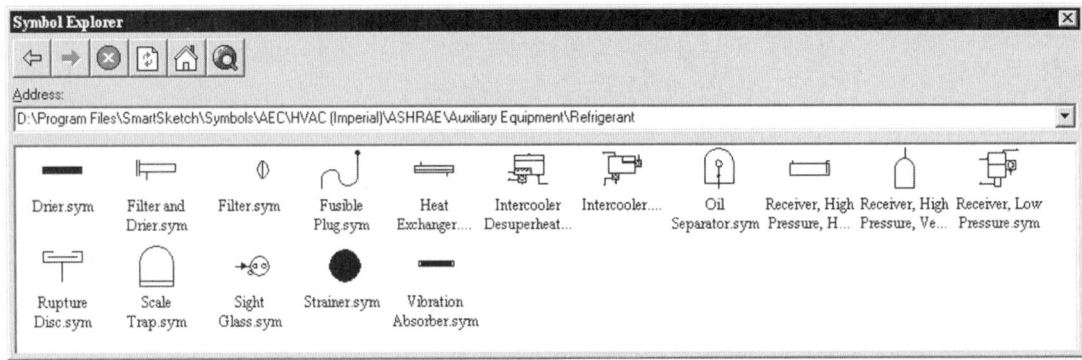

Figure 6–53. Refrigerant symbols.

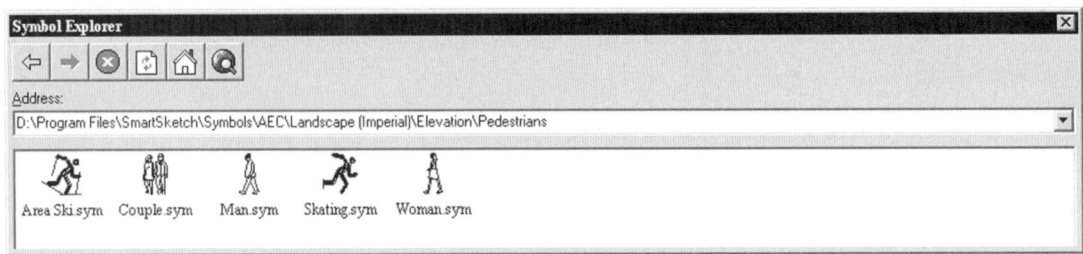

Figure 6–54. Pedestrian elevation symbols.

*Figure 6–55.
Vegetation
symbols.*

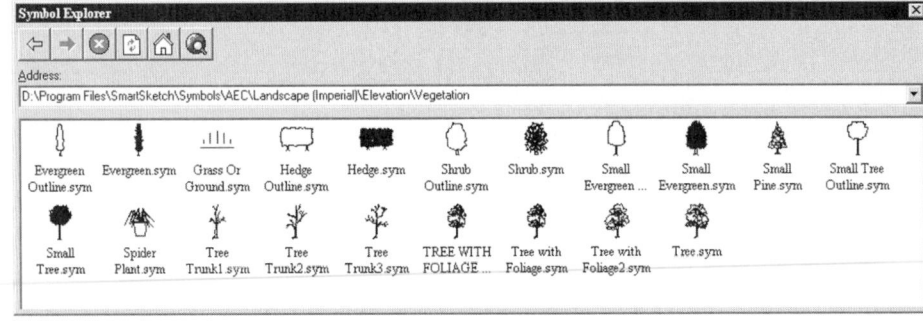

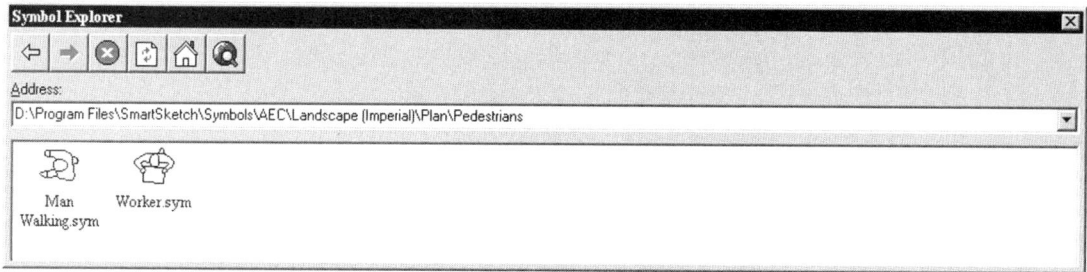

Figure 6–56. Pedestrian plan symbols.

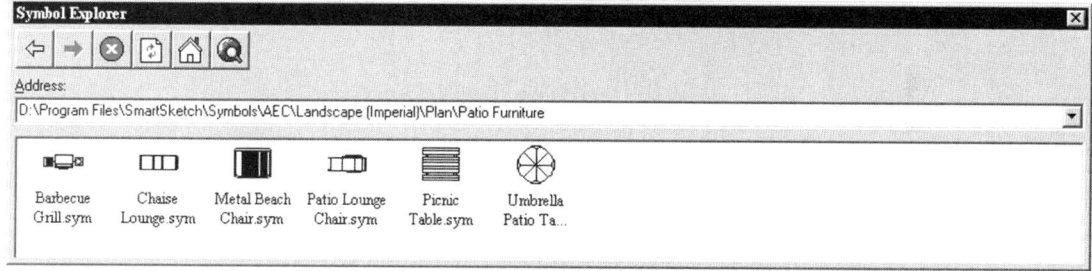

Figure 6–57. Furniture symbols.

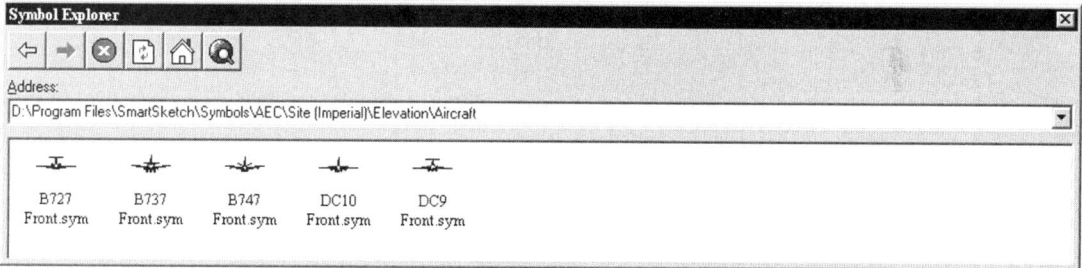

Figure 6–58. Aircraft symbols.

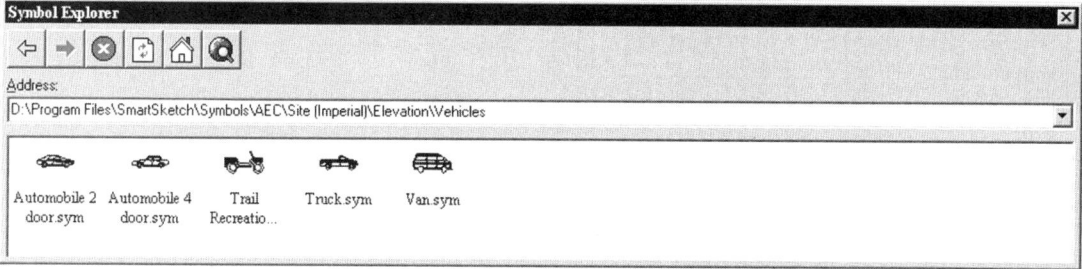

Figure 6–59. Vehicle symbols.

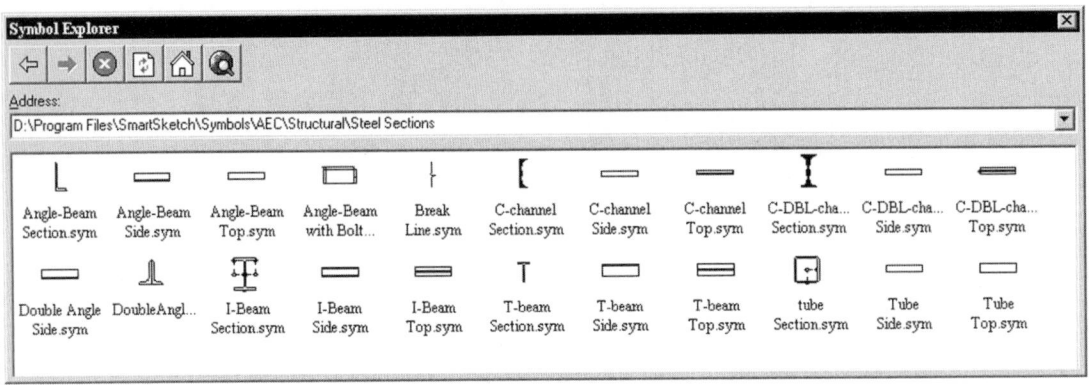

Figure 6–60. Steel section symbols.

Using the templates and symbols, a large variety of AEC drawings can be produced. Figures 6–61 through 6–64 shows four types of Smart-Sketch AEC sample drawings.

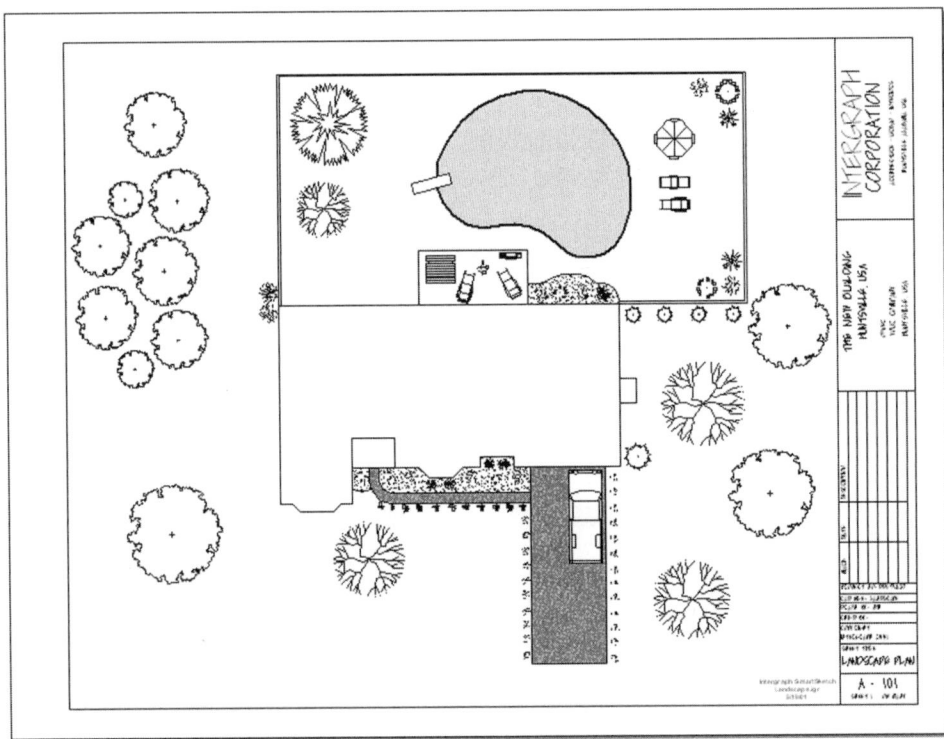

Figure 6–61. SmartSketch sample landscape plan AEC drawing.

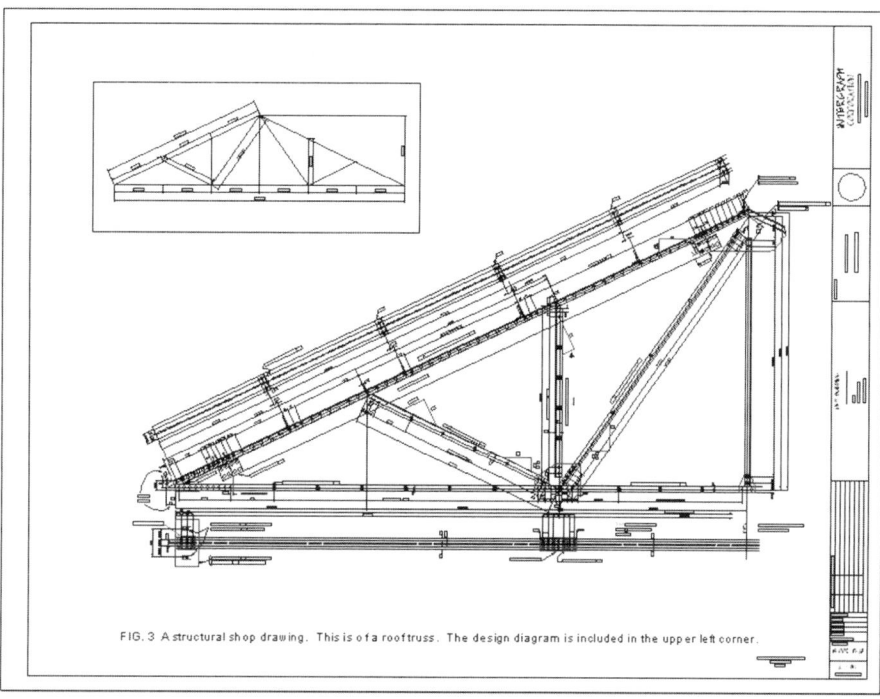

Figure 6–62. SmartSketch sample metal truss AEC drawing.

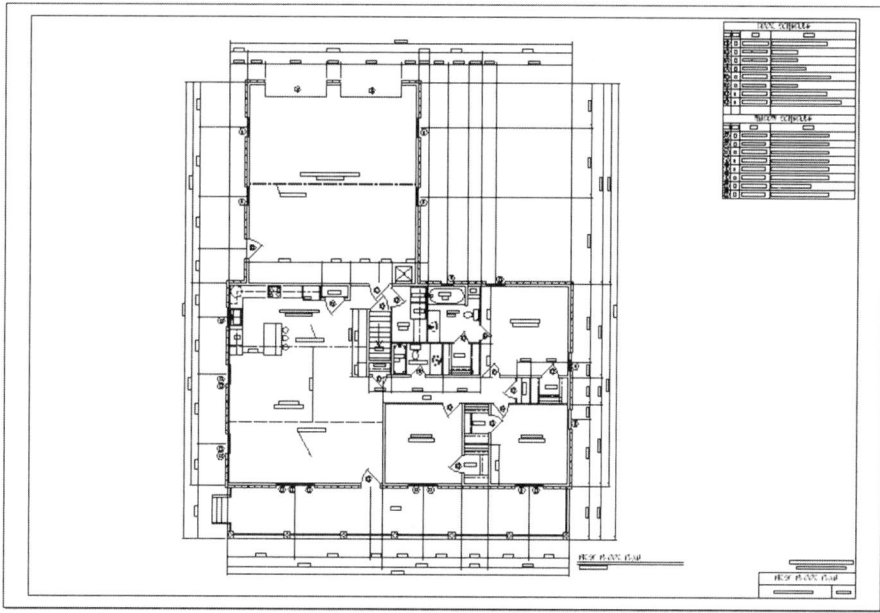

Figure 6–63. SmartSketch sample residential house plan AEC drawing.

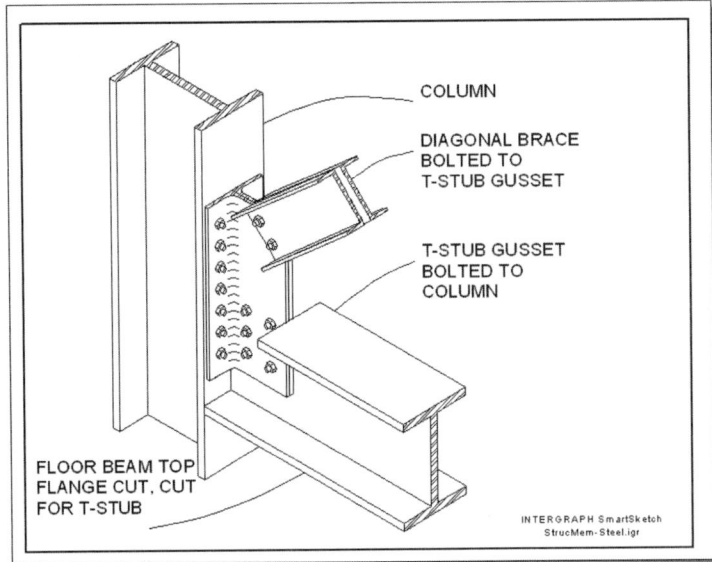

*Figure 6–64.
SmartSketch sample
isometric structure
steel AEC drawing.*

Mechanical Engineering

Mechanical engineering drawings are 2D orthographic projection drawings that depict a 3D object by using orthogonal projection views. There are two types of mechanical drawings, technical drawings and technical illustrations. To produce mechanical engineering drawings, you use the templates illustrated in Figure 6–65.

*Figure 6–65.
Mechanical
engineering
drawing
templates.*

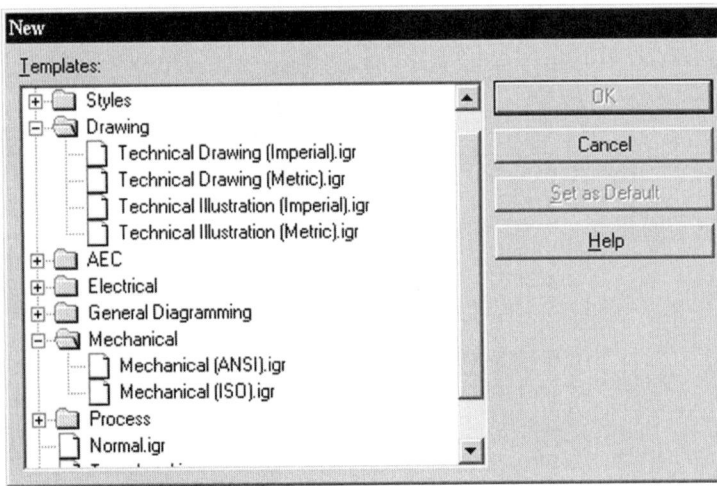

Among the templates shown in Figure 6–65, the technical illustration templates can be deployed for making 2D isometric drawings. Along with this template, there is an Isometric toolbar. Associated with the mechanical engineering drawing templates is a symbol folder shown in Figure 6–66.

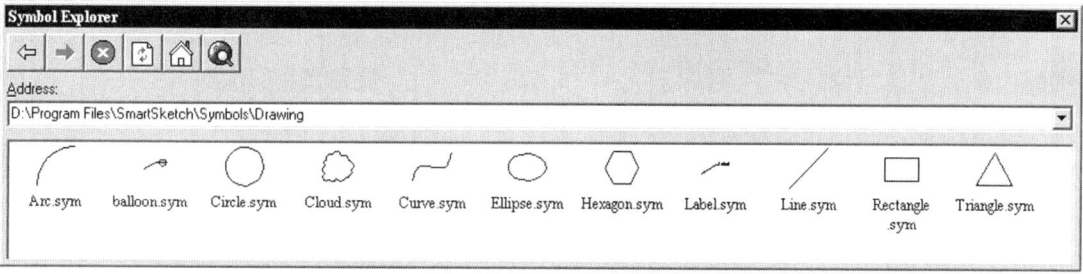

Figure 6–66. Symbols for producing mechanical engineering drawings.

SmartSketch sample drawings illustrating the use of mechanical engineering templates and their associated symbol folder are shown in Figures 6–67 through 6–70.

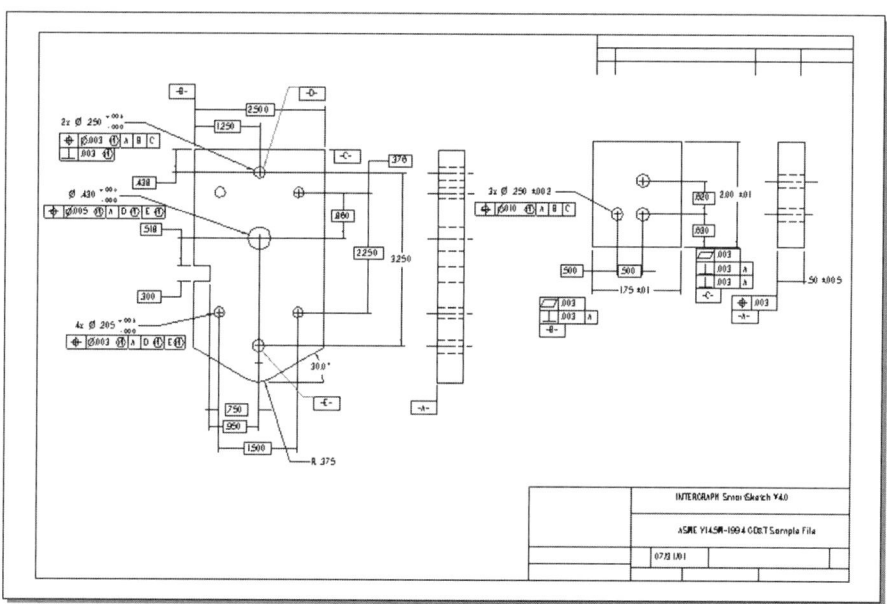

Figure 6–67. SmartSketch sample mechanical drawing showing geometric tolerances.

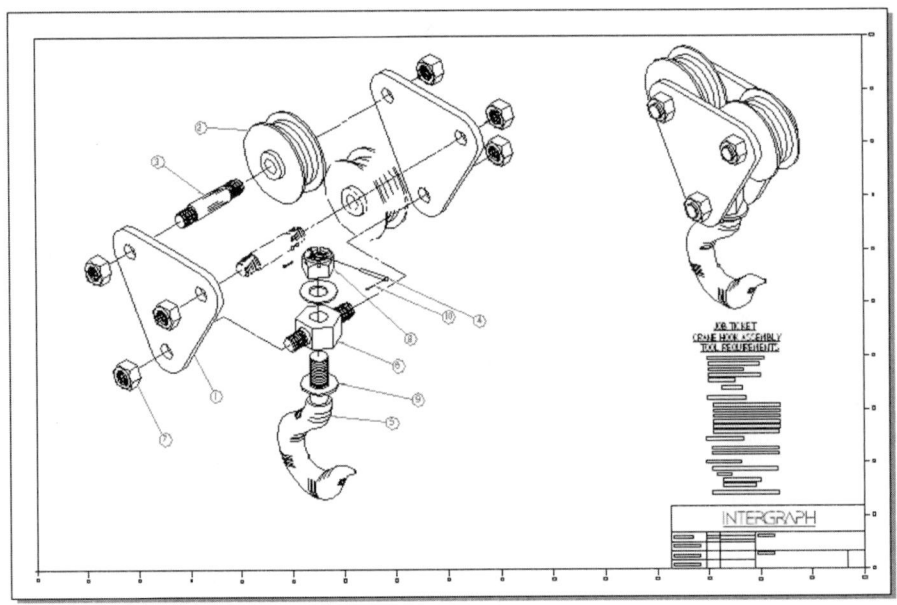

Figure 6–68. SmartSketch sample mechanical drawing showing the exploded 2D isometric views of a crane hook.

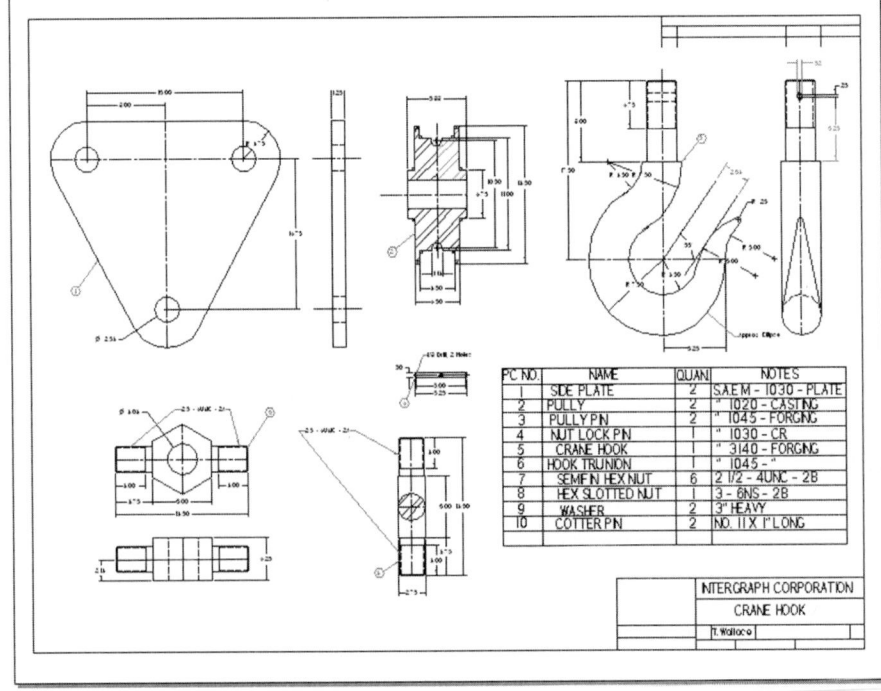

Figure 6–69. SmartSketch sample mechanical drawing showing the orthographic views of a crane hook.

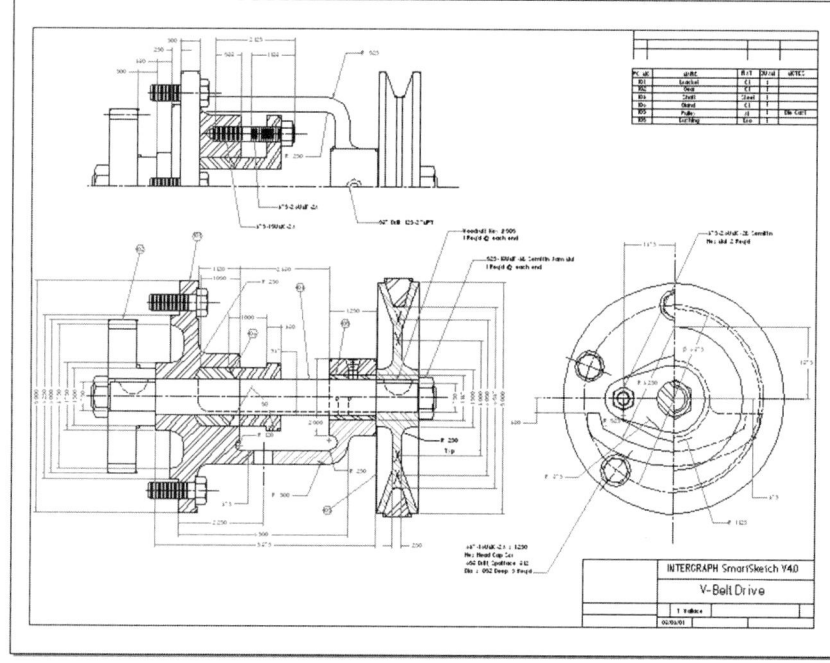

Figure 6–70. SmartSketch sample mechanical drawing showing the assembly of a V-belt drive system.

Using the parametric relationship drafting tools, you can construct mechanical drawings to simulate the motion of mechanisms. Figure 6–71 shows the mechanism of a 4-bar linkage and Figure 6–72 shows the mechanism of a pair of pliers.

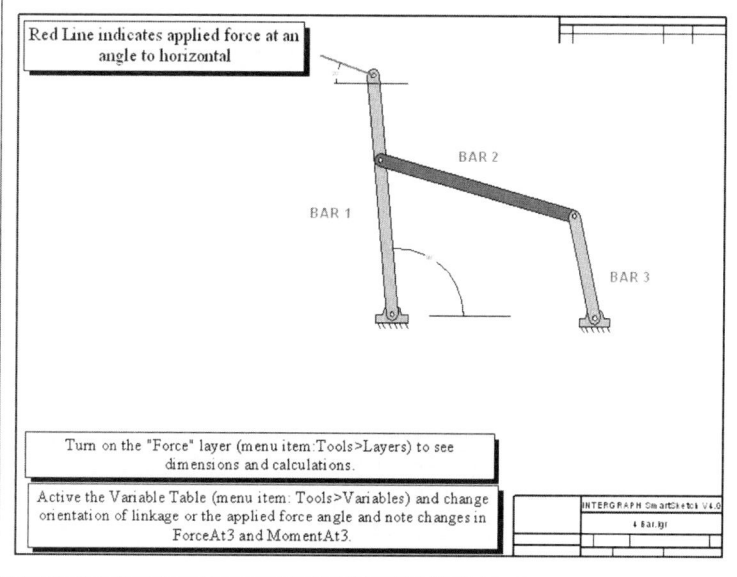

Figure 6–71. SmartSketch sample 4-bar linkage mechanism.

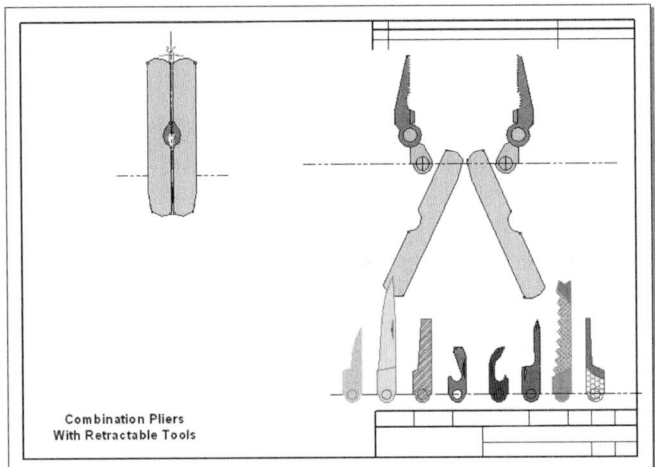

Figure 6–72. SmartSketch sample pliers.

Combination Pliers
With Retractable Tools

Summary

SmartSketch is a 2D computer-aided drafting tool for making schematic diagrams and precision 2D drawings. To produce various types of drawing documents, you use templates categorized into five groups. Three of them are schematic diagrams and two of them are precision engineering drawings. To increase drawing productivity, each template is associated with a set of symbols.

Schematic diagrams include general diagrams, electric diagrams, and process diagrams. General diagrams are atlas maps, basic diagrams, directional diagrams, flowcharts, network diagrams, office layouts, organizational charts, and workflow diagrams. Electrical diagrams are divided into control loop diagrams and electrical schematic diagrams. Process diagrams include process block diagrams, process flow diagrams, process technical illustration diagrams, ortho piping diagrams, P&ID diagrams, and plot plans.

Precision drawings are AEC solution drawings and mechanical engineering drawings. AEC drawings are architectural drawings, HVAC drawings, landscape drawings, site drawings, and structural drawings. Mechanical drawings are mainly 2D orthographic drawings, including isometric and exploded views.

Review Questions

1   What types of schematic diagrams can you produce?

2   List the types of precision drawings that can be produced.

Exercises

To further enhance your knowledge of using SmartSketch as a computer-aided design tool, try the following exercises.

■ ■ ■ ■ Staircase

Figure A–1 shows the elevation of a staircase. Start a SmartSketch document using the *Technical Drawing (Imperial)* template. In the document, construct three layers: *Outlines*, *Center*, and *Dim*. Set the line width of layer *Outlines* to 1 mm, the line type of layer *Center* to center line, the color of layer *Center* to red, and the color of layer *Dim* to blue. Save the document as a template for further exercises.

Figure A–1. Staircase elevation.

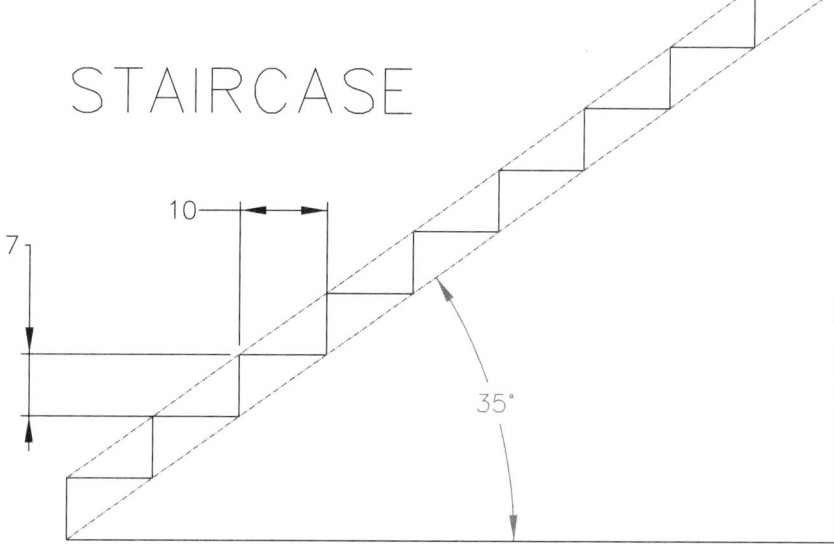

Perform the following steps to create the drawing.

1 Select File > Sheet Setup.

2 In the Size and Scale tab of the Sheet Setup dialog box, set drawing scale to 1:10 and click on the OK button.

3 In accordance to the drawing, construct the elevation with the end-points of the staircase steps coincident with the two center lines.

■ ■ ■ ■ Panel

Figure A–2 shows the top view of a panel. In accordance with the figure, construct a SmartSketch document using the template you just constructed. In your document, accept the default drawing scale of 1:1 and place the outlines on layer *Outlines*, center lines on layer *Center*, and dimensions on layer *Dim*.

Figure A–2. Panel.

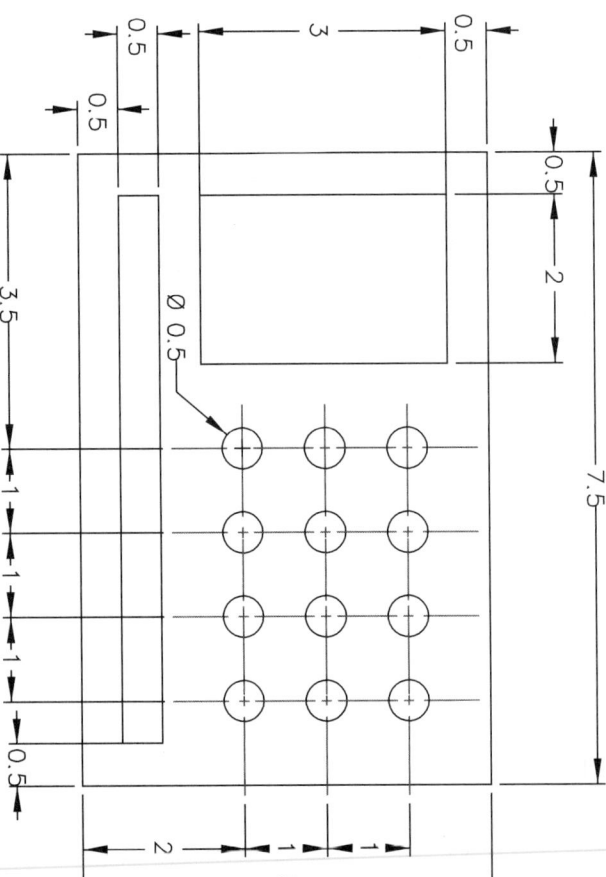

Hook

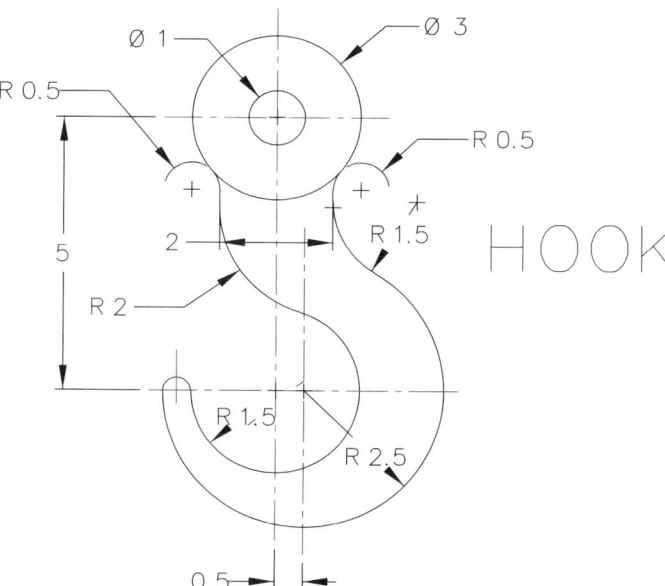

Figure A–3 shows the elevation of a hook. Using the template you constructed for the staircase, construct a SmartSketch document for the hook. Remember to place the outlines, center lines, and dimensions on three separate layers.

Figure A–3. Hook.

Bottle

Construct the elevation of a bottle in accordance with Figure A–4.

Figure A–4. Bottle.

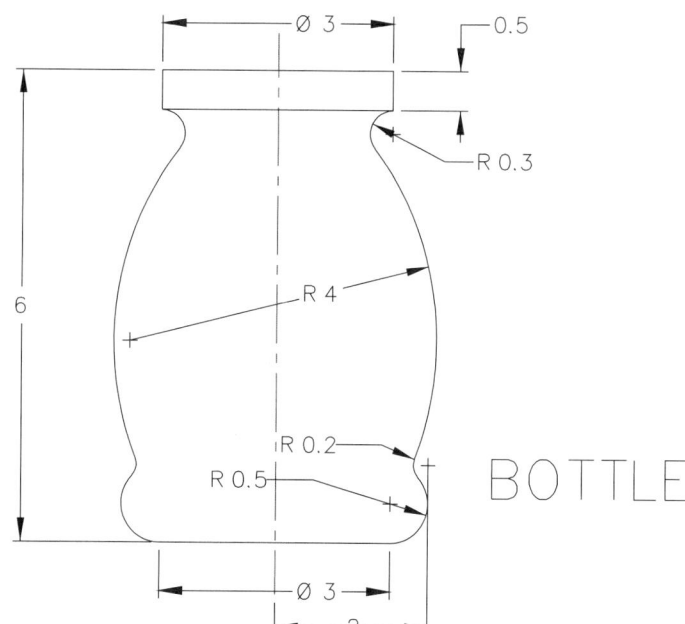

Gasket

With reference to Figure A–5, construct the drawing of a gasket.

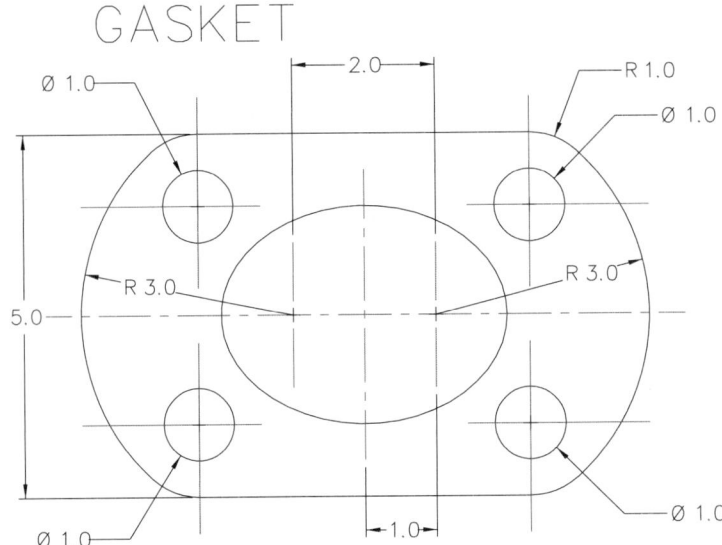

Figure A–5.
Gasket.

Block

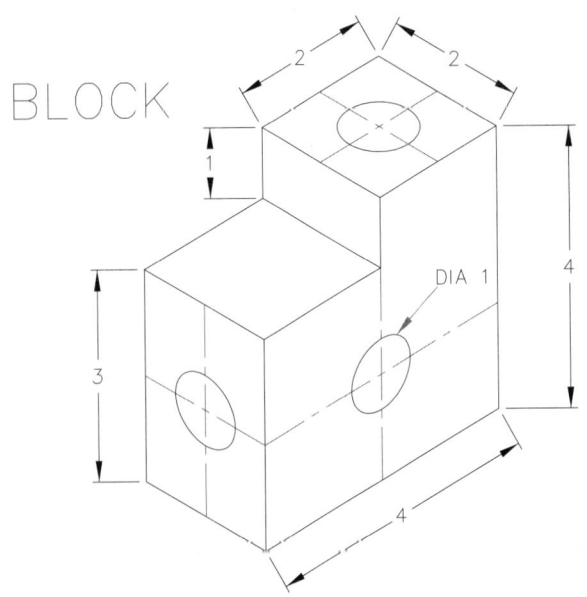

Figure A–6 shows the isometric draw-ing of a block. Use the *Technical Illus-tration (Imperial)* template and the associated Isometric Tools toolbar to construct the drawing.

Figure A–6. Block.

Flowchart

With reference to Figure A–7, use the *Technical Drawing (Imperial)* template to construct a flowchart. The blocks in the flowchart are dropped from the associated symbol folder.

Figure A–7.
Flowchart.

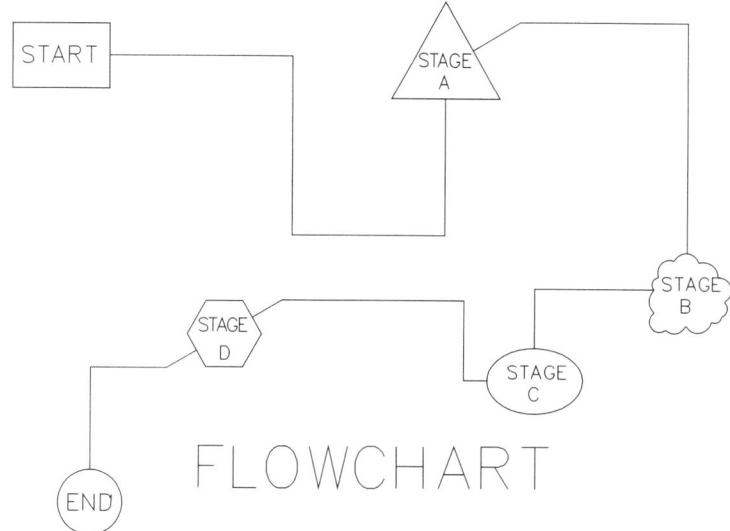

Index